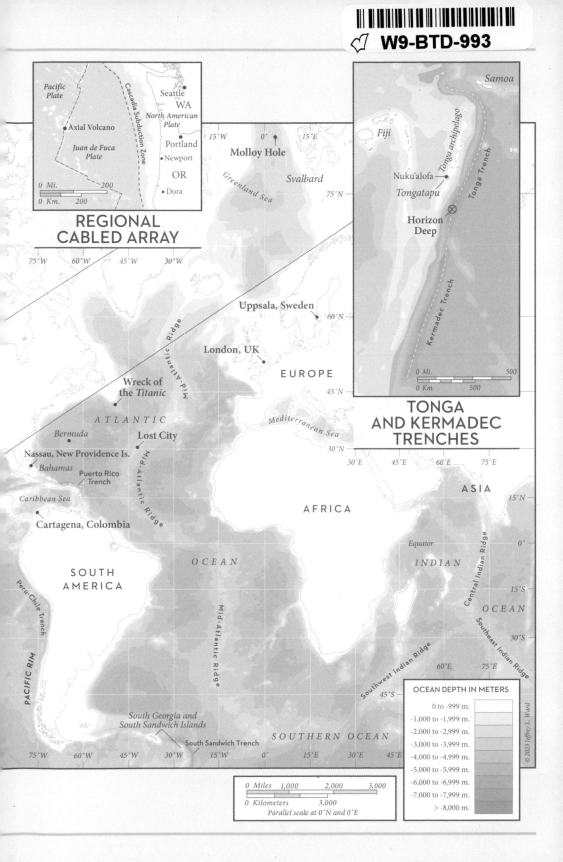

W9-BTD-993

REGIONAL CABLED ARRAY

Pacific Plate

Cascadia Subduction Zone

Seattle
WA
North American Plate

Axial Volcano

Juan de Fuca Plate

Portland
Newport

OR

Dora

0 Mi. 200
0 Km. 200

TONGA AND KERMADEC TRENCHES

Samoa

Fiji

Tonga archipelago

Nuku'alofa

Tongatapu

Tonga Trench

Horizon Deep

Kermadec Trench

0 Mi. 500
0 Km. 500

Molloy Hole

Greenland Sea

Svalbard

Uppsala, Sweden

London, UK

EUROPE

Wreck of the Titanic

Mid-Atlantic Ridge

ATLANTIC

Bermuda

Lost City

Nassau, New Providence Is.

Bahamas

Puerto Rico Trench

Caribbean Sea

Cartagena, Colombia

Mediterranean Sea

Mid-Atlantic Ridge

OCEAN

AFRICA

ASIA

Equator

INDIAN

Central Indian Ridge

SOUTH AMERICA

OCEAN

Mid-Atlantic Ridge

Peru-Chile Trench

PACIFIC RIM

Southwest Indian Ridge

Southeast Indian Ridge

SOUTHERN OCEAN

South Georgia and South Sandwich Islands

South Sandwich Trench

OCEAN DEPTH IN METERS

0 to -999 m.
-1,000 to -1,999 m.
-2,000 to -2,999 m.
-3,000 to -3,999 m.
-4,000 to -4,999 m.
-5,000 to -5,999 m.
-6,000 to -6,999 m.
-7,000 to -7,999 m.
> -8,000 m.

© 2023 Jeffrey L. Ward

0 Miles 1,000 2,000 3,000
0 Kilometers 3,000
Parallel scale at 0°N and 0°E

The Underworld

Also by Susan Casey

The Devil's Teeth
The Wave
Voices in the Ocean

THE
UNDERWORLD

*

Journeys to
the Depths of the Ocean

Susan Casey

Doubleday   *New York*

Riverhead Free Library
330 Court Street
Riverhead NY 11901

Copyright © 2023 by Susan Casey

All rights reserved.
Published in the United States by Doubleday, a division of
Penguin Random House LLC, New York, and distributed in Canada by
Penguin Random House Canada Limited, Toronto.

www.doubleday.com

DOUBLEDAY and the portrayal of an anchor with a dolphin are
registered trademarks of Penguin Random House LLC.

Pages 329–30 constitute an extension of this copyright page.

Front-of-jacket photograph: "Lost City," a frame from the Giant Screen Film
Volcanoes of the Deep Sea © The Stephen Low Company
Jacket design by John Fontana
Book design by Maria Carella
Endpaper map designed by Jeffrey L. Ward

Library of Congress Cataloging-in-Publication Data
Names: Casey, Susan, 1962– author.
Title: The underworld : journeys to the depths of the ocean / Susan Casey.
Description: First edition. | New York : Doubleday, 2023. |
Includes bibliographical references.
Identifiers: LCCN 2023002536 (print) | LCCN 2023002537 (ebook) |
ISBN 9780385545570 (hardcover) | ISBN 9780385545587 (ebook)
Subjects: LCSH: Ocean. | Oceanography—History. | Ocean—Mythology. |
Submarine topography | Ocean bottom. | Deep-sea sounding. | Marine ecology. |
Marine ecosystem health.
Classification: LCC GC21 .C36 2023 (print) | LCC GC21 (ebook) |
DDC 551.46—dc23/eng/20230518
LC record available at https://lccn.loc.gov/2023002536
LC ebook record available at https://lccn.loc.gov/2023002537

MANUFACTURED IN THE UNITED STATES OF AMERICA

1 3 5 7 9 10 8 6 4 2

First Edition

For those who love the ocean

And in memory of

Ron Casey, John Casey, Judy Casey, and Tom Walkling

We must go and see for ourselves.

—JACQUES-YVES COUSTEAU

Contents

Author's Note

When writing about the deep ocean, the first question that arises is: What is it? At what point does the ocean become the *deep* ocean? It may be surprising to learn that even deep-sea scientists don't fully agree on how to define the various depth zones. Typically, however, the deep ocean is considered to comprise the waters below two hundred meters, or approximately six hundred feet—the point at which sunlight essentially disappears. In this book, I have defined the deeper layers as the twilight zone (six hundred to thirty-three hundred feet), the midnight zone (thirty-three hundred to ten thousand feet), the abyssal zone (ten thousand to twenty thousand feet), and the hadal zone (twenty thousand to thirty-six thousand feet). These names and measurements are commonly used, although there are other ways to delineate the deep's regions, particularly within its vast mid-waters.

The abyssal zone is known in short as *the abyss*, but the word *abyss* also has a broader meaning as a deep and seemingly bottomless chasm: it's often used when referring generally to the ocean's great depths. You'll find both usages in these pages, with the distinction evident in context.

You could travel through the abyssal zone in a submersible (though few vehicles in the world can venture that deep), but if you were traveling in a submarine you wouldn't have that option. Submarines are capable of sustained, independent undersea operations, but their diving range is relatively shallow. The deep-sea vehicles I describe in these pages are all submersibles. There are two types:

manned and unmanned. Manned submersibles carry passengers, who sit within a dry, pressure-controlled atmosphere equipped with life-support systems. These compact subs can descend, ascend, and fly around beneath the surface independently, but they require a support vessel and crew for transport, launch, and recovery. Their power supplies come from batteries, so they can't remain submerged for weeks the way a submarine can.

Unmanned submersibles are robots, and include remote operated vehicles (ROVs) that are tethered to a ship and driven remotely by a human pilot; and autonomous underwater vehicles (AUVs) that may also be launched from a ship, but are preprogrammed to dive, collect data, and return to their base without real-time human input.

In casual usage the words *submarine* and *submersible* are sometimes used interchangeably, referring generally to an underwater vehicle. Both submarines and submersibles share an abbreviation: subs.

The deep ocean occupies 95 percent of the ocean's volume, and you'll notice that both terms are singular. The earth possesses one ocean, though it's traditionally recognized as having five major regions: the Pacific, Atlantic, Indian, Arctic, and Southern Oceans. Whenever possible, I refer to the ocean as a single entity.

The science of measuring the seabed terrain is known as bathymetry—the submarine version of topography. Bathymetric maps chart the depths and contours of the seafloor in three-dimensional relief, revealing its mountains, valleys, canyons, plains, rifts, trenches, and other undersea features.

American and British readers are accustomed to thinking in miles, feet, tons, degrees Fahrenheit, pounds per square inch. Science uses metric measurements. Mariners use nautical miles and fathoms. Here, I'm defaulting to the imperial system of measurement, but when it is helpful for clarity I've cited metric figures. Metric usage in quotes is verbatim.

The
Underworld

Prologue

My dear child, how did you come to this land of darkness
while still alive? It is hard for the living to get here.
—HOMER, *The Odyssey*

18.70° N, 155.17° W
THE PACIFIC OCEAN
JANUARY 31, 2021

I stood on the ship's deck in my long underwear and my fireproof jumpsuit, watching a pale silver sunrise and gauging the wind. It was blowing twenty knots, gusting to thirty, and it had been blowing this hard all night, churning the ocean into a disorderly mess. What I wanted to know—and had come outside to check—was *how* disorderly. I could see that the waves were trouble: ten-foot swells galloping from two directions, surging with whitecaps. They'd run halfway across the Pacific, gaining strength along the way. Out here, there was no land to stop them. Bracing myself against a railing, I took out my phone to check the marine forecast. Again.

Today, to me, every knot of wind speed and every inch of wave height mattered. Two hours from now, at 0800, conditions permitting, a team of engineers, technicians, and mariners would attempt to launch an eleven-ton deep-sea submersible with two people in it—and I would be one of them. If the launch was successful, the sub would bob at the surface briefly, the pilot would pump seawater into the ballast tanks, and then we would sink beneath the waves and free-fall for two and a half hours, plunging for miles and touching down in a place that human eyes had never seen. That was the plan. But right now the conditions didn't seem very permissive.

In order for the dive to go forward, the weather would have to lie down. If we didn't dive today, it was unclear if I'd get another chance—and I had been waiting for this one my entire life. In all of history, few had ventured as deep as we were going. There was a long list of reasons why, a roster of difficulties and risks, the most obvious being that at the bottom we would encounter eight thousand pounds of pressure per square inch. For oxygen, communications, navigation, shelter—survival—we'd be utterly reliant on the sub. It was a singular machine, the first submersible that could dive repeatedly with a pilot and passenger to full ocean depth (thirty-six thousand feet), so new and innovative that its creators still referred to it as a prototype. To put the deep's engineering challenges into perspective, consider that there are three Mars Rovers, while this sub was one of a kind.

But the inaccessibility of the deep, I thought, made it even more alluring. Others wanted to visit Paris, Bora Bora, the Serengeti: I wanted to go into the ocean's abyss. The idea of an unknown aquatic realm, ever present below us but invisible unless we look for it—an underworld, within our world—had always worked a sort of spell on me, an alchemical mix of wonder and fear.

It may seem as if those emotions would cancel each other out, but the opposite is true. When you add them together you get the sublime, which transcends both. "The passion caused by the great and sublime in nature . . . is Astonishment," wrote the eighteenth-century philosopher Edmund Burke, "and astonishment is the state of the soul, in which all its motions are suspended, with some degree of horror." But, he added, it was "a sort of delightful horror." The abyss might be terrifying, but you wouldn't notice because you'd be too busy gaping in awe. At least that's how I imagined it, and I wanted to see if I was right.

*

To tell you the story of how I landed on this ship, ready to climb into the world's most extreme submersible, I have to go back to the beginning. My obsession with the deep started early. In childhood, I had a recurring dream in which I floated on a moonlit sea in a small boat, while below me big fish circled ominously, or swept by like phantoms. The dream was haunting, but I never thought of it as a nightmare. Beneath the one-way mirror of the surface anything might be lurking, but even if it scared me I was determined to check it out.

In the seventies, when *The Undersea World of Jacques Cousteau* was at its zenith, I would sit in front of the TV and watch, riveted, as Cousteau and his jaunty band of aquanauts descended into coral forests and grottoes filled with hungry sharks. During the episode where the *Calypso* sailed to the South Pacific so the crew could explore sunken World War II wrecks in a lagoon, you couldn't have pried me away from the screen if the house were on fire. To roam around the globe in your own ship, diving into one mystery after another, represented a kind of idealized existence that not only seemed unattainable; it hardly seemed real. That didn't make the show any less fabulous, though: it allowed me to join the expeditions vicariously, from the safety of land. This was important because for all my desire to visit the undersea world myself, I didn't learn how to swim until I was almost ten—and even then, I was shaky about it.

I grew up in a suburb of Toronto, far from the ocean but close to hundreds of Canadian Shield lakes, gouged from Precambrian granite by lumbering glaciers. These northern lakes appear brooding, even somber. Their waters are an inky green-black, shadowed by the craggy bluffs and boreal forests that surround them. Rock shoals rise to form islands offshore, attended by lichens and pines, and serenaded by loons.

My family spent summers at a place called Port Severn, two hours north of the city. It is tucked along the southeastern shore of Georgian Bay in Lake Huron, the second largest of the Great Lakes. To my ten-year-old mind, Port Severn was a place full of dark intrigue. I would stand at the end of the dock and look down into the water and see no sign of the bottom, and envision the animals that lived in its depths.

There were northern pikes with beady eyes and torpedo bodies, their jaws lined with needle teeth; and muskies, pikes on steroids that had been known to attack children. Sturgeons, with their bony armor, looked like the offspring of a catfish and a crocodile: they could grow as big as a canoe.

Within walking distance of our cottage there was an old boat-house, unused and derelict, with a wooden door that groaned when you opened it. Inside, it was gloomy and cobwebby with a rickety U-shaped dock and a musty smell of mildew. The whole structure listed to starboard. In its cave-like, half-sunk end life, the boathouse was an ideal lair for reclusive giant fish. At dusk, I would sneak in with a flashlight and drop bits of food in the water, hoping to coax them to the surface. Most of the time there were no takers. Then one day I tossed in a scrap of hot dog and a great beast of a fish—it had to be four feet long—emerged from under the dock, raised its head, and lunged at the meat before quickly retreating. In the dim beam of the flashlight I could make out only the fish's silhouette, but that was enough. It was the coolest thing I'd ever seen, and proof of concept: the underworld was alive with surprises.

*

It was about the water, always the water. Nothing else had the same magnetism; nothing else even came close. I was irresistibly drawn to it, and I developed the skills I needed to immerse myself in the aquatic world. I became a serious competitive swimmer, then an open-water swimmer, free diver, scuba diver. The Pacific Ocean became my favorite playground, the largest waters on earth. I moved to Hawaii to be closer to my muse, and she rewarded me by letting me swim with her residents: sharks, whales, dolphins, sea turtles, eels, fish—marine creatures I'd never encountered before, each possessed of a presence that mesmerized me like nothing found on land.

The sheer *unearthliness* of everything that was happening beneath the waves rocked my world. It was plainly an empire that only the ocean could rule—an ungovernable territory that begins where the sunlight stops. It's out of sight, but once it allows you a peek at its majesty, it's never out of mind. I wanted to know more, I wanted to see more: I wanted to dive into the darkness. Somewhere in those waters I'd crossed a threshold, stepped through the doorway of the craziest boathouse of all, and now I wanted to go deeper.

When you're compelled by something I think you owe it to yourself to go out and investigate it, but my desire for deep-sea submergence came with some technical obstacles. Musing about the abyss felt like falling in love with a mirage: an image flickers into mind, then dissolves, as ungraspable as water itself. "Who has known the ocean? Neither you nor I, with our earth-bound senses," the marine biologist and author Rachel Carson lamented.

I understood that frustration—yet still I felt the deep's pull. Limited to the ocean's top hundred feet, as I was, the real abyss eluded me, as if it were an abstraction rather than a destination. It glittered in my imagination like a distant galaxy, about as tangible as a blast of radio waves. What kind of a place *was* it? What was it like to be there? *What would you see if you went?*

Even now, when every last crater on the moon has been named and interactive three-dimensional maps of Mars can be viewed on an iPhone, 80 percent of the seafloor has never been charted in any kind of sharp detail. Yet the deep ocean—defined as the waters below six hundred feet—covers 65 percent of the earth's surface and occupies 95 percent of its living space. (The Pacific alone could swallow every landmass, every continent and island, and still have room for another South America.) The deep isn't merely a part of our planet—it *is* our planet. You'd think we would want to be more familiar with it.

*

Our awareness of the ocean typically stops at its uppermost layer, known as the epipelagic (or sunlight) zone. If you see marine life and you can name it, odds are it swims in these shallows. But the epipelagic occupies only 5 percent of the ocean's volume. For all its loveliness, it's merely a ceiling. The real action takes place below.

That's where you'll find the twilight zone (from six hundred to thirty-three hundred feet), with its menagerie of creatures that twinkle and glow with bioluminescence; followed by the midnight zone (from thirty-three hundred to ten thousand feet) and the abyssal zone (ten thousand to twenty thousand feet), wraparound layers of eternal night, populated by marvelous weirdos. The intersection where these waters meet the seafloor is known as the abyssal plain, a sediment-cloaked flatland that hosts extraordinary and subtle life. Although these plains may seem quiet, they're often interrupted by outbursts of geological drama.

This is the bottom of the world, but the descent isn't over yet. Beneath the abyss lies the hadal zone, named after Hades, the Greek god of the underworld (and brother of Poseidon), and his mystical realm of the dead. Hadal waters start at twenty thousand feet and pitch down into dozens of ultradeep trenches and troughs—the vast majority of which are located in the Pacific—shaped like the inverted summits of Himalayan peaks. The deepest of them is the Mariana Trench, a fifteen-hundred-mile-long, forty-four-mile-wide gash in the seabed, near Guam. It's home to the 35,876-foot Challenger Deep, the ocean's absolute nadir. By comparison, Mount Everest is 29,029 feet high.

These great depths are a shadow kingdom. In the past, they were the blank spots on the map that carried the old warning: *Here Be Dragons*. Although now we're quite sure the abyss is dragon-free, it still has a mythical aura. Even the slightest glimpse of it is fascinating, because it doesn't reveal itself easily. Actually it doesn't reveal itself at all, unless we approach it with serious technology.

The only way to get a clear picture through thousands of feet of liquid darkness is with sound, which travels faster and farther through water than it does through air. This technology is known as acoustic imaging, or sonar: bouncing sound waves off the bottom and measuring the speed of their return. By processing these data, we can build a refined three-dimensional model of the seafloor. This can't be done from afar: to map an area the sonar array must pass directly over it.

Making precise bathymetric maps is a highly technical affair, and in most places in the ocean no one has even tried.

Our lack of a seafloor atlas lends itself to some startling discoveries, and in 2017, one of them ignited my curiosity. In my efforts to learn more about the abyss, I had been tracking the science, reading books, watching documentaries. I pored over any deep-sea news, so I was primed to notice when this headline came along: "Search for MH370 Unveils a Lost World Deep Beneath the Ocean."

*

Like millions of others, I'd followed the story of Malaysia Airlines Flight 370's disappearance with a sinking heart and a long list of questions, and I believed that, eventually, those questions would be answered. How could they not be? To simply *lose* a jumbo jet and its 239 passengers was inconceivable. But as the years ticked by without a resolution, and pieces of MH370's torn wings washed up on East African beaches, it was evident that the Indian Ocean had claimed it—and she doesn't answer to us.

To make matters more complicated, the plane was thought to have gone down in the Indian Ocean's remote southern reaches, where gale winds rake across waters up to twenty-three thousand feet deep. On March 8, 2014, the night MH370 vanished, nobody knew much about that area—except that it wasn't friendly. The only hints of what the seafloor might look like were models derived from satellite altimetry: by measuring gravity's effects on the sea surface, you can infer the depth and contours of the underlying terrain. (Above a large seamount, for instance, the surface bulges perceptibly; above a trench or other depression, it's subtly lower.) Blurry and low-resolution, the gravity models were more like estimates than facts. To have any hope of pinpointing the jet's wreckage, better information was needed.

What followed was the biggest, deepest, hardest, longest, most technically ambitious and expensive deep-sea search ever conducted. For 1,046 days the searchers scoured the abyss with robots and high-resolution sonar, creating crisply detailed three-dimensional maps of the deep across an area the size of New Zealand. They found four shipwrecks, and identified one as the *West Ridge,* a 250-foot British barque that was lost with all hands in 1883. Now it lay thirteen thousand feet down.

Although they didn't locate the plane, the search had been a true *katabasis*—a recovery mission into the underworld. Like the Greek hero Orpheus, the searchers had descended far to retrieve what they sought, and then had to return to the surface without it. But a *katabasis* is never a wasted journey: it always offers a dose of the phenomenal. According to the news article that caught my eye, this one was no exception.

The maps revealed that the Southern Indian Ocean seafloor was spectacularly, eerily beautiful. It was a symphony of extremes, a playlist of geology's greatest hits. It was as if we had discovered Tolkien's Middle-earth, four miles underwater. There were mountains taller than Swiss Alps, valleys that dwarfed Yosemite, yawning crevasses, vertical cliffs that plunged into chasms. The seabed was slashed with scars left by the supercontinent Gondwana when it broke into pieces, creating Australia, India, and Antarctica.

A hundred million years ago during Gondwana's long crackup, a gargantuan volcanic rift in the seafloor opened up like a zipper. Fiery magma from the earth's mantle poured into it, swelling into a huge mass of igneous rock that morphed and changed shape over time, rising and falling and twisting and tearing as two tectonic plates pulled apart. This split resulted in a 750-mile-long fracture zone flanked by a serrated escarpment that plummets seventeen thousand feet—passing six canyons on the way down—bottoming out in one of the Indian Ocean's deepest spots, the Diamantina Trench.

If you were to walk across the Diamantina Trench your feet would sink into soft sediment, the accumulation of millions of years of marine snow—bits and flecks of dead sea creatures, tiny skeletons and tinier shells, plankton, bacteria, organic waste, silt, and these days, microplastics—that slowly wafts down from above. Not all of the bottom is oozy, however.

In one adjacent area geologists counted 154 volcanoes, and 17 of them topped thirty-three hundred feet, which qualifies them as seamounts: full-fledged, freestanding mountains. Many were encircled by moats that had been dug by surprisingly strong currents—and as currents sweep by, they circulate food. In the deep, hunting for something to eat is a full-time job, and the seamounts serve as oases that attract a riot of unique species. Creatures would be everywhere in this terrain: wedged into crevices, blossoming on the rocks, swimming around, burrowed into sediment. And south of these peaks was another epic feature: the Geelvinck Fracture Zone, a three-thousand-foot-deep cleft in the seabed that runs so straight for so long, it looks like someone drew it with a ruler.

*

It was an undersea network of national parks for giants and until now, no one had known it existed. Reading about this fantastical seafloor, I found it impossible not to wonder: *What else are we missing?* What else is stashed in the abyss—and why aren't we dropping everything to find out? How much lost history is down there? How much knowledge? How many geological marvels? How many undescribed species? What kind of wild rumpus goes on below while we're preoccupied topside with our leveraged buyouts and political bickering and selfie apps?

Nothing would be too unbelievable. In the deep, there are creatures that breathe iron and creatures with glass skeletons and creatures that communicate through their skin. Some of its creatures can turn themselves inside out. They might have two mouths or three hearts or eight legs. Or their bodies might consist of a thousand little bodies, a coordinated army. At least one deep-sea creature squirts yellow light. Some have see-through heads. Even the most ethereal among them can handle pressures that would crush a Mack truck.

The deep's tiniest residents are the earth's mightiest biological force. These are microbes—bacteria, archaea, protists, and viruses—the single-celled organisms that power life's laboratory. They convert chemicals into energy, recycle carbon, supply oxygen, turn waste into nutrients, and chew up toxins, among countless other feats. We wouldn't be here without them. In the ocean, their numbers are so astronomical that we need to borrow a word from cosmology to estimate them: nonillion, or 10^{30}. If all 3.6 nonillion marine microbes were gathered up and put on a scale, they would account for 90 percent of the ocean's biomass. (Scientists don't know how many species they represent—possibly as many as a billion.) There were microbes gusting from seafloor vents back when multicellular life was only a glimmer in the primordial eye. They've managed to thrive in brutally harsh conditions for eons. By studying their resilience, we've discovered new antibiotic and antiviral medicines, new biomaterials, new compounds for treating cancer, and new diagnostic tests, including the one used for COVID-19.

There's nothing anthropocentric about any of this; it's a full cast of aliens. But the deep's *otherness* is the essence of its enchantment, and it deserves to be appreciated on its own terms. "To sense this world of water known to the creatures of the sea we must shed our human perceptions of length and breadth and time and place," Rachel Carson wrote, with her usual poetic clarity. It would also be helpful if we could set aside our terrestrial bias, the mistaken belief that everything important happens aboveground, because that's where we live.

In fact, our survival depends on the ocean. The more we've delved downward, the more we've had to revise our ideas about how the earth operates, how the climate behaves, what we can learn from the distant past, our place in the overall scheme of life—even our definition of life. Now it's apparent that nature runs as a massively interconnected system, with the deep sea as its motherboard. Yet even as we tinker with the machinery in potentially irreversible ways, we have only the foggiest notion of how it all works. The deep buffers our excess carbon (at least so far), drives the ocean's circulation (and thus, climate), regulates the earth's geochemistry (important, to put it mildly), and absorbs surplus heat (ditto)—to cite just a few of its services. Humming away in obscurity, it's the foundation of the planet.

This underworld is a thrilling mystery, but you'd never know it.

In general, our culture is far more interested in space. For every dollar the National Oceanic and Atmospheric Administration spends on ocean exploration and research, NASA gets a hundred and fifty. We lavish billions on the prospect of colonizing Mars, a barren dust ball. The ocean's inner space is a harder sell, because humanity has the unfortunate habit of ignoring and fearing what it can't see. The solar system is directly overhead, visible through our eyes and telescopes, but the seafloor is beyond our immediate perception. To many, it's the earth's haunted basement—sinister, shrouded in blackness, spewing molten rock and poisonous gases, a den of freaky beings and hoary specters—and they would rather stay upstairs.

The idea of a descent into anything unnerves us. We descend into madness, into grief, into chaos. We fall into disrepute, fall from grace, and even worse, fall into oblivion. We're hardwired to look upward, to head toward the light. Heaven is up there, in our opinion. "If the stars should appear one night in a thousand years, how would men believe and adore; and preserve for many generations the remembrance of the city of God which had been shown!" Ralph Waldo Emerson rhapsodized about the night sky, adding that the heavens "awaken certain reverence, because though always present, they are inaccessible."

So is the abyss, on both counts, but it doesn't get the same adoration. Mountains, forests, rivers, ponds, trees, flowers, birds, clouds, stars: they've all been exalted in literature and poetry, in art, in music, in minds, in hearts. When the ocean makes an appearance, it's usually as a soothing backdrop or a stormy surface or a medium that reflects sunlight or moonlight in beguiling ways. On the rare occasion when the abyss is considered, it looms as a threat or a cautionary tale. In a word, it's abysmal. It's as if the deep were too remote, too frightening—too ugly to be lovable.

But what if we have it upside down? What if the deeper you go, the more astonishing everything becomes? To me that seemed like a compelling possibility, but there was only one way to know. I would need to make my own *katabasis*—to descend into the depths myself, and return with the stories I found there. I knew this was easier said than done, but I also knew that technology was moving fast, enhancing our ability to study the abyss, and even to view it in person. Autonomous deep-sea robots with artificial intelligence, manned submersibles like zippy little spacecraft, seafloor observatories wired to the Internet, scanners that sequence DNA in the water, new sonars, new sensors, new science: they're all here, at last. And more breakthroughs are coming.

As with any trip to an offbeat place, you can learn from those who've gone there before you. Certainly, I wasn't the only one drawn to the deep. A formidable group of explorers and scientists had left their own aquatic trails. Over the course of five years I would seek them out—and ultimately, the journeys I took with them would lead me to the day when I would be poised to dive into the heart of the Pacific abyss.

Some of these people were famous; others were quietly knowledgeable. All of them were intrepid. And they knew firsthand that to take such a plunge is to grapple with one of humanity's most intimidating and well-ingrained beliefs: *If you dare to go down there, you may not come back.* On that January day in 2021, as I stood on deck in the blustering wind, I was worried, but not about being lost in the deep. Despite the surly ocean conditions, despite the fact that the sub had experienced its share of mechanical problems, despite my awareness that we would be diving deeper than I'd ever dreamed of—I intended to come back. I had faith in the whole operation. At that moment, my only concern was whether I would get to go.

CHAPTER 1

Magnus's Monsters

Indeed, I should also add that monsters, some long-familiar, some
unprecedented, are sighted off Norway, and this is due particularly
to the unfathomable depth of the waters . . .

—OLAUS MAGNUS

UPPSALA, SWEDEN

If you were searching for world-famous deep-sea monsters, a stately
building at the top of a hill in Uppsala, Sweden, is not the first place
you'd look. But the monsters are here, behind the butter-colored façade
and tall windows of Uppsala University's oldest library, an institution
known as Carolina Rediviva. The university was established in 1477.
Uppsala, a charming city about an hour north of Stockholm, has been
around for even longer. It was a Viking stomping ground in the first
millennium, a hub of feisty Norse pagans who worshipped the gods of
thunder and wind and war, while enjoying the odd human sacrifice,
before Christianity moved in. There's a lot of history in Uppsala, but
I had come to see one particular relic: the *Carta Marina*, a sixteenth-
century illustrated map that depicts the North Atlantic, North Sea,
and Norwegian Sea regions—and the fiendish creatures that, according to the map's author, lived in those waters.

For as long as people have gazed out at the ocean, they've shivered at the thought of its silent inhabitants. The word *abyss*, translated
from its Greek roots, means "without bottom." What kind of ungodly

beast would find such a place hospitable? It was hard to imagine what such a thing might look like, though religion and mythology offered chilling descriptions and comparisons to Satan. So when someone created a map of the sea that featured portraits of these residents, it was sure to attract attention.

The *Carta Marina* was printed in 1539, and at a glance it was no boring document. Every inch of its twenty-three-square-foot surface was covered with intricate drawings, landmarks, labels, directions, and notes written in a cramped Latin script. The map was packed with its era's latest intelligence about natural history, geography, marine life, ocean conditions, navigation, shipping routes, and local customs. It charted Scandinavia—an isolated part of the world at that time—with unprecedented accuracy. But the reason I flew to Sweden to see the *Carta Marina* is because it's a 480-year-old snapshot of the prevailing fears and beliefs about the deep ocean. Beyond its cartography, it's a map of perceptions.

In an age before science, before deep-sea exploration, before high-definition underwater cameras, what people overwhelmingly believed about the deep was that it was filled with monsters—and the *Carta Marina* made their presence official. It portrayed their malice in striking detail. From Greenland all the way to Norway, menacing creatures are shown lolling in the water, wreaking havoc on ships, devouring sailors, and in general behaving badly, having risen from the hellish pit of the abyss, and intent on dragging their victims back down there.

Adding to the map's authority, it came from a trustworthy source. Its creator, a Catholic priest and historian named Olaus Magnus, was born in Linköping, Sweden, in 1490. Magnus led a cosmopolitan life, attending university in Germany and traveling widely throughout Europe. For a while he was based in Poland. He roamed the northern countries as a papal ambassador, collecting fees for the Church and soaking up information: firsthand observations, stories from villagers, maritime insights from fishermen and sailors, and regional gossip seasoned with a dash of medieval superstition.

People have always been scared of the deep, but in Magnus's time it evoked an extra degree of terror. In the moody North, the ocean was often visibly furious, and from shore it would've appeared endless, its edges unknown. Ships would leave and never come back. Mariners vanished into its maw, sinking into an underworld that was crawling with demons like Leviathan and the Kraken. Aside from the unlucky souls who'd visited the abyss on a one-way trip, no one had ever seen it; no one knew anything about it. To the average person it was a tabula rasa of doom. But even back then, how could anyone not wonder what was hidden there?

*

We'll never know when or where the earliest efforts to learn more about the deep occurred—somewhere in ancient Oceania would be my guess—but in Western culture, the Greek philosopher Aristotle is credited as the first marine biologist. He dissected any aquatic creatures he could get his hands on in the fourth century BCE, spent years studying their behavior in a lagoon, and presented his findings in a book called *Historia Animalium* (*History of Animals*). Among other observations, Aristotle noted that cuttlefish change color when they're startled, and that female sharks tend to be larger than male sharks. He figured out how lobsters copulate. He also discerned that whales and dolphins are mammals, classifying them in a group he named *Cetacea*, from the Greek word *kētŏs*, or sea monster.

The Roman historian Pliny the Elder followed in 77 CE, publishing a thirty-seven-volume encyclopedia that contained notes, musings, and wild conjecture about marine life. Unlike Aristotle, who relied on empirical evidence, Pliny spouted sea monster lore, holding forth about three-hundred-foot-long eels, man-eating octopuses, and fish the size of islands. In fact, he declared, most of the deep's inhabitants were "of monstrous form." He wrote of submarine hurricanes that stirred up the ocean "from the very bottom, and the monsters are driven from their depths and rolled upwards on the crest of the billow." Aristotle's work was substantial and rigorous, but Pliny's tall tales were more popular. Throughout the Middle Ages, while Aristotle's anatomical diagrams were mostly forgotten, Pliny was considered the main authority.

But still the deep remained far out of reach. It was shrouded in mystery, a veiled supernatural realm. With the *Carta Marina*, Magnus intended to illuminate it. He began with a conclusion, gleaned from his research: "Inside this broad expanse of fluid Ocean, which admits the seeds of life with fertile growth, as sublime Nature ceaselessly gives birth, a conglomeration of monsters may be found."

Magnus's timing was excellent. It was the Age of Discovery and Europeans were curious about everything, especially if it was unusual or ferocious or exotic. There was a hunger for astonishment: wonders, marvels, and terrors were the order of the day. Ships sailed to far-flung shores and sent news of remarkable sights. The unexplained, the awesome, the magnificent, the freakish; rumors of a lion with a man's face in India, of babies born with inverted feet; reports of pygmies, werewolves, and an assortment of what one book, *Marvels Described*, written by a French monk named Friar Jordanus, referred to as "sundry monstrous appearances"—all of these riveted the public. ("If there were cats with wings in Malabar, as he had seen, why should there not be people with dogs' heads in the Islands of the Ocean?" Jordanus's translator asks, somewhat defensively, in the foreword.) The world was expanding—helped along by Johannes Gutenberg's invention of the printing press—and apparently it was a colorful place. In Italy, the Renaissance was in full bloom; Leonardo da Vinci had drafted designs for an underwater breathing apparatus. He was cagey about the details: "I do not wish to publish this because of the evil nature of men, who might use it for murder on the seabed."

As a cleric and a scholar, Magnus took everything he learned and filtered it through the Bible, Aristotle, Pliny, and Ptolemy's *Geographia*, a second-century CE stab at a global atlas that included calculations of latitude (not bad) and longitude (way off: it would take until the mid-eighteenth century to get that right). He also digested classical texts, with their vivid descriptions of Scylla, Charybdis, and the Hydra, before beginning his twelve years of work on the *Carta Marina*. And though Magnus couldn't have foreseen it, his map would become iconic. It would reign as the gold standard for sea monsters long after he was gone, reflecting our dread of the deep back at us in the most delightfully frightful way.

*

I walked up the hill to Carolina Rediviva on a flawless September morning, to view one of the two original prints of the *Carta Marina* that have survived the centuries. (The other one is in Munich.) On the way there, I wound through cobblestone streets and weaved among students wearing Fjällräven backpacks and passed by rune stones that were taller than me, carved with mystical shapes and inscriptions from the eleventh century. "The stone in memory of . . . soul," explained one translation placard. Trees with silvery leaves rustled and shushed in the breeze.

I was excited to see the map in person so I'd arrived at the library early, ten minutes before it opened. I stood on the front steps and looked around. In this low-slung town you couldn't miss the Gothic spires of the Uppsala Cathedral: they speared four hundred feet into the sky. This hulking redbrick building had played a bitter role in Magnus's life. It was the seat of the archbishop of Uppsala, a post bestowed on him by Pope Paul III in 1544. But it was merely an honorary title: Magnus would never occupy the office. By then he had been exiled in Italy for seven years. The Reformation had turned Sweden into a Protestant country.

He'd finished the *Carta Marina* in Venice. It was a supersize map, measuring four feet tall by five and a half feet wide, printed crisply on nine woodcut blocks. Nobody knows how many first-edition prints were made—probably not many, because after 1574 the originals had all but vanished. Fortunately, the map's popularity had spawned a smaller-scale copy in 1572, and that version stayed in circulation. (No original was seen again until 1886, when one turned up in a German library's cache of old maps. In 1961, a second original was found in Switzerland and acquired by Uppsala University.)

Magnus had also written lengthy treatises in German and Italian

about the *Carta Marina*'s contents, and a handy reference key to its
nine panels—but he still had more to say. He spent the next sixteen
years completing a 778-chapter doorstop of a book called *A Description of the Northern Peoples* that elaborated on the map's drawings,
and included a long, exhaustively footnoted section devoted to its sea
monsters. Magnus died in Rome in 1557, two years after the book's
publication, so he never witnessed its afterlife: twenty-two editions
were published in six languages. It was a hit, a Renaissance bestseller
that's still in print today.

Carolina Rediviva's doors opened, and I made my way into the
entrance hall. Inside, there was a quiet grandeur: soaring ceilings
and windows, catwalks linked by spiral staircases, crown moldings
done in Swedish neoclassical style. The reading rooms were spacious
and warmly lit, their shelves lined with hardcover volumes bound in
muted colors. The effect was sophisticated, as though an art director
had stopped by to organize the books.

The *Carta Marina* is on permanent display in a rare manuscripts
exhibit off to the side of the entrance. Sunshine is no friend to antique
paper, so the room was as dark as a cave. Spotlights were trained on
the individual pieces: Mozart's handwritten musical scores, first editions of Newton's *Principia Mathematica* and Darwin's *On the Origin
of Species*, Galileo's letters defending his theory of sunspots. I stopped
for a moment to let my eyes adjust. Then, at the back of the room, I
saw Magnus's monsters.

Mounted in its frame and sheathed by protective glass, the *Carta
Marina* takes up an entire wall. I'd seen reproductions of it, but I was
still stunned by the intensity of its detail. The map is printed in black
ink on thick ivory paper without any addition of color, and its lines are
so finely rendered they seem to have been drawn with a pin. Everywhere you look the map is buzzing with activity, but Magnus clearly
wanted to make a point. On land the action is orderly: tiny figures are
farming, hunting, skiing, playing the violin. By contrast, the ocean is
in chaos, awash in dangers and tragedies, livid with waves and currents flowing, swirling, pooling, seething. Amid the tumult, twenty-
five sea monsters make their appearance.

At eye level near the Faroe Islands, and almost as big as them, a monster with a round face, a daggerlike dorsal fin, and spiky claws is frowning as it gobbles a seal. Magnus had labeled it "the Ziphius." "It has a ghastly head, like that of an owl; a very deep mouth, like some vast chasm, with which it terrifies and puts to flight anything that sees it; dreadful eyes; a back that is tapering, or rather, raised into the form of a sword; and a sharply-pointed beak," he explained in his book. Beside it, another monster with a prominent snout and what looks like a bad case of acne is sinking its hooked teeth into the Ziphius's flanks. "These creatures frequently approach northern coasts like sea-robbers or ill-disposed visitors, and purpose harm to all who cross their path," Magnus warned.

Off the coast of Norway, a grizzled beast with a hump on its back grapples with a mutant lobster, while above them a hapless ship is being squeezed to splinters by the "Sea Orm," one of Magnus's most daunting specters. This is "a serpent of gigantic bulk, at least two hundred feet long, and twenty feet thick," he'd noted, and in case that doesn't sound awful enough: "It has hairs eighteen inches long hanging from its neck, sharp, black scales, and flaming red eyes."

Magnus stresses repeatedly that he's relaying eyewitness accounts, and that one of his sources, another archbishop, had even obtained a sea monster's head, preserved it in salt, and sent it to the pope: "The whole head of this creature, which is like very hard leather studded round with horns, is extremely heavy, perhaps because Nature has so designed it for quicker submersion." Throughout his writings Magnus comes across as meticulous, and his book is stuffed with citations to underpin his claims, so it's likely that the head of *some* large marine beast was delivered to the papal residence. On the other hand, Magnus also reported that in Iceland the streams run with beer.

I was standing so close to the map that I'd fogged the glass, so I stepped back and regarded it from a distance, only to be drawn right back in. Magnus's descriptions may be outlandish, but the *Carta Marina* is no cartoon. Its monsters were rooted in reality—and then subjected to an epic version of the telephone game during which they morphed into bigger, meaner, and more fantastical versions of themselves, until finally the garbled, exaggerated story reached Magnus, who committed it to print: "Paulus Orosius declares in his life of Caligula that in the fifth year of his reign a monster extending to a length of nearly four miles sprang from the sea's abyss . . ."

Many of Magnus's monsters resemble whales—albeit more diabolical looking—and it's obvious from his texts that he was familiar with the animals. He knew they gave birth to live young, breathed air "through tubes," and contained oil "most abundant in the head." But there seems to have been some confusion about which creatures were whales and which were another breed of monster. Magnus believed, for instance, that the male narwhal—a shy, deep-diving Arctic whale with a protruding spiral tooth—was the "Unicorn-Fish . . . a sea-monster with a huge horn on its forehead; with this it can pierce and wreck oncoming ships and kill a large body of men."

During his travels, Magnus would have met coastal villagers who'd encountered stranded whales and other denizens cast up from the deep, bloated and bedraggled and bitten and, in all fairness, looking pretty sinister. What was a medieval farmer supposed to think when he stumbled across the body of a fifty-foot sperm whale, with its boxcar head full of seven-inch teeth? Without context or knowledge there would've been horror. Magnus cited one onlooker's impression of a stranded baleen whale: The "monster" was ninety feet long, with "thirty throats" and "stupendously large" genitals, the man reported. "Attached to its palate were what looked like countless horny plates, hairy on one side . . . and no teeth, a fact which leads people to conclude that it was *not* a whale."

*

Fear can take strange forms, so it's not surprising that our fear of the ocean abyss—a realm that Carl Jung compared to the madhouse of the subconscious mind—has produced some of the strangest forms of all. From the start, we've imagined the worst. In the absence of any personal experience with the deep, we've stocked it with apparitions from our minds' darkest corners. It's the oldest archetype in the human storybook: the monsters, the freaks, the *others*—the beings we don't recognize, so we react to them by recoiling. Aristotle set out to demystify them; Pliny was determined to amplify them. Magnus's genius was to get them all down on paper.

It was hard for me to tear myself away from the *Carta Marina*, even though I'd been staring at it, in the dark, for two hours. What I loved most about the map was its exuberance. Even as he was adding horns, fangs, and scowling expressions to make his sea monsters look more fearsome, Magnus couldn't disguise his enthusiasm for them. "Such great wonders have their position in the huge extent of Ocean that even a person with surpassing talent can hardly describe them," he raved. It was that ageless mix of fear and fascination; that clash of attraction and repulsion. The biologist Edward O. Wilson summed it up in a sentence: "Even the deadliest and most repugnant creatures bring an endowment of magic to the human mind."

I left Carolina Rediviva with a handful of *Carta Marina* post-cards and a *Carta Marina* coffee mug and the *Carta Marina* seared into memory, and I wished I could tell Magnus that his visions were still mesmerizing, almost five centuries later. For decades after Magnus's death, his monsters were copied extensively on other maps. But as seafaring became more common, and colonies cropped up on foreign shores, it occurred to European monarchies that the ocean was some-thing more than a den of trouble—it was a seemingly limitless place to make money. By the mid-seventeenth century, drawings of hor-rible whalelike predators threatening ships were replaced by images of ships engaged in whaling. Sperm whales no longer had tusks and blaz-ing eyes: they had harpoons stuck in their backs. The monsters were gone, reduced to candle wax and lamp oil. The ocean still commanded respect, but our relationship with it was changing.

To time-travel through the seventeenth and eighteenth centuries—and their seismic shifts in our understanding of the natural world—all I had to do was walk a few blocks around Uppsala. From Carolina Rediviva's front lawn I could see the Gustavianum, a four-hundred-year-old museum that's a quirky shrine to the Enlighten-ment. Among its collections are the version 1.0 tools of the Scientific

Revolution: microscopes and telescopes, medical instruments, Anders Celsius's first thermometer. There are barometers (invented in 1643), and sextants (1731), and chronometers (1735), and beneath a domed roof on the top floor, the world's second-oldest anatomical operating theater, a standing-room-only gallery where medical students and steel-stomached citizens could watch surgeries performed on convicted murderers.

If you were a budding natural scientist, eighteenth-century Uppsala was a lively place to be. In 1728, there was a new arrival at the university: Carl Linnaeus. Aside from the fact that he was broke, Linnaeus wasn't your average student. At the university's botanical garden—right around the corner from the Gustavianum—he distinguished himself as a prodigy of plants. In childhood he'd memorized their Latin names, which at the time were long, baffling mouthfuls based on subjective descriptions. (A cactus, for example, was "the large melon-thistle with fifteen angles and broad recurved thorns, which are of a red color"—and it was even more convoluted in Latin.) Linnaeus became a professor of botany and medicine, and then vaulted into history by establishing the taxonomic system that is still used today to classify living things—a universal format that allows anyone, anywhere, to pinpoint where an organism belongs on the tree of life, and to identify its origin by genus and species. Over the years, taxonomy has been honed into ever more exact distinctions, and genomic sequencing has advanced the whole game, but thanks to Linnaeus we now know the orca, for instance, as *Orcinus orca*, from the family Delphinidae in the suborder Odontoceti of the order Cetacea, and not by Magnus's preferred name, "the Grampus," or Pliny's description: "an enormous mass of flesh with savage teeth."

All of this represented a great leap forward, a sea change that shunted hearsay and wacky ramblings about monsters off to the side, and replaced them with cool, rational assessment. Although science was still entangled with religion—nasty battles with the Church lay ahead—and the word *scientist* wouldn't even come into use until 1834, the post-Magnus era was a time of measuring, debating, questioning, and reasoning; of developing theories and proving them. It was a heady time of figuring things out.

As the afternoon sun glinted through a scrim of clouds, I left what is now called the Linnaeus Garden—an assemblage of plants identified by his handwritten labels—and headed back into town to check out a Viking-themed pub where you could drink a horn of mead and eat roasted deer meat with your bare hands. As I passed the Uppsala Cathedral, I realized that I'd come full circle. When Linnaeus died in 1778 he was buried here, and his tombstone, set into the church's floor, marks the spot where Magnus's career was thwarted. The last archbishop of Uppsala had never taken the helm in this cathedral, and ultimately it would be scientists armed with new technologies—not priests toting sketchbooks—who were destined to uncover the ocean's secrets. Magnus, it seems, had been aware of his own limitations. "In the deeps of the sea there are species of fish which never, or very rarely, reveal themselves to men's eyes," he wrote, sounding disappointed.

The next wave of ocean investigators had no intention of waiting for these shadowy deep-sea life-forms to reveal themselves. They wanted to hunt through the depths and find them. It wasn't enough to gawk at marvelous, monstrous creatures that happened to expire on a beach. Nineteenth-century naturalists wanted to collect specimens from their habitats, place them under a microscope, understand their life histories, how they functioned, what they ate. And they wanted to conduct physical experiments that would answer some basic yet confounding questions. How deep was the ocean anyway? What was the seafloor made of? What were conditions like at the very bottom? Was it even possible to methodically examine a realm that existed entirely in the dark? Under miles of water? The abyss was unknown—but was it unknowable?

*

There was no question that studying the deep would be vexingly difficult. Anyone hoping to pry knowledge from the ocean's underworld would need to spend months, even years, at sea, eating hardtack biscuits and drinking lime juice to avoid scurvy, wedging themselves into tight bunks aboard ships that pitched and rolled and occasionally sank. They would have to endure cabin fever, tropical diseases, heavy weather, and everything else fate could dish out. Accidents and injuries were a given; deaths were to be expected. On his voyage aboard the HMS *Beagle*, Charles Darwin suffered through five years of seasickness. "It is no trifling evil which may be cured in a week," he advised. Darwin disliked maritime life for other reasons, too. He complained of "the want of room, of seclusions, of rest; the jading feeling of constant hurry; the privation of small luxuries, the loss of domestic society and even of music . . ." The ocean, he concluded, was "a tedious waste."

There was also the issue of equipment: easy-to-use, reliable gear was nonexistent. Soundings to gauge depths were taken regularly from ships, but the devices were primitive and the method was inexact. Mariners would drop a weight attached to a long line, watch as it plummeted into the darkness, and try to detect when it hit the bottom. Then they'd slowly recover it, marking the distance in fathoms—a unit of measurement based on the length of a man's outstretched arms (about six feet). Some sounding devices were designed to grab sediment samples when they touched down, but the only way to obtain a larger hunk of the seafloor—and have any chance at catching animals—was by dredging with an iron-jawed scoop or trawling with a weighted net, both of which also had to be lowered on a line from an unsteady ship in an environment that was constantly moving. These contraptions couldn't be maneuvered with any kind of finesse. They failed often: lines got snagged and tangled and snapped. Their operations demanded hours of monotonous labor even in shallower waters. Working below three hundred fathoms would be a backbreaking production for anyone masochistic enough to try. In its 1823 edition, the *Encyclopaedia Britannica* threw in the towel, declaring that: "Through want of instruments, the sea beyond a certain depth has been found unfathomable."

But the story of human ingenuity always demands a next chapter, and by the mid-nineteenth century there was a pressing reason to learn more about the seafloor: so we could lay telegraph cables across it. Once again the deep was commercially important; sending messages by homing pigeon and packet ship wouldn't do in the age of electricity. Faster communications would also mean military advantage, not to mention bragging rights, for any nation that managed to install a live electrical wire under fifteen thousand feet of salt water. What followed was a decade-long technological experiment at the bottom of the Atlantic Ocean, as Great Britain and the United States collaborated (and competed) to run a two-thousand-mile cable from Europe to America. Where to put it? How to secure it? What might gnaw through it in the depths? Would currents sweep it away? Expertise was urgently needed.

As deep-sea research evolved from a pipe dream into a reality, the typical person who would light out on a seafloor surveying expedition was a European man—women need not apply—from a privileged background, with a polished education, academic ambitions, and connections to similarly pedigreed men. The fledgling field of marine science was a gentlemen's club, and unlike sailors, its members were not known for their high tolerance of misery. But the abyss was a wide-open frontier, a bonanza of opportunity. *Everything* in the deep was new to science; careers could be made on a single voyage. It was, in one naturalist's words, "the only remaining region where there were endless novelties of extraordinary interest ready to the hand which had the means of gathering them."

Meanwhile in nature-crazed Victorian England, people were enthralled by any aquatic discovery. Collecting seashells was considered a stylish hobby. Home aquariums, known as "ocean gardens," were a status symbol among the upper classes. "The wonders of the ocean floor do not reveal themselves to vulgar eyes," sniffed a how-to book for setting one up. Marine life transcended the ordinary, even when confined to a tank. Sea anemones were like "rare exotic flowers," the book's author waxed, before injecting a darker note: "Yet they are not flowers, but animals—sea monsters, whose seeming delicate petals . . . seize the unconscious victim as he passes near the beautiful form—fatal to him as the crater of a volcano, in which he is soon engulfed by the closing tentacles." The last chapter was an impassioned pitch for a new form of entertainment, "a Titanic Aquarium" measuring "hundreds of feet in depth," in which sharks would be "exhibited in deadly conflict with human divers, armed with net and trident."

In this go-go time of ocean enterprise, there were some notable ironies. For one thing, even as marine science was flourishing, sea monsters made a strong comeback. Suddenly, it seemed, everybody had seen one—especially in Gloucester, Massachusetts, where more than a hundred people claimed to have watched a "sea-serpent" capering in the town's harbor. Boaters and fishermen also encountered it offshore. Sightings were so frequent that the Linnaean Society of New England launched an official inquiry, taking sworn depositions. The creature was dark brown, snakelike, about eighty feet long, and as thick as a barrel, witnesses agreed. It swam with an up-and-down motion like a caterpillar, and may or may not have had a stinger on its forehead. A naturalist by the fabulous name of Constantine Samuel Rafinesque-Schmaltz even proposed a Linnaean genus and species for the serpent: *Megophias monstrosus.* "Great efforts will be made to take it dead or alive," vowed one local man. But the *monstrosus* evaded capture.

No matter though, because the deep was coughing up monsters everywhere. British mariners observed one with a giraffe's neck and terrifying teeth off the Cape of Good Hope, and a fifty-footer that resembled a frog in the Malacca Strait, and a sleek beast with a bullet-shaped head near Sicily. There was a run of Scandinavian sightings that would've made Magnus proud. "During my late passage from London I saw no less than three sea-serpents," a man named J. Cobbin wrote in a journal called *Annals and Magazine of Natural History.* The most impressive of them, he recalled, was "at least 1,000 yards long," with a head that flared like a cobra's, and glowing eyes. Reading his account, you have to wonder if someone spiked the sherry on Cobbin's crossing. The monster "swam with great rapidity and lashed the sea into foam, like breakers dashing over jagged rocks," the man insisted. "The sun shone brightly upon him, and with a good glass I saw his overlapping scales open and shut with every arch of his sinuous back, colored like the rainbow."

Even some eminent scientists were believers. Their support was largely due to the unearthing of fossils (another Victorian sensation) from ancient marine creatures, including an ichthyosaur and two plesiosaurs found at an English seaside town. If these finned dragons had once prowled the deep, who could say they weren't still down there? "There is no a priori reason that I know of why snake bodied reptiles from 50 feet long and upward should not disport themselves in our seas as did those of the Cretaceous epoch, which geologically speaking is a mere yesterday," the British biologist Thomas Henry Huxley argued in defense of sea monsters. "I can no longer doubt the existence of some large marine reptile allied to Ichthyosaurus and Plesiosaurus yet unknown to naturalists," the Swiss zoologist and Harvard professor Louis Agassiz agreed.

British paleontologist Richard Owen, the man who coined the word *dinosaur*, was a loud dissenting voice. "The sea saurians of the secondary periods of geology have been replaced in the tertiary and actual seas by marine mammals," he countered testily, pointing to the "utter absence" of any recent monster remains. Owen dismissed the sightings; he was convinced the hysterical public had misidentified whales, sharks, eels—possibly even a gigantic seal. "A larger body of evidence from eyewitnesses might be got together in proof of ghosts than of the sea-serpent," he huffed.

But the idea that the abyss harbored prehistoric creatures wouldn't be swept aside so easily. Owen himself had examined a chambered nautilus—a deep-dwelling ancestor of octopus and squid—and pronounced it "the living, and perhaps sole living, archetype of a tribe of organized beings, whose fossilized remains testify their existence at a remote period, and in another order of things." With their striped spiral shells, button eyes, and mop of tentacles, nautiluses were exhibit A: proof that a species from the age of the dinosaurs could survive to the present. The deep was such a strange lost world, so seemingly removed from the extinction events and evolutionary pressures that were known to have occurred on land, that it seemed logical for throwbacks akin to living fossils to be skulking down there. We just hadn't found them yet.

Then in Norway, somebody did. In 1864, naturalist Michael Sars and his son, Georg Ossian Sars, dredging at three hundred fathoms (eighteen hundred feet) in a fjord, hauled up an archaic animal known as a stalked crinoid. It was hardly a monster; in fact, it resembled a delicate flower, with a slender stem and feathery petals (which are actually its arms). Crinoids, also called sea lilies, were abundant in the fossil record and presumed to have died out a hundred million years ago—but here was a thriving specimen that seemed to have defied time.

The crinoid's appearance was startling. And furthermore, it shouldn't have been found at that depth. *No* creatures should have been found at that depth. Because in another irony, just as the abyss started to win people's attention, one of the world's most respected naturalists, Professor Edward Forbes from the University of Edinburgh, declared that at three hundred fathoms and below, the deep was "azoic," or lifeless. There was nothing down there at all.

*

Forbes wasn't the first to present the abyss as a wasteland. The Greek philosopher Socrates had delivered this unfavorable verdict: "Everything is corroded by the brine, and there is no vegetation worth mentioning and scarcely any degree of perfect formation, but only caverns and sand and measureless mud, and tracts of slime . . . and nothing is in the least worthy to be judged beautiful by our standards." A French naturalist named François Péron supposed that temperatures were so cold at the bottom that the seafloor was sealed by a thick layer of ice. Others believed that the deep's immense pressures compacted the seafloor into a cement-like block—and that the water above it was so stagnant and ultradense that no object was heavy enough to sink all the way down. Everything that dropped into the ocean would end up suspended at some intermediate depth. Shipwrecks, dead bodies, monster remains (which is why we couldn't find them), telegraph cables: they would all float through the abyss for eternity, as if lost in space. At great depths sounding weights would hit a false bottom, unable to fall any farther.

Of course there was no evidence for any of this, but coming from Forbes the azoic theory took on the air of fact. Forbes was smart, energetic, connected. He attended meetings, chaired committees, founded associations. He had money but he wasn't snobbish. He liked to joke around. At age twelve he published his first book and opened a natural history museum—curated from his own collections of bugs, rocks, and fossils—in his family's home. At university Forbes set out to study medicine, but quickly switched to zoology, geology, and paleontology. In his thirties he became a fellow of the Royal Society, Britain's most prestigious scientific academy, and served as the president of the Geological Society of London, and the head of natural history at the University of Edinburgh. His reputation was unrivaled.

Forbes was gung ho about ocean exploration, and something of a dredging expert, though his experience was limited to coastal waters. When he got the chance in 1841 to join a surveying expedition in the Mediterranean Sea, he jumped at it. He was keenly interested in the distribution of marine life across various depths: what lived where, and why. It was vital, he wrote, "that Britons, whether scientific or unscientific, who boast at all fitting occasions of their aptitude to rule the waves, should know something of the population of their saline empire."

During his eighteen months at sea, Forbes dredged at depths up to 230 fathoms (1,380 feet)—with frustrating results. Mostly, what surfaced was mud. Later it became obvious that his dredge was a big part of the problem. Its scoop was too small, with a narrow mouth and poor drainage, which caused it to clog with sediment. Also, the area he was working in was an aquatic desert, although nobody knew that at the time. To Forbes, who loved dredging so much that he wrote a song about it—"Down in the deep, where the mermen sleep, our gallant dredge is sinking!"—it was a blow to end up with such a skimpy haul.

But then Forbes the brilliant scientist made a wild extrapolation. If he hadn't found life at depth, perhaps that was because there was none to find. He put forth a theory that ensured his place in ocean history—although not in the way he would've hoped. "As we descend deeper and deeper in [the ocean]," Forbes famously wrote, "its inhabitants become more and more modified, and fewer and fewer, indicating our approach towards an abyss where life is either extinguished, or exhibits but a few sparks to mark its lingering presence."

This was news to Michael and Georg Sars in Norway, who had found at least nineteen species in the supposedly lifeless abyss. And to the polar explorer Sir John Ross, who as early as 1818 had brought up worms and a spectacular starfish from deep in Baffin Bay. And to Ross's nephew, Sir James Clark Ross, who'd sailed to Antarctica on an expedition in 1839, dredged up corals, jellyfish, worms, and crustaceans from four hundred fathoms, and concluded that the seabed was "teeming with animal life."

All of these discoveries were ignored or discounted. It didn't help that John Ross had kept sloppy records, and that some of James Clark Ross's specimens were eaten by his ship's cat. The Sarses' results were harder to explain away, but the scientific community embraced Forbes's theory because it made perfect sense. The deep was hostile, impenetrable. As far as anyone knew, there was no sunlight, no oxygen, no food, and no habitable environment due to the colossal pressures. To write off the abyss as a dead zone was a tidy solution to a mind-bending mystery: how life could thrive without any sign of those ingredients.

*

Forbes passed away in 1854, but his azoic theory didn't. It hung around like a musty odor, even as animals kept appearing at inconvenient depths. In 1860, a ship called the HMS *Bulldog* dropped a sounding line to 1,260 fathoms (7,560 feet) in the North Atlantic; the line came up with thirteen brittle stars clinging to it. Brittle stars are relatives of starfish, with attenuated spidery arms that allow them to scamper across the bottom and wrap themselves around objects. Often, creatures yanked from the seafloor didn't survive the journey to daylight, leading to uncertainty about whether they'd died in shallower waters and merely sunk to the bottom. But these brittle stars were alive, and they wriggled their long arms on deck. The *Bulldog*'s naturalist, an ambitious man named George Wallich, was elated. He'd set out to make a name for himself by proving Forbes wrong—and to his mind, he'd succeeded. "This sounding far exceeds in importance any previous sounding on record," he wrote in self-congratulation.

Wallich published his findings in 1862 and then waited for the accolades to roll in, the invitation to speak at the Royal Society, admission into the scientific elite. Instead, his brittle stars were met with skepticism. Wallich hadn't caught them with a dredge, people griped, so there was no guarantee they'd come from the deep. They could have been snagged anywhere. Wallich argued and fumed and wrote long rebuttals in florid Victorian prose, but no one was buying it. (He had also dented his own credibility by claiming to believe in a mythical undersea kingdom known as "the Sunken Land of Buss.") Deep-sea creatures with a better provenance had emerged off Sardinia, where a broken telegraph cable was retrieved from twelve hundred fathoms down. It had been lying on the seafloor for years. The cable was spackled with sediment and a menagerie of corals, clams, starfish, snails, and other animals that had colonized it like a reef. There was no denying this one.

Yet doubts remained: the azoic theory wasn't quite dead yet. Maybe the threshold just had to be lowered? "As to the point at which animal life ceases, it must be somewhere," wrote a puzzled academic. Other investigators, looking at seafloor mud under a microscope, noticed a filmy goo that appeared to be moving. It was everywhere, creeping through every sample, and nobody had ever seen anything like it. They named it *Bathybius* and wondered if it was a protoplasm that served as the foundation of life—the original primordial ooze. Or maybe it was a food source? Or both?

So was there zero life in the deep, or was the deep essentially a life soup? A Scottish naturalist named Charles Wyville Thomson was following the debate closely. Thomson was in his thirties and influential. He was a professor of zoology and botany at Queen's University Belfast, but he would soon follow Forbes's path to a natural history post at the University of Edinburgh. He'd traveled to Oslo to see the Sarses' crinoid and a sponge that was collected at three thousand feet off Japan, a specimen so bizarre that people suspected it was some kind of hoax. It looked like a milky-white tulip trailing long filaments of glass coiled into a thin rope. Any marine scientist examining it now would recognize it as a *hexactinellid,* or glass sponge—an intriguing, primitive animal that can live for more than ten thousand years. Today's materials engineers study glass sponges and take inspiration from them. Their fractal-like skeletons are made of silica; they can transmit light with more efficiency and durability than a fiber-optic cable. Back then, of course, no one knew any of this. In the annals of deep-sea science, it was a confusing time.

Thomson was convinced that more dredging needed to be done, at greater depths, with better equipment. In 1868, he and a colleague, Royal Society vice president William Carpenter, asked the British Admiralty to provide support for a six-week expedition. An old steamer called the HMS *Lightning* was available, mainly because it was falling apart. But it had a small auxiliary engine that could be used to lower and raise the dredge, and a captain and crew who were game to shoot for the abyss. Thomson and Carpenter loaded the ship with specimen jars, vats of alcohol to use as a preservative, a new type of sounding device called "the Fitzgerald," and some state-of-the-art dredges, and sailed north toward Greenland in August 1868.

Cranky weather plagued them throughout the voyage, including a storm that ripped off part of the ship's rigging. Leaks spurted everywhere. They took soundings and other measurements, but on most days the waves upended their deep-dredging plans. At one point the *Lightning* took shelter in the Faroe Islands, docking near a fishing boat that was transporting live Atlantic cod back to England. The cod were corralled in a saltwater well, and Thomson went over to take a look. "It is curious to see the great creatures moving gracefully about in the tank like gold-fish in a glass globe," he wrote in his journal. What struck him as even more curious, though, was how curious the cod were about the humans who had entrapped them. Whenever people approached the enclosure, the fish would swim up to their audience, staring quizzically. They seemed to enjoy being petted. The sailors were especially fond of one cod that had been injured in the net; they swore the fish recognized them. "Certainly it was always the first to come to the top for its chance of a crab or a bit of biscuit," Thomson observed, "and it rubbed its head and shoulders against my hand quite lovingly."

Perhaps in some faint way, Thomson clocked this as a hint that life in the ocean was far more nuanced than anyone had dreamed. Fatefully, it was the start of a series of events that would make him one of history's most renowned deep-sea scientists. The journey that lay ahead would be profound, but at that moment in 1868, things didn't look promising. The *Lightning* was limping through the North Atlantic with its mast hanging by a thread, two dredges had been lost in the depths, and the trip was shaping up to be a bust.

But the weather settled down near the end of the expedition, and they were able to dredge below five hundred fathoms, and those few dips brought up animals that were worth all the trouble. A fiery-orange sea star with curlicue arms. A primitive mollusk called a brachiopod. A ruby-red squat lobster with coppery eyes. Several glass sponges like the specimen from Japan, one of which was woven into a nest as fine as filigreed lace. Thomson and Carpenter found a universe of microorganisms in the gray seafloor ooze, single-celled protists with shells shaped like globules, commas, tubes, pinwheels, starbursts. (There was no sign of *Bathybius*, but you can't have everything.) The results were encouraging enough to earn them a hardier ship, more months of dredging, and the chance to go deeper. The following spring, they returned to sea in the HMS *Porcupine*.

Their luck changed aboard this new vessel. The weather calmed, the equipment worked, and they dredged as deep as 2,435 fathoms (14,610 feet). Week after week, the dredge rose from the abyss loaded with splendid goblins. Thomson sieved through the muck, picking out new species, including a skeleton shrimp—a glistening, mantis-like creature that looked more like a stick than an animal—a pink starfish as pristine as a jeweled brooch, a giant sea spider with spindly legs as long as a human's. One day, as the dredge was being recovered, Thomson spotted a scarlet sea urchin jammed against its side. He feared that its shell would be crushed, but the urchin remained intact. When he examined it, he was startled to see that it was panting like a dog. He stared at its rows of tube feet, its sharp blue teeth, its array of tiny spines, its round body heaving in and out. "I had to summon up some resolution before taking the weird little monster in my hand," he admitted.

The deep released a kaleidoscope of beings, many of which shone with bioluminescence. "In some places nearly everything brought up seemed to emit light," Thomson wrote. Even the mud sparkled. Worms glittered like gems. Gorgonian corals radiated a soft white glow. Plumy sea pens flickered like lilac flames. Brittle stars were lit with neon green. Far from being a black hole, the depths were ablaze with fireworks.

Using new pressure-tolerant thermometers, Thomson and Carpenter took temperature readings near the seafloor, and found that they fluctuated. The abyss was not frozen solid, and it was not inert— currents were moving down there, independent of the surface conditions. By now they'd completely disproved the azoic theory, though amazingly it persisted. "We concluded that probably in no part of the ocean were the conditions so altered by depth as to preclude the existence of animal life—that life had no bathymetrical limit," Thomson wrote. "Still we could not consider the question completely settled."

In the aftermath, Thomson and Carpenter ended up with the fame that George Wallich had sought with his brittle stars, much to Wallich's fury. He spent the rest of his career hurling accusations of plagiarism at the two men, referring to them as "the conspirators." By the end of his life, he'd become a downright crank. In Wallich's papers, historians would later come across manic doodlings of Carpenter hanging from a noose.

Carpenter probably didn't notice Wallich smoldering in the background. He was too busy; he had another expedition to plan. He and

Thomson had been given the approval for a massive exploration: a three-and-a-half-year global circumnavigation with every resource imaginable, courtesy of Her Majesty Queen Victoria's government. The goal: to survey the deep thoroughly. To search for new species, living fossils, reclusive monsters, and anything else that might be concealed in the abyss. To decipher the cryptic *Bathybius*. And to hammer the last nail into the azoic theory's coffin.

Carpenter, who was sixty years old, bowed out due to the length of the journey, but Thomson was slated to lead a five-man scientific team, supported by one of the best captains in the Royal Navy, twenty-three officers, a crew of 240, and a two-hundred-foot warship powered by steam engine and sail, set up specifically for dredging, kitted out with research laboratories, and refitted to carry 180 miles of hemp sounding and dredging rope instead of its usual seventeen cannons. The ship's name was the HMS *Challenger*.

*

Mariners are a superstitious bunch, always on watch for bad omens, so it wasn't seen as an auspicious start when one of the *Challenger*'s crewmen drowned before the ship had even left the dock. He'd returned to the vessel at night and accidentally stepped off the pier, and he sank like a rock and never resurfaced. Divers retrieved his body the next day. And then it was time to cast off on a bold journey that seemed to the sailors like a blind scavenger hunt. They were accustomed to serving on ships that steamed across the ocean in pursuit of territory or trade or war—but the *Challenger*'s only grail was knowledge about the invisible world below.

The first two weeks were unruly. They'd left Portsmouth, England, on December 21, 1872, and run straight into the teeth of a gale. The *Challenger*'s sails were torn, its jib boom washed over the side, one of its lifeboats was smashed. The scientists moaned with seasickness in their cabins, as furniture and dishes flew around. Along with Thomson, there was Scottish-Canadian naturalist John Murray, and a young German biologist, Rudolf von Willemoes-Suhm. Naturalist

Henry Moseley was English, chemist John Buchanan was Scottish, and the expedition's artist, John James Wild, was Swiss. They'd all been handpicked for the voyage by Thomson and the Royal Society, but as the ship bucked in the angry seas, they may have regretted accepting the invitation.

Initial attempts at dredging resulted in thousands of fathoms of rope being lost, but eventually they brought up a smattering of crustaceans, which the sailors referred to as "insects." The *Challenger* steamed on—past the bobbing wrecks of other vessels capsized by the storm—and pulled into Lisbon, its first foreign port. The ship was greeted by the king of Portugal, the British ambassador, and other dignitaries; it was the height of the empire and diplomatic niceties had to be observed. The sailors ran for the nearest bar, quaffing large quantities of Madeira and griping about the herculean workload. Now they had experienced the reality of dredging: the endless hours and grueling physicality of it, the tedium, the mess, the lack of commensurate pay. The grumbling would continue. Dredging soon became known as "drudging."

Next the ship headed to Gibraltar, drudging and sounding along the way. The captain suggested dragging the trawl net, rather than the dredge, across the bottom. The trawl had a twenty-foot mouth, so the equipment change was effective (if destructive), and deep-sea fish began to appear in the net. In a letter to his family, the assistant steward, eighteen-year-old Joseph Matkin, described the fish as having "heads larger than their bodies, and eyes in the middle of their backs and several other peculiarities." Sublieutenant Lord George Campbell—who happened to be the son of a duke—proclaimed them "ghastly objects." One of the fish was a rattail, a goggle-eyed distant cousin of the cod, with a brief snip of a tail fin and a body that tapers sharply to a point. Rattails are regular customers in the abyss, so successfully adapted they come in hundreds of species, but they also carry the deep's imprimatur of weirdness. They could never be mistaken for a plain, garden-variety fish, and it's likely that upon seeing one, the sailors woke up to the fact that this was no routine fishing trip.

Off the Canary Islands Thomson hit pay dirt, pulling up a slinky, eel-like fish called a halosaur, along with a shrimp as transparent as crystal, a violet-hued brittle star, a dazzling glass sponge, obscure species of corals and urchins—and a stalked crinoid. He celebrated by popping champagne for the entire crew. They had started to find their rhythm. Like clockwork, day after day, instruments went over the side: thermometers, trawls, dredges, sounding devices, and sampling bottles with valves that popped open at various depths so Buchanan could analyze the water chemistry. Measurements were plotted on graphs and entered into logbooks; specimens were pickled in alcohol and scrutinized in the lab. As they crossed the Atlantic, their soundings revealed a vast mountain range that ran down the center of the ocean basin like the backbone of a gargantuan armored beast. Slowly the abyss unfolded beneath them.

Then, in the West Indies: tragedy. The dredge got stuck on the bottom and the line tightened, shearing off an iron block that swung wildly, fatally striking a deckhand. He was buried at sea, the captain intoning, "We commit his body to the deep." And it *was* deep: 3,875 fathoms (23,250 feet). In Bermuda, another shipmate died from a stroke. A twenty-foot tiger shark followed the *Challenger* for days, freaking out the sailors, who believed that its appearance portended more deaths. (And they were right.)

They surfed the Gulf Stream north, studying that majestic artery of warm water, its turbulent eddies and cobalt colors and quicksilver moods. The trawl brought up a worm the size of a snake. When they arrived at Halifax, the town was busy recovering the bodies of more than five hundred passengers from the White Star Line's RMS *Atlantic*, a luxurious ocean liner that had rammed into a submerged rock. (In 1912, another White Star ship, the *Titanic*, would follow it to the seafloor.) The weather turned dreary and drippy and cold. In what would become a recurring problem, several *Challenger* sailors deserted.

South again, dropping off a gravestone in Bermuda for their two men, and recrossing the Atlantic to the Azores. "Dredge, sound, trawl; sound, trawl, dredge; and so on all the way," one officer complained in his diary. At night, the waves winked with bioluminescence. "There was light enough to read the smallest print with ease," Campbell observed. "It was as if the 'milky way,' as seen through a telescope, scattered in millions like glittering dust, had dropped down on the ocean, and we were sailing through it." They hauled up a black seadevil anglerfish with a luminous lure dangling from its forehead; and an eyeless crab with sea anemones growing on its back; and a four-foot-long pyrosome, a fat, glowing, gelatinous tube that's actually a colony of organisms known as zooids. The sailors called it a "blubber-fish."

Geological novelties surfaced as well. The dredge often returned full of black lumps that, in deeper waters, apparently carpeted the seafloor. They ranged in size from gumballs to potatoes to grapefruits. Murray cut them open and inspected their composition. Inside, they were layered like tree rings, with concentric circles of manganese, iron, cobalt, copper, nickel, and other metals that had accreted around a scrap of seafloor debris, like a fossilized shark's tooth or a whale's ear bone or a nub of coral. The genesis of these nodules was a mystery, but the process must have been infinitesimally slow, like the growth of a pearl. Compared to the deep's flamboyant life-forms, the lumps looked dull. It wouldn't have occurred to anyone aboard the *Challenger* that in the not-too-distant future, nations would be seeking the mining rights to them.

*

Over three and a half years they sailed 68,890 nautical miles. Their route took them to the edge of Antarctica's Great Ice Barrier, into Indonesia and the Philippines, and along both coasts of Africa and South America. In Australia, a place the sailors adored, there was a stampede of desertions. During a rough passage through New Zealand's Cook Strait, they lost a man overboard. They dodged cyclones, smallpox, icebergs, cannibals. They stopped in Hong Kong and Japan (where they admired the geishas) and wound through the lush islands of the South Pacific (where ten men contracted syphilis). In

Papua New Guinea they were chased by natives paddling war canoes. They brought on board two giant Galápagos tortoises, which trundled around the deck eating pineapples. Between Hawaii and Tahiti, biologist Rudolf von Willemoes-Suhm died of a bacterial infection.

On one memorable occasion, three hundred miles off Japan, they sounded a depth of 4,575 fathoms (27,450 feet). The measurement was such an outlier that they repeated the sounding. It was correct: what gaped below them gave fresh meaning to the word *abyss*. "This is the greatest depression as yet ascertained—the deepest wrinkle, so to speak, on the face of old Mother Earth," Wild, the artist, wrote in his memoir. These days we know it as the Mariana Trench.

With four months left in the expedition, they sailed through Chile's Strait of Magellan and found themselves back in the Atlantic, on their last leg home. "So, hey for the *Challenger*, and may we soon see the last of her!" Campbell hurrahed. You have to suspect the others shared his opinion. "It is possible even for a naturalist to get weary of deep-sea dredging," noted Moseley, who was also fed up with another aspect of shipboard life. In his cabin, he'd been tormented by insects—moths, ants, flies, crickets, spiders, mosquitoes, and in particular, an aggressive cockroach. "He bothered me much," the naturalist recounted, "because when my light was out, he had a familiar habit of coming to sip the moisture from my face and lips, which was decidedly unpleasant." Moseley took his revenge by shooting the cockroach with an air gun. "At last I triumphed," he wrote.

The ship returned to England on May 24, 1876, crowds lining the docks to cheer them. Throughout the years, Thomson had dispatched magazine articles so the public could follow the team's progress in serialized form. After Jules Verne's blockbuster novel *Twenty Thousand Leagues Under the Sea* was released in 1870, the abyss was all the rage. "The sea is merely the medium which supports the most fantastic, prodigious forms of existence," Verne's antagonist, the obsessive Captain Nemo, declared—and now here was the *Challenger*, pulling into port with the receipts: nearly five thousand never-before-seen species, sediment samples from across the seafloor, and enough raw data to keep scientists busy for the next twenty years. "Let us hail the work of those who lifted a piece of the thick veil that covered the abyssal depths of the ocean," toasted the scientific journal *Nature*.

But the real work was only beginning. Thomson and Murray set up an office in Edinburgh to compile the expedition's results. They enlisted seventy-six specialists from around the globe to help create what would eventually become a lavishly illustrated fifty-volume report encompassing everything known to date about the deep ocean. The *Challenger* Report made all that came before it obsolete. So much conventional wisdom and so many preconceived notions had turned out to be wrong.

Now it could be said emphatically: the deep sure as hell wasn't azoic. And it wasn't a hideout for prehistoric creatures that had managed to avoid evolution. They'd found a few anachronistic characters—including a live nautilus and a *Spirula:* an archaic squid with an internal shell—but these were few and far between. Also, nothing they'd captured in the abyss qualified as a monster, although some of the fish wouldn't win a beauty contest.

As for the distribution of life, it was everywhere. While Olaus Magnus had pictured marine beasts gliding into their personal undersea lairs like the Batmobile returning to its cave, and Edward Forbes believed that each species had its own headquarters, which he called its "center of creation"—the *Challenger* team found similar animals roaming the depths no matter where in the world they looked, even when geological features formed natural barriers that should have kept them apart. And far from dwindling away to "a few sparks," to quote Forbes, in some cases abyssal life-forms grew to gigantic proportions.

Bathybius, however, was a dud—and a bit of an embarrassment. Two years into the expedition, Buchanan had cracked the code of this "organism." It wasn't life's matrix, burbling on the seafloor. In truth, it wasn't alive at all. It was merely the residue of a chemical reaction between salt water and the alcohol used to preserve specimens. Left bottled up for long enough, those two components precipitate a film of calcium sulfate.

Thomson died suddenly in 1882, probably due to exhaustion,

so like Magnus he didn't live long enough to guess how enduring his deep-sea contribution would be. Murray steered the project to its completion in 1895. Reviewers praised the *Challenger* Report as a "momentous enterprise," "a scientific drama," and "a stupendous undertaking." The journal *Science* called it "the most magnificent contribution to natural science, and monument of enlightened research, which has ever been given to the world in any age or by any country." In the torrent of compliments I found only a single criticism, with one reviewer carping that the expedition "seems to have needed a live harpooner, for it got no porpoises during the whole voyage, though many played about the ship."

Even today it's hard to overstate the *Challenger*'s success. What a legacy, what an odyssey, what a mother lode of work. It transformed our understanding of the deep, and left tantalizing clues for future investigators. Instead of a map filled with monsters, Thomson and his team had given them a compass—and fifty volumes of reasons to keep seeking knowledge in the depths.

Thomson described the deep as an "enchanted region," but that very enchantment had perplexed him. The elegance of a glass sponge, the symmetry of corals, the intricacy of a lobster's stalked eye: to him, they all raised the same question. *Why?* Why would nature create something so exquisite, and then put it in a place where nobody would ever see it? Thomson found himself amazed at "the recklessness of beauty, which produces such structures to live and die forever invisible, in the mud and darkness of the abysses of the sea."

Of course, he was forgetting that the deep's beauty was fully accessible to the creatures that lived there. To them its pulsing lights and phantasmagoric shapes and jewel colors are as present as flowers, birds, and stars are to us. Thomson's lament was more specific: that the abyss was "forever invisible" to human eyes. He was half right. People couldn't witness the glories of the deep. But our blindness was only temporary.

CHAPTER 2

Aquanauts

These descents of mine beneath the sea seemed to
partake of a real cosmic character.
—WILLIAM BEEBE

WAIMANALO, OAHU

S hould we go see the sharks?" Terry Kerby asked, treading water
beneath the Makai Research Pier. This was a rhetorical question.
Of course we were going to see the sharks. Before I could answer he
was gone in a hail of bubbles, weaving through wooden pilings and
arrowing twenty feet down to the seafloor. I adjusted my goggles,
took a deep breath, and followed him. Kerby was close to seventy
years old, but to watch him free dive you'd never guess it.

We popped up about fifty yards away, clear of the gauntlet of fishing lines hanging from the pier. To our left, volcanic cliffs framed
Oahu's eastern shore. To our right, the Pacific Ocean ran uninter-
rupted to Baja California. By Hawaiian standards it was a drab day,
with stern clouds overhead and a brisk wind giving the water a bouncy
chop. I knew that didn't matter much to Kerby. Rain or shine, in perfect calm conditions or in the face of approaching hurricanes, he swam
the same two-mile circuit every day at lunchtime—a routine he'd
observed for the past forty years. To commute from his desk to the
ocean, all he had to do was climb down a ladder: Kerby's workplace,
the Hawaii Undersea Research Lab (HURL), occupied most of the

pier. While other people stepped out for sandwiches, Kerby was traversing Waimanalo Bay, clad in a black shorty wet suit, scuba mask, and fins. "It's a spiritual thing," he told me.

It's also unsurprising: Kerby is one of the most aquatic souls I've ever met. In his role as operations director and chief pilot of the *Pisces IV* and *Pisces V*, HURL's two deep-sea submersibles, he had spent thousands of hours roving the Pacific depths. On Kerby's résumé there were no stints in an office building, no gigs that involved clock punching, nothing that remotely resembled an average job. In fact, throughout his career, none of his employment had even occurred on land.

In his twenties he served as a U.S. Coast Guard navigator, stationed on ships in Alaska, the Caribbean, and the Gulf of Mexico. He'd also worked as a salvage diver in the cold, bleak waters of Northern California, fending off great white sharks and prying metal from shipwrecks: "That was my suicide period." He was a stuntman in James Cameron's classic deep-sea movie, *The Abyss*, for which he spent twelve hours a day immersed in a three-story water tank; and a shark wrangler on the James Bond film *For Your Eyes Only*, a post that required Kerby to crouch inside a sunken shipwreck gripping a tiger shark, and then shove the animal into the frame when the director called for action. "That was an experience," he recalled wryly. "Tiger sharks aren't too interested in being in the movies."

Hawaii had always figured large in his imagination, and in 1976 Kerby landed on Oahu and stumbled across his destiny. "I was driving around the island, past the pier," he recounted. "There was a crane picking a submersible out of the water, and I thought, *Oh, there's my dream.*" As a boy, he'd inhaled science and adventure books, especially tales about the ocean. He'd thought about how cool it would be to have something like the starship *Enterprise*, but underwater.

The submersible gleamed in the sunlight. It was the *Star II*, a goofy-looking yellow vessel that resembled a kid's oversize bath toy, but could dive to twelve hundred feet. Kerby pulled over to watch and ended up talking to the crew. They were shorthanded, they told him, because two members of their team had been killed in a recent accident. Kerby joined on the spot.

He started out operating the launch system; before long, he was training as a pilot. Kerby had grown up in the High Sierra, where his family ran a construction company. He'd learned to drive heavy machinery as a teenager; the *Star II*'s complexities didn't faze him. One day another pilot announced that he was terrified of the sub and never wanted to set foot in it again, and Kerby got his shot at the controls.

Now, some nine hundred dives later, he was a part of a rarefied group that I'd come to think of as the "aquanauts"—human emissaries to the deep. They were upside-down astronauts, mavericks who had plunged into inner space at the first opportunity. In his decades of submersible piloting, Kerby had witnessed underwater scenes the HMS *Challenger* scientists couldn't have conjured in their dizziest fantasies. He'd logged dives that would've made Jacques Cousteau keel over in a fit of envy.

When I learned about Kerby and HURL and the *Pisces,* I'd reached out immediately. Their base on Oahu was only a thirty-minute flight from Maui, where I lived. It felt like serendipity to have deep-sea submersibles and their pilots so close by. Aquanauts aren't exactly ubiquitous; as far as vocations go, it would be fair to call it an outlier. Earlier, when I'd interviewed a marine engineer named Will Kohnen who manufactured subs and ran a committee to oversee their safety, he stressed that it was a tight-knit, eccentric field. The industry's annual meeting, Kohnen told me, was "like a dysfunctional Thanksgiving dinner." "So what draws people to it?" I asked. "Oh," he replied with a chuckle, "because they're just demented."

*

We stroked across the bay for another few minutes, in the lee of an islet that was shaped, appropriately, like a shark's dorsal fin, and then Kerby stopped and pointed down at a rocky rise on the seafloor. "That's their cave," he said. "Let's see how many are home." (Kerby's record for shark encounters on this swim was ten; he drew little shark symbols in his calendar to keep a tally.) I watched as he lay on top of the rocks, pulling himself over the edge and peering into the cave's mouth from above. I wasn't sure if the best way to view sharks was to back them into a closed space and then position your head in front of the exit, but Kerby stayed down there for a while and then resurfaced, smiling. "There are three of them," he said. "Take a look." I dived, and saw a trio of whitetip reef sharks piled beneath the overhanging lip, sleeping on the sand with their eyes open.

We swam on, past another group of fishermen angling from shore, and a homeless encampment that sprawled along the beach. Often Kerby came upon fish, sea turtles, and other creatures ensnared in gill nets and nylon lines that people had cast and abandoned. He carried a pair of wire cutters in a waist pack and would stop, hovering underwater, to snip the animals free. On two occasions he'd removed a hook from a shark's mouth. He also picked up trash along his route. "I discovered what looked like a whole campsite dumped in the ocean," he told me, shaking his head in dismay. "Clothes, towels, blankets, rugs, hammocks, tarps—all scattered over a live coral colony."

Kerby took such despoilment personally, the way you would feel if someone turned your garden into a junkyard. "I remember the first time I saw the ocean it had such an impact on me," he said. "It was as if I'd come home." Like others drawn to a frontier—in Kerby's case, a saltwater one—he'd gravitated west and arrived in the right place, at the right time. In 1980, when HURL began its operations as a joint venture between the University of Hawaii and the National Oceanic and Atmospheric Administration (NOAA), his piloting experience made him an obvious hire. HURL's mission was broadly ambitious: to study the depths of the Pacific Ocean.

For Kerby, this was nirvana. It was also far too much work for a lone submersible. He found the *Pisces V* in Scotland and the *Pisces IV* in Canada, and had them shipped to Hawaii. After refurbishment, the subs were certified to dive to sixty-six hundred feet. The university also bought an oil industry ship at a bankruptcy auction, customized it as a platform for the subs, and named it the *Ka'imikai-o-Kanaloa*. Kanaloa is the Hawaiian god of the ocean; *Ka'imikai*, translated, means "Heavenly Searcher of the Seas." The only question that remained was where, in the Pacific's sixty-four million square miles, to start searching.

*

After our swim, Kerby gave me a tour of HURL's headquarters, a weatherworn building that resembled a small airplane hangar. The front of the structure was open and I could see the two *Pisces* hunkered inside, thirteen-ton sea creatures temporarily stuck on land. They were twenty feet long, roughly the size of a minibus, set atop skids that enabled them to land on any type of seabed terrain. Their front and back ends were rounded; their tops were flat, with a fire-engine-red hatch tower poking up. The passenger compartment, known as the pressure hull, was a white sphere positioned up front. A viewport gazed from the center of each sphere like the pupil of a Cyclopean eyeball.

On the outside the subs bristled with high-definition cameras and sonars, lights, altimeters, laser-measuring devices, acoustic tracking systems, long banks of batteries. On their front bumpers they carried plastic crates stocked with sampling containers for water, gases, rocks, sediment, and marine life. "We have two hydraulic manipulators on each sub," Kerby explained. He pointed to one of them, a robotic appendage with multiple joints and a clawlike hand: "This is like an extension of your arm, it's so fluid." Working the manipulators in concert, a skilled pilot could pluck even the most delicate organisms and secure them in a jar.

Under their hoods, the *Pisces* contain ballast tanks that can take in or pump out air and water, as the pilot adjusts buoyancy throughout the dive. The goal, as with scuba, is to be able to rise and fall as needed through the water column, but to be neutrally buoyant on the bottom so it's easy to cruise around. Thrusters positioned on both sides of the pressure hull can propel the subs in any direction; the *Pisces* glide gracefully underwater despite their size and weight. Most of their bulk comes from blocks of syntactic foam—a buoyant, crush-resistant material made of glass microspheres in epoxy resin—that are padded around the frame. Each sub also carries four hundred pounds of steel shot. This ballast weight aids the descent; on the bottom, half of it is dropped. The remainder is released at the end of the dive. (The steel oxidizes on the seafloor, helped along by metal-eating bacteria.) In an emergency, the pilot can jettison all the weight to rise to the surface more quickly.

We had changed into our street clothes and Kerby was wearing his usual uniform: a HURL T-shirt, khaki shorts, hiking shoes, and white ankle socks. He is north of six feet and athletically built, with Nordic blond hair that flops lightly over his forehead and a perpetual tan. As he walked around the subs pointing out their features, he gave the impression of a doting parent. "There aren't many of these around," he said, resting his hand on the *Pisces V.* "It's a real coup to have them."

To be precise, in spring 2018, there were five other manned subs that could dive below six thousand feet. They were all owned by nations or institutions, rather than individuals. There was the *Nautile* from France, the *Shenhai Yongshi* (Deep-Sea Warrior) and the *Jiaolong* (Sea Dragon) from China, the *Shinkai* (Deep Sea) from Japan, and the U.S. Navy's research sub, the *Alvin*. Each of these vessels could go deeper than the *Pisces*. Their maximum depths ranged from 14,600 feet to 23,000 feet, which allowed them to fly throughout the abyssal zone. But the hadal zone, with its 35,000-foot trenches, remained largely out of reach. This was no minor omission. The hadal zone occupies less than 2 percent of the seafloor, but it accounts for 45 percent of the ocean's full depth.

Despite this limitation, these subs were impressive. They'd come a long way in a short time, given that manned deep-sea exploration was pioneered in 1930, and wasn't pursued with any rigor until the 1960s (and even then it was sporadic). This slow start wasn't for lack of interest: the itch for submergence can be traced back to antiquity. According to dubious legend, Alexander the Great was an aspiring aquanaut, descending three hundred feet in a glass cage and guided by an angel who asked him, "Dost thou wish me to show thee some of the wonderful things which are in the sea?" (After allegedly spending eighty days underwater, Alexander returned triumphantly and reported that he'd seen "many fishes.")

Inventors had long dabbled with designs for diving bells, and submarines like the *Drebbel,* a seventeenth-century craft that was essentially a sealed rowboat, bristling with a dozen oars. It was demonstrated in London in 1620, gliding fifteen feet beneath the surface of the Thames, but no plans or drawings of the *Drebbel* remain so it's unclear how the rowers were able to breathe. In 1775, an American sub called the *Turtle* made its debut, a one-man wooden barrel that was propelled by foot pedals and a hand crank, submerging and resurfacing with a crude system of weights and ballast tanks. To call the *Turtle* a prototype would be generous, but it became the first submarine ever used in combat. During the American Revolution it was deployed in New York Harbor in an attempt to attach a bomb to a British warship, but the mission was aborted when the *Turtle*'s pilot succumbed to carbon monoxide poisoning.

The early history of manned subs isn't short on precocious failures and elaborate contraptions. Most of them barely functioned in the shallows; none were even vaguely suitable for the deep. Jules Verne's fictional visions aside, before World War I the materials simply didn't exist. And even once the industrial technologies needed to build a deep submersible became available, nobody knew what sort of vessel would work. The abyss was hardly the place to risk human lives in a trial-and-error experiment. And who in their right mind would want to go *first?*

*

William Beebe wanted to go first.

He was a guy who loved birds, jungles, cocktails, expeditions, costume parties, museums, marine life, writing, science, and women—and not necessarily in that order. Beebe was an unlikely celebrity in the glamorous Roaring Twenties: a lanky, balding man from a middle-class New Jersey family who'd dropped out of Columbia University to work as a curator at the newly built Bronx Zoological Park. As a kid, his hobby was taxidermy. Basically he was a nerd, but an energetic one with a flair for storytelling. From his earliest days, Beebe was driven by a consuming passion: he was crazy about the natural world.

He wanted to study every animal. He wanted to learn everything about how every species lived and how they interacted. He kept lists of the creatures he'd seen, noting the subtlest details about their behavior, and made lists of the ones he wanted to see, charting his progress in voluminous personal journals. He liked to end his journal entries with the sign-off: "I have spoken!"

Beebe had a lot to say. He was a prolific writer, and his first magazine article, a vignette about bird-watching, was published in 1895 when he was seventeen. Even then, his curiosity and zeal were infectious. Beebe kept writing, his audience steadily growing. His job at the Bronx Zoo brought him into contact with wealthy patrons in New York City, where money was sloshing around as the stock market boomed, and Beebe found that he also had a knack for fundraising. The zoo needed animals, and he was the logical person to go out and get them. Backed by flush industrialists and sportsmen, Beebe embarked on a series of collecting expeditions, scouting for exotic fauna.

He sailed through the Caribbean and South America, scooping up parrots, iguanas, and tree snakes. He rode on horseback through Mexico, and trekked into the Himalayas, and wound through Malaysia by houseboat. In Borneo, hacking through tropical rain forests, he obtained flying lemurs, a pangolin, and a banded civet. He spent years globetrotting, gathering reams of material along the way for articles, research papers, and books. Beebe's voice was appealing. He'd created his own literary niche, a mix of travelogue, science, and adventure. By the time he turned forty he was one of America's most popular natural history authors, a respected (if self-taught) biologist, and a swashbuckling figure in New York society.

In the Galápagos he'd experimented with helmet diving, clomping along the seafloor in heavy copper headgear, tethered to the surface by a hose. (Scuba diving was still decades away.) Beebe reeled at the fantasia of colors and creatures found on a coral reef; his longtime interest in the sea flared into an obsession. He spent so much time underwater, pushing the limits of his improvised equipment, that his ears began to bleed.

Back in Manhattan, he charmed some financiers into giving him a research ship. Not long after that, Britain's Prince George offered him the use of a private island in Bermuda. It was the ideal site for Beebe's next project: he wanted to survey all life-forms within a single cube of ocean, measuring eight miles square and two miles deep. He'd dive in the shallows making observations, and trawl through the depths netting specimens. In the spring of 1928, Beebe decamped to Bermuda with his pet monkey and an entourage of assistants, and set up a research station.

The location was stunning, but Beebe quickly became frustrated by the limitations of trawling. Most of the creatures he netted came up mangled beyond recognition. Other animals, he suspected, were too fast or too savvy to get caught in the first place. This was a half-assed method, Beebe concluded; to get a true picture of deep-sea life, he'd have to go down there himself. Years earlier, he and a friend, President Theodore Roosevelt, had sketched submersible designs on a cocktail napkin, debating the mechanics of such a craft. Roosevelt had since died, but now Beebe wanted to move beyond hypotheticals. He envisioned being lowered a mile into the ocean in a cylindrical capsule that would encase his body like a phone booth. Knowing that media attention would help raise the cash to build it, Beebe had announced his plans: he would become the first man ever to study the deep in person.

On Manhattan's Upper East Side, a twenty-six-year-old engineering student named Otis Barton read the *New York Times* headline— "Beebe to Explore Ocean Bed in Tank"—with a jolt. Barton was a fan of Beebe's, but he took this news as a setback. Barton, an avid sailor and diver, had his own plans to become the first man into the deep. He, too, had devised a submersible. Maybe Beebe would agree to a partnership? Given Beebe's fame and Barton's obscurity, this was a big ask. But Barton had a major advantage: he was rich. He'd already engaged a top marine architect. Plus, he knew that Beebe's design invited disaster: it was the wrong shape. Only a perfect sphere could withstand the deep's pressures, because the force would be distributed equally across all angles. A cylinder would get crushed like a tin can. He reached out to Beebe to request a meeting.

When Beebe saw the blueprints for Barton's vessel, he was impressed. Beebe was nowhere in terms of execution, of a cylindrical sub or anything else, and he recognized the wisdom of building a sphere. Barton's design was brutal in its simplicity: a hollow steel ball with walls an inch and a half thick. It was just shy of five feet in diameter, with three viewports made of fused quartz, a superior-strength glass. If comfort was not a priority, two people could fit inside it. Entry and exit were through a fourteen-inch top hatch that would be sealed from the outside with bolts. A fitting on the hatch connected it to a steel cable so the sphere could be winched into the depths, where it would dangle from a ship like a two-ton Christmas tree ornament. Breathable air came from oxygen tanks; trays of soda lime and calcium carbonate powder would absorb carbon dioxide and moisture. *It could work,* Beebe thought.

They agreed that Barton would foot the bill and supervise construction of what Beebe had christened the *Bathysphere*, from the Greek word *bathys*, or deep. Beebe would provide a winch, a ship, a crew, and maximum publicity. Their descent would take place offshore from his island in Bermuda. The two men shook hands. They would both go first.

*

Kerby invited me to climb into the *Pisces V*, which meant ascending a tall ladder up the side of the sub and then shinnying down through the hatch tower. Inside, the ambience was spartan. Except for the rows of switches and dials and buttons and gauges that lined the control panels, the pressure hull itself hadn't changed much since Beebe and Barton's day: it was still a hollow steel ball. Each *Pisces* can hold three people. As pilot, Kerby kneels in the middle of the sphere, navigating with his forehead pressed to the center viewport. Two passengers—typically scientists—squash in on either side of him to observe through their own viewports. These days the viewports are spherical wedges of acrylic rather than plates of fused quartz—a big safety improvement. Acrylic can take far more abuse before shattering, and the wedge flexes inward under pressure, acting like a cork: the deeper the descent, the more impregnable the seal. The shape also acts like a fish-eye lens, providing a wider scope of vision.

Even with the hatch open it felt stuffy inside the sub, like there might not be enough air to go around. The life support system was still pretty retro. Oxygen is pumped in from tanks; carbon dioxide is scrubbed by canisters of crystals that turn purple when saturated. "We're always monitoring the oxygen content and the CO_2 content," Kerby explained. "Yeah, we can tell by the crystals."

Obviously there is no bathroom, and the dress code is simple: wear layers. The water temperature drops as the submersible descends, chilling the orb like a meat locker. "You start off in shorts and a T-shirt," Kerby said, "and by the time you're at six thousand feet you're wearing thermal underwear, socks, and a knit cap." I looked around the seven-foot sphere and tried to imagine three adults crammed into it for eight hours, the average length of a deep-sea excursion. Diving in the *Pisces* would be an unappealing experience for anyone claustrophobic, easily frightened, or weak-bladdered. You would have to pray for diving companions with nondescript body odor, and that nobody brought egg salad for lunch. For Kerby, any minor discomforts were swept aside by adrenaline. "Every time I climb in and close the hatch it's like the first time," he said. "It's still the same rush. It never goes away."

Sitting in the tiny chamber I got a sense of the fizzing anxiety Beebe and Barton must have felt on June 6, 1930, the day they made their inaugural dive. Bolted into the *Bathysphere*, swung out over the Atlantic, their six-foot-tall bodies folded like origami, the men could only hope they hadn't overlooked some crucial detail. There were so many things that could go wrong. Things could break, fail, implode. Any leak could be fatal: under the deep's pressure, water would shoot through even a pin-size hole with the force of a bullet. And if, God forbid, the cable became separated from the hatch or the winch, the *Bathysphere* would cannonball to the seafloor, and Beebe and Barton would find themselves two miles down in a ready-made coffin.

What if they ran out of air? What if the viewports caved in? Of the five quartz plates that Barton had commissioned, two cracked as they were being installed, and a third failed a relatively gentle pressure test. Only the two remaining plates made it onto the *Bathysphere;* the third viewport was plugged with a metal cover. The viewports were caulked around the edges with white lead paste—hopefully that would keep them watertight. It wasn't like anyone could consult an instruction manual.

Beebe and Barton were painfully aware that any openings in the sphere were weak spots that would be relentlessly probed by the water pressure—and now there were four of them. Along with the three viewports, Barton had drilled a one-inch hole where a power line and a telephone line entered the *Bathysphere,* sheathed in a rubber hose. The hose, which would wind into the depths alongside the steel cable, passed into the sphere through a stuffing box that would (fingers crossed) create a seal. Having electrical wires running through water into a small space that contained potentially explosive oxygen tanks was, to put it mildly, less than ideal. But without the lines there would be no communication with the surface, and no way to run their searchlight in the abyss. Neither of those items seemed negotiable.

The *Bathysphere* was lowered into the ocean. For the first sixty feet the scenery was familiar, the turquoise aquarium that Beebe and Barton knew from their helmet dives. On the topside end of the phone line Beebe's assistant, a willowy blonde named Gloria Hollister, relayed depth measurements as the cable slowly uncoiled from the winch. In turn, Beebe called up observations for Hollister to write down. At two hundred feet the red and yellow light wavelengths had ebbed, leaving only green, blue, and violet. When a crimson-colored shrimp darted by, it appeared to be jet-black.

At three hundred feet they stopped abruptly when Barton noticed water trickling in through the hatch. This cannot have been a pleasant moment, but rather than panicking or aborting the dive, Beebe asked for the *Bathysphere* to be lowered more quickly. He guessed, correctly, that increased pressure would tighten the seal.

Their descent continued, past four hundred, then five hundred, feet. Around six hundred feet, the electrical wire erupted with sparks. Barton lunged for the hose, jiggling it. The fireworks stopped. At seven hundred feet Beebe called for a pause to absorb his surroundings, calm his nerves, and reflect on the gravitas of where he was. "Only dead men have sunk below this," he noted to Hollister.

What really struck him was the uncanny quality of the light. Only the last pinch of the visible spectrum remained, its indigo embers brilliant in their purity. On the brink of utter darkness, the final traces of light appeared eerily radiant. "We were the first living men to look out at the strange illumination: And it was stranger than any imagination could have conceived," Beebe would later write. "It was of an indefinable translucent blue quite unlike anything I have ever seen in the upper world, and it excited our optic nerves in a most confusing manner." The blue was so piercingly vivid, he felt, that language couldn't describe it. It was more like an emotion than a color, one that "seemed to pass materially through the eye and into our very beings."

They had planned to continue to a thousand feet, but Beebe's inner alarm bells started jangling—"some mental warning which I have had at half a dozen critical times in my life"—and at eight hundred feet he called for the descent to stop.

On the way back up, with everything seemingly in order, Beebe turned his focus to the animals that flashed by his viewport. Little glinting fish and jellies floated and tumbled and jetted past, flitting in and out of the searchlight. Beebe resolved that on subsequent plunges, with the terrifying novelty of the first dive behind him, he would get serious about identifying species. When the *Bathysphere* was safely back on the ship's deck, and the bolts were unscrewed and the four-hundred-pound hatch cover was lifted off, the men crawled into the sunshine—which now paled in comparison to the deep's divine noctilucent blue—and Beebe pronounced the milestone: "The window to a wholly new world has been opened at last to human eyes."

*

Over the next four years Beebe and Barton made thirty-three more dives, eventually reaching 3,028 feet—more than a half mile down. The realm below six hundred feet, Beebe learned, was a pulsing night-club of life. Even Dr. Seuss couldn't have dreamed up the parade of creatures that danced by the *Bathysphere,* checking out the dead squid that Hollister had hung by the viewports as bait. "I sat crouched with mouth and nose wrapped in a handkerchief, and my forehead pressed close to the cold glass—that transparent bit of old earth which so stur-dily held back nine tons of water from my face," Beebe wrote.

He watched, mesmerized—"with every available rod and cone of both eyes"—as sawtooth eels with pincer jaws twisted their snaky bodies to get at the bait, and troupes of winged pelagic snails fluttered by like pixies, and six-inch dragonfish with dagger teeth and lumi-nous barbels swinging from their chins emerged as tiny frights in the inky water. A gang of silvery hatchetfish—palm-size predators with bulbous eyes and gaping mouths—glittered like tinsel. Little green-glowing lanternfish twinkled past the *Bathysphere* in a chorus line.

The deeper they went, the more dashes and streaks and arcs of bioluminescence flared like intergalactic signals. The *Challenger* expe-dition had established that many abyssal animals can illuminate them-selves, and Beebe had examined fish with photophores speckling their bodies—light-producing organs that look like neon dots—but he and Barton were the first to witness the deep's pyrotechnics. "I watched one gorgeous light as big as a ten-cent piece coming steadily toward me," Beebe recalled, "until, without the slightest warning it seemed to explode, so that I jerked my head backward away from the window."

Even more haunting were the stealth visitors that didn't adver-tise their presence with lights, advancing under cover of darkness. Beebe sensed the ghostly forms of larger beasts moving just beyond the scope of the searchlight. One time, at 2,450 feet, he caught a glimpse of a behemoth: "Twenty feet is the least possible estimate I can give to its full length, and it was deep in proportion." He had no idea what it was. And there were other cryptic fish, resembling no known species. Beebe was elated by these sightings and poured out stream-of-consciousness riffs about them, which Hollister dutifully transcribed.

He reported seeing a quartet of skinny fish—with vermilion heads, yellow bodies, peacock-blue backs, and the attenuated jaws of an alligator—hanging vertically and moving through the water upright, as though they were out for a stroll; and a creepy-looking dun-colored fish with oversize pectoral fins and no tail. Another fish was impossibly pretty, striped with purple and yellow photophores: "In my memory it will live throughout the rest of my life as one of the loveliest things I have ever seen," Beebe enthused. But the real show-stoppers were a pair of six-foot-long, lilac-lighted, barracuda-shaped fish that cruised by the *Bathysphere* with their mouths open, baring "numerous fangs." Beebe appointed these newcomers with fanciful names like the Abyssal Rainbow Gar, the Pallid Sailfin, the Five-Lined Constellation Fish, and the Untouchable Bathysphere Fish. In proper Linnaean form, he also translated the names into Latin.

*

Overall the dives went smoothly, but two near tragedies demon-strated that luck played a greater role than anyone would've cared to admit. Twice during unmanned test dives, the caulking around a viewport blew out and the *Bathysphere* was flooded—events that prompted Beebe to reflect how "in the icy blackness, we would have been crushed into shapeless tissues." By a spin of life's roulette wheel, the sphere had been empty—no harm done—and the viewport was repaired. Harrowing anecdotes like this made for spicy news copy, however, and Beebe and Barton were deluged with attention. They even did a radio broadcast from twenty-two hundred feet down. In August 1933, the *Bathysphere* was exhibited at the Chicago World's Fair, displayed next to the aluminum sphere built by the Swiss physi-cist and inventor Auguste Piccard, who'd sailed it 53,153 feet into the stratosphere, attached to a towering hydrogen-filled balloon. And they seemed to belong together, the upward-soaring sphere and the downward-diving sphere, as avatars of humanity's bold pioneering spirit.

The most vexing part of Beebe and Barton's project turned out to be the partnership itself. The two men were in a shotgun marriage, and neither of them liked it. Beebe was in his fifties and complicated, and his purpose for venturing undersea was intensely scientific. Barton was in his thirties and mostly wanted to be famous, like Beebe. He resented being called Beebe's "aide" by the media, and couldn't have cared less about the genus of an eel. Barton viewed Beebe as a bossy gadfly; Beebe considered Barton to be a passive-aggressive stiff. When they weren't jammed together in the *Bathysphere*, they avoided each other. After their last dive, they never spoke again.

Barton tried unsuccessfully to parlay his undersea adventures into a filmmaking career; Beebe went off to write another book, titled *Half Mile Down*, which became an instant bestseller. In *The New York Times*, a reviewer gushed: "This book ought to create a boom for the bathysphere. Henry Ford, or some other high priest of mass production, might wisely mobilize his plant." But there was no run on steel spheres. Even the original one was retired. Beebe soured on marine research after other scientists, especially ichthyologists, dismissed his *Bathysphere* dives as a stunt. They scorned his reports of strange fish. Beebe had no right, they argued, to name species based on a quick drive-by in the murky depths. The Five-Lined Constellation Fish— what the hell was *that*? Stung by the criticism, Beebe turned his attention back to the jungle.

The abyss carried on, unobserved, in its timeless fashion. But then in 1948, Auguste Piccard unveiled a new machine he'd been working on: a submersible. It was a successor to the *Bathysphere*, and a radical upgrade. To Piccard, the idea of being yoked to a cable was absurd. His vessel was autonomous, and he intended to take it far deeper than Beebe and Barton had gone. It was a hulking sub that, more than anything, resembled a zeppelin. Piccard called it a bathyscaphe, or "deep ship."

*

Kerby and I left the *Pisces V* and walked through the hangar to his office in a loft above the subs. HURL's décor could be described as man-cave chic, minus the chic. It was the ultimate garage—thousands of square feet of machinery, tools, workbenches, diving equipment, spare parts, and men in surf shorts tinkering with gear. Zodiacs were stacked on trailers. Dog-eared manuals were piled on shelves. An outboard motor hung from the ceiling. Ocean posters and magazine articles featuring the *Pisces* were thumbtacked to plywood walls. A fridge was plastered in stickers with a distinct undersea theme: the Deep Submersible Pilots Association, the Schmidt Ocean Institute, Poseidon USA, Micronesia Aquatics of Truk Lagoon. One bumper sticker of a rampaging shark boasted: "I've Been Seen by the Great White."

Kerby ushered me into his office, a well-loved space that was lined with mementos: pictures, awards, battered leather chests, scraps of coral and driftwood. He went to the kitchen to get us some coffee, and I settled in on a couch that was actually the torn-out bench seat of a car. I had about a million questions, and wanted to spend the rest of the day talking. Or rather, what I hoped was that Kerby would talk and I would listen, because I wanted to hear every last detail about his deep-sea experiences. To ask Kerby what he has seen in the abyss is to unleash a torrent of recollections, historical accounts, names of remote seamounts, GPS coordinates, facts, figures, dates, locations. He seemed to have total recall of every dive he had ever made. On top of that, he had thousands of photos taken from the subs, hours of video, and logbooks dating back to the eighties. Kerby, a talented artist, had even made paintings of his favorite undersea spots.

One site he knew intimately was Lōʻihi, the submarine volcano that is currently building itself into Hawaii's next island. It rises about thirteen thousand feet from its base to its summit, which lies nearly a mile beneath the ocean's surface. Like the other Hawaiian Islands, Lōʻihi was created by a hot spot: a plume of magma welling beneath the seafloor, and eventually bursting through. It's the handiwork of Pele, goddess of volcanoes and fire, one of the fiercest and most revered deities in the Hawaiian culture. At four hundred thousand years old, Lōʻihi is her youngest child, a little sister sitting at the feet of Mauna Loa, the world's largest volcano; Kilauea, one of the world's most active volcanoes; and Mauna Kea, another mammoth volcano that ranks as the world's tallest mountain (if you measure it from the seafloor). Scientists don't know exactly when Lōʻihi will grow tall enough to poke its head above the waves. Maybe a hundred thousand years from now—maybe more, maybe less.

"I did my first dive on Lōʻihi in 1987," Kerby said, handing me a mug and sitting down in his desk chair. "And I'm dropping down there in *Pisces V* going, *What am I doing diving on an active submarine volcano?*" No one knew if this was a good idea. There was no map to follow, no best practices to avoid getting buried by eruption debris or crushed beneath unstable lava shelves. Live undersea volcanoes are uneasy places, and Kerby was aware that he needed to approach this one with caution.

The submersible sank into the darkness on its white-knuckle reconnaissance, drifting down and down until it reached Lōʻihi's highest point, which would later become known as Pisces Peak. Trimming the sub, Kerby began to orient himself. He could see mounds of black pillow lava, and rust-colored mineral deposits that signaled the presence of iron, and strands of bacteria waving lazily in the current. Jumbles of rocks glistened with volcanic glass. It was a landscape of stark Plutonian beauty.

Suddenly, an immense pinnacle reared up in his viewport. It had to be a hundred feet tall. Chimneys sprouted from its sides, pumping translucent fluid. Kerby knew what to call the strange formation—a hydrothermal vent system—but vents had been discovered only a decade earlier, on the Galápagos's deep seafloor. Scientists were just beginning to study them and marvel at their weirdness. Like hot springs on land, hydrothermal vents pop up in volcanically active areas, roiling out a mix of seawater, minerals, gases, and microbes from the earth's superheated plumbing. When this brew hits the cold, deep water, it precipitates minerals that form chimneys of various heights. Kerby named the giant looming before him "Pele's Vent." At that moment it seemed prudent to pay her some respect.

After that first dive scientists kept clamoring to return, and Kerby became familiar with Lō'ihi's twisted gray-green chimneys and spooky ochre-yellow rocks, its rubble-strewn craters streaked with something that looked like dried blood. There were uncommon animals down there, too. Kerby regularly ran into a toad-like fish called *Sladenia remiger* that squatted on the rocks with fins that resembled feet. It's a member of the anglerfish family, and so awful looking that it's cute. Steel-blue eels would zip by the viewports: these were synaphobranchids, nicknamed cutthroat eels because their gill slits are slashed across their necks.

Kerby also encountered chimaeras, or ghost sharks, primitive cartilaginous fish with big heads, pointy snouts, fins like airplane wings, long trailing tails, and shiny silver-dollar eyes. A sensory network of lateral lines curve around the chimaeras' bodies, making them look as though they've been stitched together, or assembled from jigsaw puzzle pieces. Sometimes a false cat shark would swish by like a runway model, sporting the elongated eyes of a gray alien and a wide jack-o'-lantern grin. It's one of the many deep-sea shark species that we barely know, because they wisely spend their lives as far from us as they can possibly get.

On one memorable dive, the *Pisces* subs were greeted by a Pacific sleeper shark—a thick-bodied deep-dweller with mottled skin and a buzz-saw mouth. It's closely related to the Greenland shark, the earth's longest-lived vertebrate, with a life span that can top four hundred years. (Researchers once thought they were the same species.) Pacific sleeper sharks are covert creatures, hefty as great whites, and the only predators besides sperm whales that are known to hunt giant squid. One Greenland shark was found to have ingested a polar bear.

Kerby showed me a video of the sleeper gliding by in dramatic chiaroscuro and closely approaching both subs, one after the other, while excited scientists shouted in the background. The shark had an oddly gentle vibe, a body as brindled as old granite, and blind-white eyes thanks to a parasite that eats its corneas. It wasn't like any shark I had ever seen. It seemed to have come from deep time rather than the deep ocean, like a visitor from a vanished era. "Look at her," Kerby said, gesturing to the screen. "If ever there was an ancient Hawaiian spirit wandering Lōʻihi, that was it."

*

In 1996, the seafloor around Lōʻihi rattled with a swarm of four thousand earthquakes, the largest seismic event ever recorded in Hawaii. "Nobody had any idea what was happening," Kerby recalled, raising his eyebrows for emphasis. "It just sounded like something major was going on." A *Pisces* expedition was quickly mounted. Descending into a deep-sea eruption is not on the average person's to-do list, but this was an event scientists couldn't afford to miss. That didn't mean it wasn't wildly dangerous.

Submarine volcanoes don't always present themselves politely. During one notorious tantrum in September 1952, the U.S. Navy's deep-sea hydrophones detected a series of loud explosions in the Pacific Ocean, 230 miles south of Tokyo. It was a known spot for frisky tectonics, part of a longer arc at the seam where two oceanic plates collide. Active volcanoes had been charted on the nearby seafloor.

Over the next week the blasts continued, becoming so convulsive they generated multiple tsunamis. Often these outbursts were accompanied by thunder and lightning that lasted for hours. "Great sparks rose into the sky," one fisherman noted. Someone else called in a "pillar of fire." Marine observers watched a two-hundred-foot dome of water swell up on the surface like a colossal bubble, its edges running with waterfalls. They heard roaring and moaning noises that seemed to come from the ocean itself, which had turned a sickly green color and was puking up dead fish. When U.S. Air Force pilots flew over the site, they saw spiky black rocks emerge in a boil of whitewater, and then sink back into the depths.

For marine geologists this was blockbuster stuff, so when the explosions stopped—momentarily, as it turned out—a group of thirty-one Japanese scientists and crew motored out on a research ship, the *Kaiyo Maru 5*, to document the action firsthand. We'll never know what they witnessed that day, for the ship was never seen again. A few days later, scraps of it were found floating nearby. The wreckage was shot through with lava shrapnel.

It's hard to imagine the force that's needed to propel hundreds of tons of volcanic mayhem upward through a mile of water, but it's safe to say that you don't want to be near it in a submersible. And the Hawaiian Islands have hosted a lot of turbulent rocks. On a wall outside Kerby's office, I'd noticed a bathymetric map of Hawaii that revealed vast debris fields on the seafloor. Rocks the size of bungalows, buildings, and city blocks had, at some point, careened across thirty-eight thousand square miles of submarine real estate, an area five times larger than the combined landmass of the islands.

I felt humbled by the sight of the map because I knew what it meant: monumental violence had occurred here in the past, when the volcanoes rose to a point where they shuddered and partially collapsed, generating mighty submarine landslides. (Some of the slides would have caused mega-tsunamis, which explains why coral fragments have been found high on the slopes of the Big Island.) During a massive earthquake swarm, anyone familiar with this submerged carnage would've instantly wondered: Was Lōʻihi shifting and sliding and shedding its skin in that same way now?

"It was nerve-racking," Kerby confirmed. "We got out to the site and there was still activity coming off the bottom. The ship was getting hit with these shock waves, just—*BANG!* I was supposed to go down there to see what was going on." He laughed. "I never would have done a dive like that if I hadn't been exploring that volcano for nine years already."

Kerby descended in the *Pisces V*, easing the sub down warily. The water in the depths was turbid, and it gave off an unsettling, almost electric, vibe. Visibility worsened. "I worked my way up to where Pele's Vent should've been. We came to the edge of this huge drop, and we just sat there staring at it." It took a moment to grasp what had happened. Pele's Vent was gone: in its place was a thousand-foot-deep crater. The volcano's magma reservoir, its molten heart, had drained, flowing down the rift zone and causing the peak to implode. Later, scientists would discover vent fluids emitting from the new pit crater at temperatures up to 392 degrees Fahrenheit.

Creeping forward, Kerby dropped into the maw: "It got to the point where I couldn't see anything." Orange bacterial floc and white flecks of sediment whirled around the sub like a blizzard. On his sonar, Kerby saw that the *Pisces* had flown perilously close to the crater's wall. He reversed with one of the thrusters, triggering an avalanche of loose rock. "The thruster started all this stuff moving, so I got out of there," Kerby said with a grin. "After that I was completely addicted to volcano diving."

*

Deep-sea submersibles are an inherently risky proposition, but the field has a sterling safety record. No one has died in a manned submersible since 1974, when an electrical fire inside a Japanese craft caused the two-man crew to be overcome by toxic fumes. But there have been many close calls in the depths, and the *Pisces* were no exception. Kerby admits to moments of jeopardy, referring to them as "inconveniences." One time, a pilot he was training flew the *Pisces IV* into a narrow canyon and got wedged between two rock walls. "He freaked out," Kerby recalled. "In his brain, it was like the door closed and he was trapped." It took some cajoling and soothing to calm the trainee down, and eventually the sub was freed. "Panic, that's the killer," Kerby noted.

Occasionally the *Pisces* have been swept up in undersea currents. These internal waves—created by tidal movements and differences in water temperature and density—rush through the ocean like rivers. Submersibles aren't built for quick evasive action. Three knots is their upper speed limit; beyond that, they're out of control. Internal waves, meanwhile, are among the most overwhelming forces in the deep. They're enormous, some measuring fifteen hundred feet high and running for hundreds of miles, and they're largely invisible until you're caught in them. In 2014, a Chinese submarine was engulfed by an internal wave and dragged down into an abyssal trough. It barely survived the encounter. The sub's captain later compared the feeling to "being in a fast car that suddenly runs off a cliff." "I got a taste of one at French Frigate Shoals," Kerby told me. "I was flying along the bottom trying to avoid crashing into anything."

Of the many ways a pilot can find himself in trouble beneath the ocean's surface, the biggest risk is entanglement: getting snagged on fishing gear, cables, debris, or even ropes. There is no plan B when you're stuck a mile down. Somehow the sub must be freed. "I've been hung up twice," Kerby said, looking sober. The first time happened early in his career. "I was a green pilot, and I got caught in the lines from a shrimp trap. And I knew that if I couldn't get out of it, I would die." The second entanglement was more recent: Kerby ran into a pile of cable that had been dumped by a tugboat. "The worst thing you can do is thrash around," he said. "You have to stop and take stock of your heading, and try to project yourself outside of the sub." Both times, he added, it took hours of dressage-style maneuvering to escape. "Self-help is the only help, that's our motto."

Having the two subs dive in tandem provides a safety advantage, Kerby explained, because one can potentially come to the other's rescue. But depending on circumstances, that aid isn't guaranteed—and it also adds the risk of collision: "You have to be really disciplined when you're down there, because they're not bumper cars." The *Pisces* are equipped with five days of life support, so anyone marooned underwater would have plenty of time to consider their fragile mortality. This wasn't a theoretical situation. In 1973, another *Pisces* submersible, the *Pisces III*, suffered a hatch failure that flooded its external machinery tank, overburdening the sub and causing it to plummet to the North Atlantic seafloor with its two pilots trapped inside the pressure hull.

Thankfully, the *Pisces III* crashed on soft bottom mud without catastrophic damage, at a depth of 1,575 feet—shallow enough to attempt a rescue. (It narrowly avoided falling over a shelf that would've carried it beyond reach.) That was the good news. The bad news was that submarine rescues are logistically complicated scrambles that often fail to beat the mercilessly ticking clock.

The *Pisces V*—at the time owned by a Canadian company and working in British Columbia—was airlifted to the scene, along with the *Pisces II*, which had been laying cable in the North Sea. While the stranded pilots, Roger Chapman and Roger Mallinson, endured this long wait, they watched the *Pisces III*'s oxygen supply dwindling like an hourglass. Hypothermia set in, then dehydration, then delirium from breathing too much carbon dioxide. Terrible weather and plain bad luck slowed the rescue, but finally the other subs were able to attach lines to the *Pisces III*. By the time Chapman and Mallinson were craned to the surface, they'd been in the sphere for eighty-four hours. Only twenty minutes of oxygen had remained in their tanks.

*

With all this in mind, you have to believe that it took more than a little courage to agree to dive 35,800 feet down in Auguste Piccard's bathyscaphe. This would be far deeper than anyone had gone before. It was also as deep as anyone *could* go. On January 23, 1960, the bathyscaphe—rebuilt, named the *Trieste*, and now the property of the U.S. Navy, which had bought it from Piccard—was scheduled to descend into the Mariana Trench and touch down in a thirty-mile-long, five-mile-wide slot known as the Challenger Deep: the ocean's most profound destination.

Not to take anything away from Beebe and Barton, but this was a different ball game. This was no dunking in Bermuda; this was a trip to the bottom of a yawning crevasse in the bowels of the Pacific Ocean. Was the *Trieste* up for it? Quite possibly, although nobody knew for sure. From the original bathyscaphe's first test dive to forty-six hundred feet in 1948, Piccard's design had been validated (albeit with some hiccups along the way). The vessel worked like a hot-air balloon, if it were headed down instead of up. The steel passenger sphere hung like a gondola beneath a sixty-foot-long, blimp-shaped float encased with a thin metal shell. Piccard's *aha* moment had been to fill the float with twenty-eight thousand gallons of gasoline—a fairly incompressible fluid that's lighter than water—and attach ten tons (or more) of ballast weights. To descend, the pilot would vent some of the gas, allowing heavier seawater to flow in for negative buoyancy. To ascend, he'd drop the ballast weights.

Piccard had approached the U.S. Navy in 1956 because he needed a partner to cover the *Trieste*'s operating costs. He and his son Jacques, a marine engineer, had scraped together the funding to build it, but a submersible isn't going anywhere without a ship, a support crew, and a hefty budget. The Navy was all ears: the Cold War was raging, and the deep had compelling military prospects. The United States and the Soviet Union were already competing to send men into space; the ocean was another arena. After a series of test dives, the Navy proposed to buy the *Trieste* outright. Piccard agreed, with a caveat: on any mission presenting "special problems," his son Jacques could opt to be one of the two pilots.

The Challenger Deep is as special and problematic as it gets, so the Navy had only one seat on the historic dive. It went to a twenty-eight-year-old submarine lieutenant named Don Walsh. A likable ace from Berkeley, California, Walsh had volunteered for *Trieste* duty before he knew the full scope of what it entailed. "I just thought it would be fun," he told a reporter. "Something different, you know?"

Fun, sure—if your idea of fun is test piloting an experimental vehicle into an unexplored realm so savagely unaccommodating to human life that a mistake might result in your bones being mashed into liquid. No one was out to be a daredevil; it was all about calculated risk. The last thing the U.S. Navy wanted was a high-profile disaster. But there was no fail-proofing a dive into the Mariana Trench—there were too many unknowns. Walsh and Piccard didn't even know what kind of terrain they'd be landing on. One Navy scientist was concerned that the trench floor might be a gooey soup that would behave like quicksand. "Could we sink and disappear into this material before being aware that we had contacted the bottom?" Piccard worried.

On launch day the Pacific was uncooperative, kicking up twenty-five-foot waves that backhanded the *Trieste* as it rocked on the surface. Some exterior instruments were torn off, prompting Piccard to reflect: "Was it sheer madness to dive seven miles into the sea under such conditions?" Apparently not. At 0815 hours, he and Walsh climbed into the passenger sphere through a tunnel that connected it to the float, closed the hatch, and flooded the tunnel so it wouldn't collapse in the depths. They arranged themselves on two footstools, stowed the "lunch" Walsh had brought (fifteen Hershey's bars), vented some gasoline, and began their five-hour trip to the bottom.

They fell through the twilight zone, the midnight zone, the abyssal zone, and into the hadal zone, monitoring their gauges under the dim glow of the interior lights. Condensation dripped down the sphere's walls, and the air was clammy and cold. The *Trieste* had a single three-inch viewport so the men shared it, trying not to dwell on the fact that it was battling seven tons of pressure per square inch. Outside the window there was no sign of Beebe's Untouchable Bathysphere Fish or other finned extravaganzas; nothing to see beyond a few bioluminescent sparks and the ceaseless fall of marine snow.

At thirty-one thousand feet that quietude was ended by a moment of sphincter-clenching terror, when a dull bang, like a muffled explosion, shook the sphere. "It got our attention," Walsh noted with understatement. But the *Trieste* continued its descent and nothing calamitous ensued, so they shrugged nervously and kept going. They knew, as all aquanauts do, that if something serious happens in the depths, you won't be around to troubleshoot. As one submersible expert put it, "You might just have time to hear a click."

If they were still alive, in other words, it wasn't that bad. Eventually they figured out that the entry tunnel above the sphere had cracked. This wasn't ideal, but it also wasn't fatal. At worst, it would delay their exit when they returned to the surface. For the moment the sub was sound, but it was protesting the hadal depths with a cacophony of creaks and squeals and groans. "We were venturing beyond the tested capabilities of the *Trieste*," Piccard would later write. "Above me, in the float, icy water was streaming in as the gasoline contracted, making the craft ever heavier and heavier. It was as if this icy water were coursing through my own veins."

When the seafloor finally came into view, their external lights reflected off what Piccard described as "a waste of snuff-colored ooze." "Good—we've made it," Walsh said, his eye pressed to the viewport. He called up to the support ship: "This is the *Trieste*. We are at the bottom of the Challenger Deep at sixty-three hundred fathoms. Over." To everyone's amazement the communications system worked.

The trench floor was smooth and flat, blanketed with sediment so fine that when the *Trieste* landed, it stirred up a milky fog. They had planned to take photos but the sediment billowed around them, wrecking the visibility. It was ironic to arrive at such an exotic destination and then fail to see it. But they took temperature readings (the water was 36.5 degrees Fahrenheit) and checked for currents (none detected). In their twenty minutes of bottom time, Walsh and Piccard did spy something long and flat swimming away—our first clue that Hades wasn't an empty tomb.

*

As the afternoon passed, Kerby told me story after story, but hours of talking couldn't begin to exhaust his repertoire. Even his offhand comments had stories trailing behind them like party streamers. He'd throw out a phrase like "After we left Eel City . . ." and I'd cut in with "Wait a minute—*what is Eel City?*" and then he would backtrack to explain that while exploring Vailulu'u, a submarine volcano near Samoa, he'd come across a craggy hump of lava. At first it had looked pretty ordinary. But when he tapped it with the sub's manipulator arm, it exploded with purple eels. "There were hundreds of them," he said, pulling up a video to show me. "At first we couldn't figure out where they were coming from. But then we realized they were *in* the rock."

It was like the eels were being hatched out of lava. The scientists in the *Pisces* gaped; the discovery made international news.

At that same location he found a thousand-foot-tall volcanic cone growing from the floor of the caldera—a new volcano sprouting up inside the existing volcano. The cone was ringed by a layer of bubbly acidic water that had killed the creatures unfortunate enough to swim through it, their corpses entombed in the fluid. "There was everything you could imagine down there," Kerby recalled. "Fish, squid, shrimp, eels—fossils in the making. We called it the Moat of Death."

There was the time in the Kermadec Islands when Kerby descended into a volcanic crater and accidentally landed the *Pisces V* on the thin crust above a lake of molten sulfur, melting the sub's sample basket. And the dive on the Giggenbach Volcano near New Zealand, where a family of giant groupers sidled up to the sub and proceeded to examine it with the intensity of forensic accountants. "I gotta say, they're my favorite animals," Kerby noted. "Because they're *so* curious. They're the only fish that'll swim right up and look you in the face. Anything you're working on with the manipulators, they want to see it, and if you grab something, they want it. Yeah, they're characters."

Swordfish were less enamored of the sub and more likely to ram it at top speed—altercations that didn't end well for the swordfish. And there was another gang of highly interactive fish on the flat-topped summit of the Cross Seamount in the southwestern Hawaiian Islands, a spot that Kerby had named Jurassic Park because it was a hangout for dozens of *Hexanchus griseus,* a shark species also known as the bluntnose sixgill. Sixgills would seem to qualify as living fossils: they could've swum right out of the Cretaceous period. (One researcher equated them to "a *T. rex* in the water.") They've got six gill slits instead of the five that modern sharks have evolved, and one dorsal fin instead of two. Their eyes are fluorescent green. "*Really* large sharks," Kerby stressed. "I've sat there and felt the sub moving—they'll bump into it and push it around."

For all the spectacles Kerby has beheld, all the breathtaking natural history, many of his dives have involved man-made things. He'd flown his sub into nuclear blast craters around tiny Enewetak Atoll in the Marshall Islands, documenting the aftermath of forty-two atomic tests conducted by the United States from 1948 to 1958. But that wasn't the worst human-caused havoc he had encountered.

On countless dives, Kerby seethed at the damage caused by reckless commercial fishing. Bottom trawlers had demolished the seabed; reefs of gold and black corals thousands of years old were pummeled into crumbs. Seamounts, with their awesome biodiversity, were smashed and scraped into ruins—vandalism that will take millennia for nature to repair. Discarded tangle nets, gill nets, trawl nets, drift nets, drag lines, and long lines were draped everywhere, killing what little marine life remained. "You would not believe how much gear we find on the bottom," Kerby said with disgust. "I've come across it even when we're in the middle of nowhere."

There was also abundant garbage. Plastic was rampant, and so was every other kind of trash. Among the most common sights on the seafloor, he told me, were Budweiser beer cans: "It's the preferred beverage of the environmentally unconscious."

In his own underwater backyard on Oahu, Kerby had found detritus from various wars—and more than a hundred wrecks. He and two other *Pisces* pilots, Steven Price and Max Cremer, had become experts on Hawaii's submerged military history. The depths around Pearl Harbor were scattered with hulking skeletons like the USS *Baltimore*, a late nineteenth-century cruiser that was the flagship of the North Atlantic Squadron; the USS *Chittenden County*, a tank landing ship that fought in World War II and the Korean War; and the USS *Bennington*, a gunboat that served in the Pacific War. There was a squadron of eight 1920s flying boats—amphibious biplanes with open cockpits—and a Japanese fast-attack submarine, and two American S-class submarines, among many others.

Kerby had also come across two Japanese *I-400*-class submarines sitting upright on the seafloor, as though temporarily parked. These four-hundred-foot marauders had been cutting-edge technology during World War II, able to slink through the ocean for months without refueling. Each contained three bomber seaplanes tucked into an interior hangar. "The subs could surface, open a giant watertight door, move the aircraft out, and catapult them off," Kerby said. "The Allies hadn't even known they existed."

The behemoth Japanese subs were central to a long-standing World War II mystery that Kerby and his team had solved. Japan had spent years developing a top-secret "midget submarine"—a kamikaze craft that carried two men and two torpedoes, designed to sneak into harbors undetected. On December 6, 1941, the night before Pearl Harbor was attacked, five Japanese megasubs had approached Oahu, each ferrying a midget sub on its back. The midgets were released ten miles offshore. Their plan was to slip into Pearl Harbor and torpedo the American battleships as the aerial bombardment began.

In the end, none of the midgets succeeded in destroying anything but themselves. They had buoyancy problems, battery failures, and in one case, a chlorine gas leak, on their defective mission. One got stuck on a coral reef. Nine of the ten pilots died; one man survived when he washed up on the beach. (He became U.S. POW No. 1.) Sixty years later, the U.S. Navy could account for four of the midgets—but the fifth was still out there. And it was the most historically important one.

In the prelude to the attack on Pearl Harbor, the U.S. forces had been aware that enemy submarines might be lurking, but they weren't expecting them to be pipsqueaks. So when the sailors on the USS *Ward*, a destroyer patrolling the harbor entrance, claimed to have fired on a tiny conning tower at 6:45 a.m. on December 7, 1941—an hour and ten minutes before the Japanese mounted their full assault—no one at naval headquarters paid much attention. There had been false reports before.

The missing midget was the sub the *Ward*'s crew claimed to have sunk. If it were found, it would prove their account—in effect, they had fired the opening shots in the Pacific War—which was still disputed. Over the years, Kerby, Price, and Cremer hunted for the fifth midget at every opportunity. "We never had dedicated time or funding," Kerby explained. "It was more of a personal obsession."

By 2002, they had eliminated thirty-eight sonar targets. One promising target remained, but it was two miles outside their search zone. Kerby's gut told him it was worth a look, and that turned out to be correct. The midget lay twelve hundred feet down with a hole in its conning tower, exactly as the USS *Ward*'s men had described. Sediment coated its sides and corals studded its hull, but it was otherwise frozen in time. The excitement of finding it instantly gave way to the sober realization that it was a war grave. The two pilots' remains were still inside.

*

Amid all this sunken history, the *Pisces* often encountered more menacing artifacts. Bombs of every shape and vintage litter Oahu's seafloor, as though tossed with abandon into a dumpster. And Hawaii isn't the only place where you can find a submerged arsenal. Between 1919 and 1972, what the military blithely called "sea disposal" was standard practice. Globally, millions of tons of live munitions were sunk.

There are hard questions to be asked about a civilization that produces lethal weapons in such gleeful surplus, and with such cavalier disregard for the well-being of living creatures that no one even bothered to keep track of *where* in the deep the worst cluster bombs and poisons and radioactive materials were dumped. But if you cruise around Oahu's seafloor, Kerby said, these remnants of war are easy to locate. "There are trails of ordnance that just crisscross each other. It's an insane amount of stuff."

Some of it was relatively harmless, inert after decades underwater. But there were chemical weapons down there, too, and those could still inflict major harm. "We came across a mother lode of Mark 47 mustard gas bombs," Kerby recalled, adding: "That prompted a visit from the deputy assistant secretary of the Army."

No one knew if the bombs were leaching their contents, if chemical agents might one day waft through the waters along Waikiki Beach, or end up in somebody's sushi. When exposed to seawater, mustard gas forms a concentrated gel: it remains active in the ocean for a tenaciously long time. (Fishermen are regularly burned by contact with mustard shells that surface in their nets.) According to a 2007

U.S. congressional report, more than sixty-one thousand mustard bombs and mortar shells and 1,038 tons of sulfur-based chemical warfare agents were tipped into Hawaiian waters. They were joined by 4,220 tons of "unspecified toxics and hydrogen cyanide," 1,100 half-ton cyanogen chloride bombs, and 20 half-ton hydrogen cyanide bombs.

In 2012, the *Pisces* were asked to return to the site and run tests for contamination. "We were down there with a mass spectrometer, sniffing right at these bombs," Kerby recalled. "We had a whole Army team out there in their space suits."

So far the chemical weapons were intact. The Army opted to leave them on the seafloor, deeming it more hazardous to remove them. But another type of weapon, a World War II anti-submarine device called a Hedgehog, was harder to ignore. "I've got a whole Hedgehog file," Kerby said, opening a desk drawer and rummaging through it. He held up a photo: "Here's a big one lying on the bottom." The Hedgehog is a cruel-looking metal cylinder with protruding spikes that will detonate at the slightest contact, unleashing hydrostatic shock waves. "All it would take is for one of them to hit the subs . . . ," Kerby said, his voice trailing off. "Let's just say we don't want to go banging up against them."

*

I flew back to Maui in the early evening, on a ten-seat Cessna that hopped between the islands. As the sky ripened to navy, a last strip of sunset bronzed the horizon. I thought of Walsh and Piccard, surfacing in triumph as dusk descended on the Pacific back in 1960. As they'd climbed out of the *Trieste*'s hatch through the cracked—but blessedly still functioning—entry tunnel, two U.S. Navy jets screamed by and dipped their wings in salute. Their nine-hour dive had been made in near secrecy out of fear that it would fail. Now it was time to broadcast the mission's success. "The purpose of today's dive is to demonstrate that the United States now possesses the capability for manned exploration of the sea down to the deepest part of its floor," read the Navy's announcement. ("And the Russians don't" was implied.) Walsh and Piccard had launched a new age of undersea exploration.

I'd left HURL halfheartedly, trudging to my rental car as though I were leaving a fabulous party. Soon, Kerby and his team would embark on a four-week expedition to the Clarion-Clipperton Fracture Zone, a vast region of the Pacific between Hawaii and Mexico that is, Kerby told me, "destined for destruction from deep-sea mining." The industry hadn't begun in earnest yet, but it was imminent. Mining exploration contracts had been granted, and now scientists were scrambling to determine how bad the damage would be. "We'll be doing environmental studies on ecosystems below five thousand meters to find out what's down there before it's all wiped out," he said tartly.

Kerby wasn't excited about the trip. Instead of the *Pisces*, he'd be diving with the University of Hawaii's remote operated robot, Luʻukai. University bureaucrats considered the robot to be more efficient than a manned sub, an opinion that made Kerby roll his eyes. "It's *awful*," he said, grimacing. "It's *painfully* agonizing. It creeps along at a quarter of a knot, and you're basically dragging an entire ship along with you."

Even so, going to sea for a month sounded good to me—better than any other option I'd heard of lately. In part because it was 2018 and the world had turned sullen and mean, and hardened by all the sullen meanness to a degree that felt foreboding. Things on land were in shambles. So maybe I was desperate for a bit of escapism, some respite from the hailstorm of troubles, the acceleration of everything, the sociopathic society we'd built. And as always, what drew me to the ocean was the fact that enchantment was still available underwater.

This longing to go into the deep wasn't new, but part of its power came from the knowledge that the deep would be new to me—it was somewhere I'd never been. I wasn't alone, of course. Most people have never been to the deep and will never go there, and it is incredible to me that anyone would be fine with that. If you tell me a place is off-limits or too difficult or otherwise forbidden, I can guarantee you I'll want to see it. But in this case—*how?* Scientists waited years for a seat on the *Pisces* or the *Alvin*. It wasn't as though you could buy a ticket to descend into a submarine volcano.

I imagined the sleeper shark, the spirit of Lōʻihi, Pele's sentry at the bottom of the world. The vents humming with microbial life, the ghost sharks and the purple eels and the whole cast of characters. *I want to meet you,* I thought. I was aware the odds were stacked against me getting to dive miles into the abyss. It was a hope that was frankly unreasonable to hope for. It was audacious. It was unlikely. It was something that I absolutely had to do.

CHAPTER 3

Poseidon's Lair

There are environmental worlds on earth every bit as weird as
what we may imagine to revolve by far-off suns.
——LOREN EISELEY

NEWPORT, OREGON

The morning was soft and gauzy, overcast in a palette of grays.
In the harbor, the ocean was flat as glass, a temporary state of
calm that suited the fifty-two scientists, engineers, and crew aboard
the RV *Atlantis*. The 274-foot U.S. Navy vessel—run by the Woods
Hole Oceanographic Institution and used by marine researchers nationwide—was loaded with two hundred tons of gear for the
expedition ahead. Perhaps the most essential item was the espresso
machine: for the next ten days at sea, operations would be running
around the clock.

Everyone on the ship knew the mission would be anything but
easy sailing. It was the fourth and final leg of the 2019 maintenance
survey of the world's most advanced deep-sea observatory, known as
the Regional Cabled Array (RCA). The RCA is a sprawling and complicated affair, with six hundred miles of submarine fiber-optic cable
powering more than 150 seafloor instruments, many of which needed
to be cleaned, examined, adjusted, or replaced. New instruments—
including a fifteen-foot-tall sonar platform—were also being installed.

These would be tough jobs even on land, but the degree of difficulty soars when you're ten thousand feet underwater. In turn, these instruments stream images and data from the abyss to the Internet: this half-billion-dollar network has wired a swath of the deep so that we can monitor it in real time. Why does this matter? Why would anybody care? And why was this corner of the northeastern Pacific selected for the most ambitious oceanographic project in history? These are questions with long and intriguing answers, I'd discovered—and the best way to understand them was to go out and see the RCA in action.

That's how I came to be standing on the bow of the *Atlantis* as its engines rumbled to life and it pulled away from the pier, guided by a pilot boat. We motored slowly through the channel, under a suspension bridge, past the funky seaside town of Newport, Oregon, and out into the open Pacific. Seagulls banked overhead, and the air smelled like brine and diesel, with a top note of fish.

It was a peaceful scene, at least on the surface. But whoever came up with the saying "As above, so below" had never seen this region's seafloor. The bathymetry from Cape Mendocino all the way up to Vancouver Island is a crazy quilt of extremes, with many of the deep's most dramatic features clustered in a relatively small area. It's a riotous underworld, heaving and seething: Hieronymus Bosch would be impressed. And it's a place where the ocean's secrets are more accessible, more visible—if you've got eyes down there. Or a high-tech deep-sea observatory.

The RCA spans a slab of oceanic crust called the Juan de Fuca tectonic plate. Geologically speaking it's a cute little thing, about half the size of California. Like all tectonic plates it moves around, buoyed by heat generated beneath it in the earth's mantle. Seven major plates, eight minor plates, and about sixty microplates—like the Juan de Fuca—form a patchwork shell around the planet's blast-furnace interior. On their perambulations, these plates can separate, converge, or slide past one another. It's like a puzzle where the pieces are stealthily and endlessly rearranging themselves.

The Juan de Fuca plate is migrating northeastward at approximately two inches per year. As a result, it's colliding with the older, bigger, thicker North American continental plate. Where the two plates butt heads something has to give, so the Juan de Fuca plate is driven downward, beneath the North American plate, in a process called subduction. As the subducting plate descends, it's exposed to such scorching heat and pressure that it ultimately melts and sinks back into the mantle. (The Juan de Fuca plate's eastern edge is melting beneath the continent now, and this magma fuels the Cascade volcanoes—including Mount St. Helens, which was responsible, in 1980, for the most destructive eruption in U.S. history.) At the same time, along its western edge, the Juan de Fuca plate is moving away from the Pacific oceanic plate—a colossus that spans forty million square miles—in a process called seafloor spreading.

Neither of these plate boundaries is a gentle place. In subduction zones, rock deforms into mountains and volcanoes, faults appear as the ocean crust fractures, sediment is scraped off as one plate grinds beneath the other, generating undersea landslides. As the stresses of subduction build, the battling plates can abruptly slip, causing violent megathrust earthquakes. And while any seismic spasm can cause damage, megathrust quakes are the ones that produce devastating tsunamis, like the 2004 nightmare in Indonesia and the 2011 horror show in Japan. If an earthquake exceeds magnitude 8.5, you can be sure that it sprung from a subduction zone. Sinking tectonic plates do not go quietly into the night.

Spreading centers are even livelier. Where the plates are separating, magma rises from the mantle to fill in the gap, birthing new seafloor real estate. The older crust is forced outward; ridges and peaks are squeezed into existence on either side of the rift. Spreading centers are molten cauldrons that shudder with earthquakes, crackle with faults, gush with microbes, and produce 75 percent of the planet's volcanism. They're home to fields of hydrothermal vents belching superheated fluids like something you'd imagine coming out of a dragon's nostrils.

Spreading centers cleave the seafloor wherever tectonic plates are moving apart, forming the axis of a forty-thousand-mile-long system of mountains and rift valleys known as the mid-ocean ridge. The earth's dominant geological feature, the mid-ocean ridge encircles the globe like a jagged wound. (The Juan de Fuca's spreading center, called the Juan de Fuca Ridge, is a small section of that.)

In the 1950s and '60s, when oceanographers began to unravel the facts about the mid-ocean ridge and its prodigious volcanic activities, they were perplexed: Wouldn't the constant creation of fresh ocean crust mean the earth was expanding like a balloon? What they didn't account for, at first, was subduction. It's an elegant system, with new crust arriving and old crust departing in perfect equilibrium, as the planet recycles itself.

The mid-ocean ridge is responsible for the paradoxical truth that on this 4.5-billion-year-old earth, the longest-lived parts of the sea-floor are a spry three hundred and forty million years old. The young-est rocks popped up today. Over eons they will inch to their demise in a subduction zone, creeping away from their spreading center like an epically slow conveyor belt.

For a long time, geologists had been baffled by the fact that South America's east coast and Africa's west coast would fit together as neatly as jigsaw pieces. How could *continents* move across the seafloor, for God's sake? The revelation that it was the seafloor itself that was doing the moving—engaged in an eternal repaving project—solved the mystery. And then finally, it was official: far from being a sleepy, hoary, uninteresting place, the deep is the red-hot center of creation.

*

All of this is to explain why, despite its petite size, the Juan de Fuca plate is a hell-raiser, thrumming with unrest within a stone's throw of the West Coast, and why it's a prime candidate for full-time deep-sea surveillance. At its spreading center, the Axial Volcano—the feistiest volcano in the northeastern Pacific—sports a caldera as big as Man-hattan. Between 1998 and 2015 Axial had erupted three times, and it was swelling with magma again, making a fourth eruption imminent. (In 2015, its lava flow was four hundred feet thick.) The volcano's exuberance is a scientific jackpot—whenever the earth's innards come blasting out, there are plenty of things to study—but Axial isn't a threat to anyone living onshore.

The Juan de Fuca's real hazard, the source of its truly fearsome potential, lies coiled in its subduction zone. At the margin where it dives beneath the continent—a region known as the Cascadia subduction zone—a seven-hundred-mile-long megathrust fault is locked against the North American plate, accumulating so much strain that it has caused Vancouver Island to buckle. In fact, the whole Northwest Coast is being shoved so hard that it's bulging upward by a few millimeters each year. When the fault ruptures—and it will—all of that pent-up elastic strain will release in a king-size earthquake. The compressed edge of the continental plate will suddenly rebound, snapping downward by as much as six feet; the seafloor will lurch vertically, shaking the ocean the way you'd shake out a blanket, causing submarine landslides and displacing billions of tons of salt water. Tsunami waves, which move at jet speed, will crash onto land in minutes.

It wasn't until the 1980s that scientists realized this assassin was hiding on Washington and Oregon's front porch. Unlike other subduction zones that grumble with tremors, the Cascadia barely whispered. In the regional seismic records, there was no evidence it had ever uncorked a megathrust earthquake. Actually, it didn't seem to have done much shaking at all. At first, geologists wondered if it was dormant. Maybe it was nothing to worry about.

But the undersea stillness turned out to be ominous. We now know that Cascadia's megathrust fault isn't moving because it's locked along virtually its entire length: every bit of the Juan de Fuca plate's mighty force is rammed against the continent. The Cascadia subduction zone is all in. That means one day it will be all out.

Studying sediment cores from the seabed, excavating smothered marshes, and radiocarbon dating "ghost forests" where trees died from saltwater flooding, geologists have found the traces of forty-one earthquakes that occurred along the Cascadia fault over the past ten thousand years. Eighteen of them were caused by a rupture along its full margin.

The most recent megathrust release happened on January 26, 1700, around nine p.m., when a magnitude 9.0 earthquake sent tsunami waves thundering ashore across the Pacific Northwest. The event didn't appear in the regional records because those records extend back only to 1774. The first seismograph in the area wasn't installed until 1898, in British Columbia.

But of course the earthquake and its aftermath had not gone unnoticed. Japan, the world's most tsunami-aware nation, has kept track of every treacherous wave since the sixth century CE, and those records show that the Cascadia tsunami raced across the Pacific basin, striking northeastern Japanese shores ten hours later—near midnight on January 27, 1700—razing houses, ruining crops, and battering the coastline until noon the next day. (The tsunami came as a shock because no one in Japan had felt the earthquake that spawned it: for nearly three centuries it was referred to as the "orphan tsunami." In 1996, geologists finally correlated it to its North American source.)

Back at the epicenter, the Native American tribes that inhabited the Northwest Coast documented in their oral histories how, on that night, "The ocean rose up and huge waves swept and surged across the land." The survivors chronicled the disaster—forests drowned, villages erased, canoes and bodies found hanging from the treetops—in stories, artworks, and songs that have been passed down through generations.

Piecing together these clues to the Cascadia subduction zone's past, scientists arrived at a sobering conclusion: its recurrence interval for large-scale destruction is approximately three hundred to five hundred years. Since we're now into the 2020s that makes for some chilling math. The next time the Cascadia fault fails it will most likely unzip completely, moving from south to north. We can expect a bone-rattling earthquake. The shaking will be of such intensity that it will cause parts of the ground to liquefy. Then a phalanx of tsunami waves up to a hundred feet tall will hit a coastline that's home to eight million people. (For comparison, consider that the 2004 Indonesian tsunami—a similarly sized event—was estimated to contain the energy of twenty-three thousand atomic bombs.) Forget about roads, buildings, and bridges; forget about utilities and power stations. And if you happen to be near the ocean, forget about escape.

*

After we'd cleared the harbor, I left the deck and went into the main lab to find out more about the schedule. The previous night had been a blur of safety drills, drinks at a dockside bar, meeting my shipmates, and in general trying to settle in. The cruise was tightly packed with dives that would be made by an ROV named Jason, a workhorse of a robot with impeccable deep-sea credentials. Jason can dive to twenty-one thousand feet and lift a two-ton payload off the seafloor and perform intricate maneuvers with its hydraulic arms. It's tethered to the ship by an armored fiber-optic cable and flown by a pilot sitting in a control van on deck. Like other abyssal robots, it's designed to handle heavy jobs that humans, even humans in submersibles, could never do.

The lab was crowded with people setting up computers, fiddling with equipment, and hustling around as though they were backstage at a Broadway show and the curtain was about to go up. It was a room the length of an Olympic swimming pool, with institutional lighting, worktables running down the center, and electrical cables snaking from the ceiling. A large video monitor hung on the wall. The expedition's chief scientist, Deborah Kelley, stood at a whiteboard writing a series of long—and to me, incomprehensible—instructions for the first three dives.

It was Kelley, known to everyone as Deb, who'd invited me onto the *Atlantis*. She is a distinguished marine geologist, a professor of oceanography at the University of Washington, and the director of the RCA. The first time I met Kelley, at her office in Seattle, she made me a cappuccino and we sat down and began to talk as though we'd known each other for a decade. It's that way with Kelley; there is nothing formal or pretentious about her. Throughout her career she had done rugged fieldwork. She had spent months on the International Ocean Discovery drillship, coring fossilized magma chambers beneath the seabed, and trekked through Cyprus and Oman examining ophiolites—chunks of ancient ocean crust that escaped subduction by being thrust up on land. She had made more than fifty dives in the *Alvin*, the U.S. Navy's deep-sea research submersible.

Now in her early sixties, she was the ideal person to manage an operation this stressful, to execute the bogglingly technical, stickily political, high-stakes work the RCA demanded. It was a NASA-scale project in the abyss, one of the biggest ocean investments the National Science Foundation had ever made. Any slipups out here would be multimillion-dollar errors. Kelley is both laid-back and laser sharp, a rare combination. It is impossible to imagine her melting down in anger or choking under pressure. Histrionics are not in her DNA.

"J2-1186," she wrote on the whiteboard. "Swap deep profiler, transit ≈ 2 hrs to Slope Base (2900m)." Behind her, a group of scientists were debating seasickness remedies. "I take the Coast Guard cocktail," a studious-looking man with wire-rimmed glasses said to a heavily bearded guy with a scopolamine anti-nausea patch on his neck. Judging from the orientation meeting, this was no idle consideration. "Please do not vomit in the hallways like people did on the last leg," Kelley had requested. "Or if you do, make sure you clean it up." Then we'd all watched a safety video titled *The Ocean Is a Hostile Environment*.

Clearly there were many colorful ways to injure yourself aboard the *Atlantis*. There were cranes swinging machinery overhead, wires under tension, chemicals that could splash into your eyes, vats of flammable liquids. Twenty-four-hour operations meant that fatigue-induced clumsiness was also a risk, especially when walking on a slippery deck at night in rough seas. The ship's infirmary carried the basics, but it was nowhere you'd want to be with a sliced-off finger or a broken femur. The pre-cruise information letter served as a reminder that once we got offshore, we were staying there: "If you need dental work, get it done before you leave."

Kelley finished the dive plan and sat down at her workstation. She is a genial woman of medium height, with kind eyes and silver hair that she wears short, with her bangs swept across her face. Like everyone on the ship she was dressed for the elements—which included the lab's overaggressive air-conditioning—in jeans, hiking shoes, and a long-sleeved shirt under a fleece vest. A hot-pink hard hat hung on a nail above her computer. I thought it might be time to admit that I had no idea where we were going or what we were doing, so I went over to ask her. Earlier, when I'd tried to figure it out myself by reading a summary, I'd run into sentences like "CTD-DOSTA-OPTAA 2016 (right) and LJ01A-2016 (center left) with LV01A-2014 in the background (left) during J2-913."

"Yeah, it's a lot," Kelley said with a chuckle. "There's no other place in the world with this much infrastructure in the water." She pulled up a map of the observatory that showed the ten-thousand-volt cable running from a shore station in Pacific City, Oregon, into the ocean, and then branching across the Juan de Fuca plate. The Cascadia subduction zone was delineated by white triangles that looked like sharks' teeth. Colored dots and squares represented the seafloor nodes where robots, sensors, cameras, and other equipment had been placed. Currently there are seven primary nodes on the RCA—four in the subduction zone, two on Axial Volcano, and one in the middle of the plate—but the network is highly expandable. Each primary node distributes power to multiple junction boxes (secondary nodes), where the instruments are plugged in with extension cables. "It's like a wall socket, right?" Kelley said. "And everything's streaming live."

Aside from the cool factor of being able to spy on the deep 24/7, this fire hose of data is exactly what scientists need to tackle the big questions about how the ocean works, and therefore, how the planet works. To stand on one ship studying one location at one time—it isn't nearly enough. The deep is a mosh pit of complexity. It is always in flux, on all scales, at all depths, in all manner of conditions. With a cabled observatory everyone can be everywhere at once, witnessing everything as it happens. It's an interactive marine laboratory in cyberspace.

Kelley told me that our first stop was the subduction zone. At a site called Slope Base, seventy miles offshore, a malfunctioning robot had to be replaced. Slope Base is striking, she added, because you can see the edge where the Juan de Fuca plate is descending, and as it does, the North American plate is bulldozing off its top layers. Two miles down on an otherwise dead-flat bottom, those heaped-up sediments form a steep cliff. "It's like driving along a floor and then suddenly hitting a wall," Kelley said. "It's that abrupt."

To humans with our blink-and-you'll-miss-it life spans, the geological gears of the planet turn slowly, so slowly that we don't even notice they're moving—until something calamitous happens and we have to react. Tectonic plates wrestling with each other is the superheavyweight clash of the Titans, but it's not much of a spectator sport. That's not to say subduction zones are boring. Subduction generates heat, so the plowed-off sediments are warm, and they vent fluids and gases, notably methane. These bubbling seeps dot the Cascadia margin. "There are probably a thousand of them off Washington and Oregon," Kelley explained. "And there are teratons of methane in the seafloor."

A rowdy ecosystem arises wherever the gas is burbling. Methane-eating microbes gorge themselves. Then other creatures flock in to dine on the microbes, all the way up the food chain. This included some unusual characters. Kelley showed me a video of a shockingly ugly fish with a bulbous head, spiky teeth, a Quasimodo hump on its back, and a diaphanous, eel-shaped body. It was known simply as "weirdfish." "There's a little herd of them, and we see them every year," Kelley said. "They're voracious, apparently." The only other place weirdfish had been sighted was Antarctica.

Any movement in the subduction zone would be significant, so there were two broadband seismometers embedded in the seafloor, ready to report the subtlest twitching. "There's some evidence that there are precursors to these large earthquakes," Kelley noted. "In Japan's 2011 earthquake, a pressure sensor moved before the big event."

But distressingly, for all its capabilities, the RCA was underequipped along the Cascadia fault. "So there's no early warning system?" I asked. "Oh God no," Kelley replied. "We're phenomenally sensor poor." More cabled seismometers and tsunami-detecting instruments were needed (many more), but the installation would cost several hundred million dollars, and to date the funding hadn't been available. Meanwhile, wave-wary Japan had poured a fortune into seabed sensors that would give its population a sliver of tsunami fore-warning, enough time to shut off power, stop the trains, get people out of buildings. "It's all about priorities," Kelley said with a sigh. "We *will* have a magnitude nine here. And it's not going to be pretty."

*

On the outside, the Jason virtual van is a shipping container like any other, a corrugated-steel box located on the deck above the ship's fantail. That is where the resemblance stops. To open the van's pressure-sealed door is to enter a pulsing electronic brain. The pilot, its cerebellum, sits front and center in a Captain Kirk chair, flying through the deep using Jason's eyes—that is to say, its fourteen high-definition cameras. Everything Jason sees is displayed on a wall of monitors that glow vibrantly in the blacked-out van. Beside the pilot is a navigator who positions the ship above Jason as it roams through the deep. Scientists, engineers, and people logging data sit at computer stations arranged in tiered rows; an observation gallery lines the back wall.

I had been assigned to work as a data logger. It was the easiest job in the van—and the only one I could feasibly do—but that's a bit like saying it's easier to assist on a heart transplant than it is to perform one. My training had consisted of watching a five-minute tutorial video and listening in bewilderment as one of Kelley's students, a junior named Katie Gonzalez, reeled off a list of commands from SeaLog, Jason's software interface. "It's better to log more things than less things," she advised, as if I knew how to log anything at all.

But there was no time to linger over questions. My shift, the graveyard stretch between midnight and four a.m., was about to start. Jason was in the water at Slope Base, diving to ten thousand feet. I wound through the ship and up to the van's deck to report for duty.

Sea conditions were placid, but the weather was changing: the wind was picking up. The stars were gone, snuffed by cloud cover. I could hear the steady drone of winches as Jason's umbilical cord unspooled.

When I opened the van door it made a whooshing noise, releasing a gust of cold air. There was so much circuitry whirring in this metal box that if it wasn't kept as frigid as an ice rink, things would overheat. ("Bring a blanket," Gonzalez had recommended.) Inside, a half-dozen people were shooting the breeze in the semidarkness, preparing for Jason's arrival on the seafloor. The pilot was Chris Lathan, a quiet guy with a mop of brown hair. A silver disco ball twirled above his head, spilling glitter onto the ceiling. Vampire Weekend wafted through speakers.

On this cruise, Jason was traveling with a crew of ten technicians from Woods Hole, all of whom wore multiple engineering hats: to fly Jason means that you also know how to maintain, repair, and program it. With few exceptions, they tended to be younger men who had arranged their lives so they could stay at sea for months at a time. "This job is hard on relationships," one of them had confided. "It just is." The Jason techs had an air of hipness without being obnoxious about it. When they weren't manning a bank of computer screens that looked like mission control on *Battlestar Galactica*, you could imagine them fitting in seamlessly at an artisanal brewery in Brooklyn.

I sat down at the data-logging station. Somebody handed me the dive plan, pages of step-by-step instructions for how to swap out the broken robot. I was supposed to log each step as it occurred, so there would be a precise record of the dive—time-stamped, web searchable, and synced with images. I scanned the instructions, a thicket of abbreviations and acronyms. To my relief, Gonzalez arrived and settled in at the video logging station next to me.

The robot we were replacing was known as a deep profiler. It was a yellow pod the size of a tall coat closet, laden with instruments that it used to measure chemistry, temperature, and currents throughout the water column. On its vertical route, it crawled up and down a wire that stretched from a seafloor docking station to a mooring buoy suspended four hundred feet beneath the ocean's surface. The deep profiler was a peevish beast with a heavy workload. It was known to be balky, a regular in the repair shop.

The video wall was alive with Jason's lights illuminating the depths below us. On the screens I could see its burly manipulator arm clutching a brush shaped like a coffee mug. As Jason descended, it was dragging the brush along the profiler's wire to clean off strands of furry bacteria. This made for a movie with a painfully slow plot, but creatures would dart into the frame for cameo appearances, livening things up. Squid, in particular, were curious about Jason. They liked to jet in on a reconnaissance, assess the situation, and express their disapproval with a burst of ink. A posse of black cod swarmed the cameras, as if posing for a group selfie. Marine snow flittered down, tinted green in the strobe lights. Jellies ghosted by like alien moons.

When Jason got to the bottom, people swiveled their chairs around and locked onto the screens. The deep profiler loomed into view looking terribly lonely, its orange extension cord trailing across the seafloor to a junction box. A mauve-colored rattail with huge eyes was curled up beside it. Nearby, a second junction box sprouted cords attached to pressure sensors, hydrophones, and a seismometer. The equipment was coated with muddy sediment and wispy bacteria, the deep doing its best to claim it. If you didn't know these were sophisticated machines performing important scientific tasks, you might think that someone had shoved some old factory parts over the side of a ship.

Whenever instruments were placed in the observatory, it wasn't long before marine life moved in. These were homes, hiding spots, surfaces to cling onto. Any perch gives the animals an advantage: it helps them catch food drifting by in the currents. Rose-pink anemones, electric-yellow sea stars, and lilac-purple octopuses had staked out the junction boxes like a luxury subdivision. At another site, a platform had been colonized by so many fluffy white anemones it was nicknamed "The Sheep."

It occurred to me that this was my first live glimpse of the abyss, and even though I was viewing it from two miles above, I found it enthralling to watch. I was struck by the thought that it was *really here*. The deep was more than an intellectual concept or a strong suspicion. Diving beneath the surface with Jason was like peeking into the box to check on Schrödinger's cat: Was there truly an animal under there? If so, was it dead or alive? Here was the abyss, going about its business in synchrony with everything above it, but on its own epochal clock. I felt delight, recognition, an enveloping sense of peace. I felt my body and mind unclenching, as if hanging out on the seafloor were as relaxing as a massage. Whenever Jason's cameras zoomed in on an area, I could pick out the half-hidden life-forms: a larvacean sailing by in its opulent mucus house, a lone slithering hagfish, sea cucumbers etching tracks across the sediment like undersea graffiti.

Jason had descended with the new deep profiler latched to its side, so its first move was to clamp this replacement robot onto the wire. Then it could focus on removing the old one. Lathan was concentrating hard, his mind projected into Jason's eleven-thousand-pound body. Lathan-as-Jason lifted an arm, shook off some debris, and unplugged the broken profiler from its docking station. The plugs were oversize with square handles that fit Jason's pincer hands. To prevent seawater from flooding in and shorting the circuits, the electrical connections were filled with oil.

After a while I felt like I was getting the hang of data logging, mainly because Gonzalez was doing most of it. Even so, if you'd told me that by the end of the cruise all the cryptic lingo would come to make sense to me, and that I would hear myself confidently saying things like "Oh, are we still on J2-1993?" and "Roger that on the monkey fist and the disengaged stabs," I wouldn't have believed you.

Jason finished at three a.m. and I logged the start of its ascent. The disco ball spun to the beat of Led Zeppelin, Lathan leaning back in his chair. The dive wouldn't be completed until Jason and its cargo— the old deep profiler—were safely back on deck. No one was particularly worried; the robot had been through a lot worse. One time in the South Pacific, it had swooped into an explosively erupting volcano. On another occasion it was attacked by a shark.

An hour passed, then ninety minutes, Jason rising through the twilight zone. "Our squid friends are back," Gonzalez noted. The second mate called over the intercom to warn: "It's going to get a little squally." I stepped outside to watch Jason's return and felt the approaching storm; its blustery winds and spray of rain were a tonic after being cocooned inside the van. Looking over the railing I saw Jason's lights blazing twenty feet underwater, a cobalt halo in the black seas. The robot surfaced in a fury of bubbles and a crane hoisted it into the air. I turned to go down to my cabin. It had been ages since I'd pulled an all-nighter, but I figured I had better get used to it. Day, night—who cared, anyway? In the abyss there was no difference between them.

*

After a seventeen-hour transit we arrived at Axial Volcano, where the seabed became a theme park of lava, studded with hydrothermal vents. Like fingerprints, no two vents were the same. Each was the unique product of its geology, fluid chemistry, temperature, age, and location, among other variables. The vents ranged in size from "OH MY GOD, LOOK AT THAT THING!" to "Geez, I almost stepped on it." Some vents roared; others fizzed. There was Inferno, which had the face of a horned ogre, crawled with worms, and spewed boiling sulfide minerals; and dainty Diva, a fragile white vent that shimmered with carbon-dioxide-rich fluids. One vent was shaped like a snail: it was named Escargot. "I love going to the vents," Kelley told me, "because they're kind of where I grew up."

From her earliest days as a scientist, Kelley had a knack for finding remarkable vents. In 1982, as a student at the University of Washington, she was part of the team that discovered the Endeavour vent field off Vancouver Island, at the northern end of the Juan de Fuca Ridge. Endeavour is a hub of giant black smoker chimneys, pinnacles with knobs and lumps and cornices that bring to mind Gaudí on an acid trip. Black smokers are the dark avengers of the vent kingdom, bellowing fiery fluids. One chimney named Godzilla was 150 feet tall.

"There's nothing like Endeavour," Kelley said. "It's one of the most active hydrothermal systems in the world."

Here on Axial Volcano, Kelley had encountered another astonishing type of vent called a snowblower. Three months after Axial's 2011 eruption, a robot called ROPOS was flying over collapse pits where lava lakes had drained and imploded, a spooky moonscape of mangled basalt. Some of the pits gaped like howling mouths encrusted with white bacteria. Passing the eerie orifices, ROPOS was engulfed in a blizzard of microbes. It was a pulse of life emitting from far beneath the seafloor, a teeming netherworld known as the deep biosphere. This unseen realm is one of the earth's oldest and most expansive biomes— and one that's largely impervious to surface trauma. We could nuke ourselves, or get walloped by a comet, or die off in any number of messy ways, and the deep biosphere would carry on. (Unless the ocean boiled away, which would be quite an event.)

The study of this intraterrestrial life is on the frontier of science. Microbes that can reside way down in the pores and fissures of the ocean crust—in extreme temperatures, under annihilating pressures, and subsisting on toxic chemicals—are the champions of adaptation. They can stay alive, or barely alive (in a kind of zombie state), for millions of years. Collectively known as extremophiles, they're precisely the types of organisms we might find enjoying the methane seas of Titan (Saturn's largest moon), or under the ice on Europa (Jupiter's smallest moon), or on other ocean worlds.

But the most intriguing thing about hydrothermal vents is that they hold clues to the origin of life on this planet. How life emerged is a tetchy scientific debate that may never be resolved, but everyone agrees the recipe is exacting. First, you need water. Next, a steady supply of raw materials—chemical elements like hydrogen, sulfur, and carbon—and a site that's just right for them to react. Then, somehow, there must be an ignition, a catalysis that helps this brew generate living cells. Finally: a stable mechanism that keeps the process going over time. This is not as straightforward as it sounds. It's far simpler to list life's components than it is to determine how, when, or why a cell sparked into existence. But some deep-sea vents appear to have all the right ingredients.

This became stunningly apparent in 1977, when a team of geologists diving in the *Alvin* encountered these seafloor hot springs for the first time. From the moment they were sighted on the Galápagos Rift, the vents were hailed as geological wonders. But their biology was the bigger story.

No one had expected to find much wildlife in a lava-choked, pitch-black volcanic gash where oxygen scarcely existed, but the Galápagos vents were a zoo of strange animals. The *Alvin*'s pilot and two passengers stared in disbelief at the six-foot-tall tubeworms waving blood-red plumes, clams the size of footballs, eyeless shrimp flitting about, albino crabs scuttling across the chimneys. Photosynthesis—the conversion of sunlight to energy that was thought to provide life's only fuel—had no role here. Instead, these creatures relied on chemosynthesis: energy coming from the earth's interior, produced by chemical reactions between fluids and rocks. The microbes were eating hydrogen sulfide—a poison to creatures on land—and oxidizing it into food for the vent animals, many of which hosted the microbes inside their bodies in a symbiotic relationship. It was a *Star Wars* bar-scene ecosystem that flouted all of our rules, born of an audacious chemistry experiment that has been under way for as long as the earth has had an ocean.

Black smokers were discovered in 1979—once again by scientists diving in the *Alvin*—during an expedition on a segment of the mid-ocean ridge called the East Pacific Rise. From the dive's start, the ambience was unsettling. The water was murky, the seafloor littered with dead clams. Flying along, eighty-five hundred feet down, the *Alvin* came upon a chimney that was pumping sooty fluid like a manic locomotive. The pilot, Dudley Foster, found himself struggling to control the sub; there was so much heat and energy shrieking out of the vent that the *Alvin* was caught in its updraft. Swept into a vortex of black clouds, unable to see worth a damn, Foster banged into the chimney, toppling it and exposing its lining of metal crystals. Smoke continued to chug from its base, so he inserted a temperature probe—

which promptly melted. (The vent fluid would later be measured at 662 degrees Fahrenheit.) Foster backed away, but not before the fluid seared off some of the *Alvin*'s fiberglass trim. For submersibles with their meltable acrylic viewports, sidling up to black smokers was a dangerous game. These vents had primordial power. The expedition's leader, French geophysicist Jean Francheteau, put it bluntly: "They seemed connected to Hell itself."

*

I fell into my night-owl data-logging rhythm, snatching naps whenever the opportunity arose. The volcano dives were more demanding, and soon even the perkiest among us were drooping. "You need some sleep, bud," Kelley told oceanographer Mitch Elend, who was disassembling a camera in the lab. "Your eyeballs are looking a little rough." Elend glanced up and smiled. He was wearing a T-shirt that said "Oceans Cover ¾ of the Earth: Majority Rules." Later that day I saw Kelley herself nodding off, arms crossed, in front of her computer.

In the van, though, I was never tired. Even when Jason was doing tedious chores like untangling a cable, the scenery was riveting. On every shift I'd see meteor showers of bioluminescence, squids having dance parties, fierce little fish flashing their lights, and big nosy fish playing chicken with Jason's propellers. "There are six thrusters, each capable of two hundred and fifty pounds of thrust," a pilot named Chris Judge explained. "Every so often you'll see a fish learn that the hard way."

The first vent dive I logged was at a site called Tiny Towers. It was exactly as its name suggests, a skyline of pocket-size spires. Tiny Towers was young and hot, rippling with clear fluid. It was splotched with colors that seemed borrowed from other cosmic spectrums: a six-dimensional Martian rust, the gray of lunar craters, ruddy supernova red, Neptune's zingy indigo.

Kelley's co–chief scientist, Orest Kawka, sat hunched in the van's hot seat—so called because the person in it is responsible for accomplishing the dive's goal—directing the pilot, Korey Verhein.

Kawka, a tall, outdoorsy man in his forties, had been up for more than twenty-four hours. His voice was hoarse. Verhein was voluble and funny, with a *Duck Dynasty* beard jutting out from under a woolen toque. They were trying to position a probe, searching for an optimal spot within the vent field. "It sucks in hydrothermal fluid and filters out DNA," Kelley had explained to me earlier. "So we're looking at organisms. We want to know who's living there, and how many— the community makeup." Curiously, each vent had its own microbial populace, a roster of critters unique to it alone: "They're like little islands."

The probe situated, Kawka left to get some rest, and Jason headed for the surface. Verhein stood up and stretched, yawning. "What's the best part of piloting Jason?" I asked. "Ummm, that's kind of hard to answer right now," he said dryly. "I'm a bit jaded. I've been out here for a while."

"Okay, what's the worst part?"

"Exhaustion."

This cruise wasn't sexy, like exploration. It was the unglamorous work of keeping machines running smoothly in an environment that excels at destroying them. And if that work yielded insights into the earth's playbook, or identified vent microbes that could become next-generation medicines, or demystified the rumblings of a subduction zone, it was because armies of people had spent years on this ship and on others like it, applying brainpower and elbow grease instead of sleeping. The observatory was a mix of techno-wizardry, deep-sea savvy, insomnia, and scopolamine-infused brow sweat.

The man who dreamed up the whole wild thing was Kelley's mentor, oceanographer John Delaney. For decades the RCA had percolated in his mind; his tenacity and connections made it a reality. His email signature is a T. S. Eliot quote: "Only those who will risk going too far can possibly find out how far one can go." Now seventy-seven and an emeritus professor at the University of Washington, Delaney had pinpointed his field's most daunting challenge: "We need to be *in* the ocean to understand it."

You could be forgiven for wondering why, in a world with a mess of problems, the deep ocean should be high on anyone's list of concerns. After all, we've made it this far while mostly ignoring it. But Delaney argues that our unfamiliarity with the ocean's inner workings is the most urgent problem we face. "It's the engine that runs the climate system," he points out. "We need to be able to anticipate any serious tipping points in ocean behavior since they will affect us so much."

It's hard to know how drastically the ocean is changing if you don't know what it's up to in the first place. Since 1970, it has gulped down 93 percent of the excess heat and 30 percent of the carbon dioxide we've generated from burning fossil fuels—an extraordinary burden. As a result, it's becoming warmer, more acidic, and less oxygen-rich. The state of ecological balance that we're dependent on isn't permanent: if we perturb it enough, the ocean could make things terminally unpleasant for us. To Delaney's mind, if we hope to survive whipsawing weather extremes, ecosystem upheavals, and storms of biblical intensity, we'd be wise to turn our attention downward, even if it delays our space tourism plans.

Listening to Delaney talk about future ocean technology is like screening a sci-fi movie inside your head. He envisions fleets of drones whizzing through the water, scanning their surroundings with artificial intelligence, and beaming back holographic data visualizations and genomic analysis from the middle of an undersea eruption or a tsunami or a hurricane. "This is not far-fetched," he stresses. "We're on the threshold of being able to do it. With the right types of robotic systems, we can do things that nobody would imagine." "John's always been ten years ahead of his time," Kelley noted.

Delaney's latest quest was to add autonomous vehicles to the RCA. Specifically, he wanted a free-swimming, remote-controlled robot that would live on Axial Volcano in its own little garage. It would cruise the vent fields daily, taking samples and streaming video, and when Axial blows its top, it would be poised to fly through the eruption plumes before they disperse. "It's very hard to sample volcanic gases because we're never in the right place at the right time," Kelley explained. "There are eruptions every single day along the mid-ocean ridge, with the equivalent of huge ash clouds going off in the water—at supersonic speeds, probably. But we have no sense of their overall impacts on the ocean."

Since the RCA was switched on in 2014, scientists worldwide had submitted proposals for ingenious instruments they wanted to plug in. "One of my next projects is to work with a German scientist who wants to install a rock crawler," Kelley told me. "It looks like a Tonka truck. It would wander around the methane seeps taking measurements. They'd control it from Germany."

NASA had recently come on board with a laser system that could analyze molecular vibrations, which would be handy for detecting any extraterrestrial organisms that wriggle by. (Somebody with a twisted sense of humor had named it INVADER—an acronym for "In-situ Vent Analysis Divebot for Exobiology Research.") Eventually, INVADER might be deployed somewhere like Saturn's ocean moon Enceladus, which is venting so vigorously that geysers shoot through cracks in its icy crust and arc into space. But first it would be tested here on Inferno, a black smoker that was presumably the most accommodating site NASA could find, short of leaving the planet.

Inferno and its neighbor, another black smoker called Mushroom, were popular spots for gear. They were both wired up like intensive-care patients. Next to Mushroom and Inferno stood Phoenix—a vent that kept getting knocked down by eruptions and building itself back up—and another vent named Hell, for obvious reasons. But there were even grander black smokers on the other side of Axial's caldera, and one giant in particular that Kelley wanted to check on. "El Guapo," she said, sounding maternal. "We'll do a little tour up there and look at the flames."

*

During the cruise I was aware of a celebrity passenger holed up on the main deck: the *Alvin*. The iconic sub wasn't diving on this expedition, but the *Atlantis* was its home. It rested in its two-story hangar with an air of nobility, red boxing gloves covering its robotic hands. From its first mission in January 1966—when it was sent to retrieve a hydrogen bomb that fell into the Mediterranean Sea after a midair collision between a U.S. B-52 bomber and an aerial fuel tanker—the *Alvin* had gone on to prowl the deep for more hours than the rest of the world's manned submersibles combined.

One afternoon I got a tour of the *Alvin* from a Woods Hole engineer named Drew Bewley. The sub loomed over us, hooked up to ventilation hoses and sporting miles of highly organized wiring. Bewley knew the purpose of every one of those wires: he oversees the sub's electrical systems. As a kid growing up in Kentucky, he'd stumbled across a library book by the oceanographer Bob Ballard, who played a key role in the *Alvin*'s history. "I would check it out obsessively," Bewley recalled. He nursed a quiet yearning to work with the sub one day. It was a worthy dream, until pragmatism kicked in. "I thought, there's only one *Alvin*, it only has three seats—there's no point in even trying to approach it." He decided to become a musician instead. But when his high school physics teacher demonstrated the principles of force distribution by lying on a bed of nails, Bewley gave science a second look.

He'd studied engineering at university, graduating with "iffy marks" and certain that he had no chance whatsoever of getting a job at Woods Hole. "I was too intimidated to apply for a long time," he said. "But eventually I did." Now, at age thirty-seven, he was training to become an *Alvin* pilot. The certification process is an extended ordeal that, since 1965, only forty-two people had ever completed. The final exam is a grilling by a board of U.S. Navy admirals. "They'll sit you down in a room and throw anything they can at you."

At a glance you can tell the *Alvin* means business. It's the *Pisces* on Schwarzenegger-level steroids, outfitted for every scientific need. Its six-foot pressure hull is made of three-inch-thick titanium, with five acrylic viewports. I listened as Bewley described the sub's systems, its triple redundancies and kitchen-sink safety features. On multiple occasions, the *Alvin* has been overhauled so thoroughly that it might as well be a new sub. In early 2020, it was scheduled to undergo a major refit that would extend its range from 14,800 feet to 21,325 feet.

Bewley wrapped up and asked if I had any questions. My only question about the *Alvin* was whether I could dive in it, but I knew he couldn't answer that. As I left the hangar, I saw Kelley sitting near the bow. It was the first sunny day we'd had, a brief shot at some vitamin D. I walked over. I'd been angling to talk to her when she wasn't doing ten things at once. Nineteen years earlier, on another expedition aboard this ship, she'd made a discovery that had haunted me ever since I'd read about it. The pictures of it were so surreal that if they hadn't been published on the cover of the prestigious scientific journal *Nature*, I would've assumed they were concocted in Photoshop. I wanted to hear more about the place known as Lost City.

*

The middle of the Atlantic Ocean is a solitary post, especially in winter. On the night of December 3, 2000, the *Atlantis*, its decks aglow, was the only speck of light for miles. Kelley was asleep in her cabin. Swiss geochemist Gretchen Früh-Green was on watch in the van as Argo—a towed camera system—traversed a million-year-old undersea mountain, the Atlantis Massif. "It's about the size of Mount Rainier," Kelley began, after I sat down and asked her to tell me the story from start to finish. "We were trying to figure out the processes that had gone into forming it."

To Kelley and her colleagues the Atlantis Massif was a fascinating case study, worthy of the five-day journey to get there. It's located south of the Azores and just ten miles west of the Mid-Atlantic Ridge—the spreading center that bisects the Atlantic basin—but it's not a volcanic peak. Instead, it's the result of the massive crumpling that occurs when seafloor spreading fractures the ocean crust. In this instance, the fracturing had uplifted a hunk of the earth's upper mantle, offering a window into an even deeper realm.

Mantle rock is solid, but it behaves like angry Silly Putty. In its hottest state down in the earth's bowels, it flows, albeit sluggishly. When dragged from its deep womb and exposed to seawater it reacts dramatically, becoming unstable. It swells and deforms, morphing into a scaly dark green rock called serpentinite; it throws off heat, hydrogen, and methane in a kind of planetary hissy fit. You can't accuse it of being dull.

The scientists were focused on the massif's south face, a steep serpentinite wall that was craggy with shattered rocks. During daylight hours Kelley and the others were diving in the *Alvin*, examining the mauled terrain. At night Argo would take over, tracing the massif's contours and relaying black-and-white images to the van.

As Früh-Green monitored the screens on December 3, Argo roved the escarpment. Two thousand feet below the summit, the wall dead-ended at a flat terrace. Argo set off across it, lights playing through the darkness. Oddly, the terrace was coated with what appeared to be smooth, pale gray cement. Früh-Green moved her chair forward. Then suddenly, a ghostly figure flashed by the camera.

"It was around midnight," Kelley recounted now, leaning against the ship's railing. "When stuff happens it's always late at night, right? So Gretchen comes flying into my cabin and says, 'I think we saw something.' And it was clear immediately—I mean, it was *very* clear: this was like nothing we'd ever seen before."

For the next five hours they toured Argo around the terrace, which was the size of a few city blocks. "It's as flat as this deck," Kelley said, gesturing. "It looked like a patio." But all around it, rearing up from below this ledge, were enormous white towers as attenuated and elegant as Giacometti sculptures. Why were they *white*? What were they made of? Excitement ran through the van like a current.

The expedition was near its end, so there was time for only a single dive. At daybreak, Kelley, geologist Jeff Karson, and pilot Pat Hickey climbed into the *Alvin*. The descent was a quick two thousand feet, but they spent precious time trying to locate the site. "We were flying blind because we didn't have a map," Kelley recalled. "So we squirreled around until we found a tiny white chimney. It was *beautiful*."

Now oriented, they floated through the field of pinnacles, white as alabaster, white as snow, white as bone. These were *vents*, but of an utterly different nature. They were made of carbonate, like the limestone in caves. Up close, Kelley could see their fine detail. Some were shaped like fluted Greek columns; others looked like Christmas trees or elaborate sandcastles. They were all ornately wrought, brimming with crystals. Some of them were more than a hundred feet tall.

Everywhere the *Alvin*'s lights shone, spectral formations appeared. One monolith dominated the field, four spires crowning its summit. It rose two hundred feet like a bouncer, but its features were ethereal. Along its walls, carbonate crystals had formed clusters of chimneys as graceful as Gothic steeples, and overhanging flanges shaped like large inverted bowls. Warm, clear fluid was weeping from the walls and pooling beneath the flanges, creating mirrored surfaces. As she described it to me, Kelley still seemed amazed: "You don't expect to see upside-down reflecting pools in the ocean." They named the magnificent structure Poseidon.

Hickey took fluid samples with a titanium syringe, plucked rocks and chunks of carbonate, but biological specimens were elusive. Aside from the beefy wreckfish (relatives of the grouper) that were following the sub like golden retrievers, the place seemed to be abandoned. There was none of the creeping crawling carnival found at black smokers.

The scientists would later learn that appearances were deceiving. These vents were home to crustaceans, worms, snails—the whole gang. "But they're like the size of your thumb and transparent," Kelley explained. This subtle fauna was cached in the chimneys' nooks and crannies, hidden behind strands of white bacteria that waved like feathery kelp. Inside the towers was another surprise: filmy layers of archaea—from the Greek word that means "ancient ones"—a group of microbes that are the earth's most primitive and enigmatic organisms. They're also the toughest and most resourceful, flourishing in the most extreme environments. If a place is especially cold, hot, acidic, alkaline, anoxic, toxic—challenging in the superlative, would kill us in an instant—you can be sure that some variety of archaea loves it.

They're ubiquitous in the microbial world, though they were only discovered in 1977 by molecular biologist Carl Woese (who was pilloried, at first, for his breakthrough), constituting a third domain of life to accompany bacteria (the first domain) and eukaryotes (the second domain: organisms whose cells have a nucleus, including fungi, plants, and animals). Archaea are so weird and so old, possessed of such unexpected and versatile capabilities, that many scientists now believe they were central to the emergence of complex life; that eukaryotes themselves (i.e., us) may have an archaeon ancestor. (Specifically, our most distant relative may be a Lokiarchaea, a jaw-droppingly odd group of microbes that was found seventy-seven hundred feet down, in a hydrothermal vent field called Loki's Castle, on the seafloor between Greenland and Norway.)

"We'd driven around so much we were nearly out of power," Kelley continued. "And after a while the carbon dioxide builds up so you get a little stupid." Before leaving, they wanted to get a sense of the field's scale. Kelley asked Hickey to fly to the top of the tallest tower, Poseidon, and then float all the way down to the seafloor: "And we just kept *going and going and going*." They fell past white vents shaped like minarets, turrets, beehives, hands. Twenty-four hours earlier, no one could have conceived of this place. On the bottom, Kelley recalls wondering what else in the deep we hadn't yet imagined.

Reluctantly, they headed for the surface. "On the way up I started thinking about all these white columns, and we're there on the ship *Atlantis,* in the Atlantis fracture zone, on the Atlantis Massif . . ." It was as though they'd flown through the skyscrapers of a long-forgotten sunken metropolis. "You named it after the lost city of Atlantis?" I asked. "Yeah," Kelley said, laughing. "I got a lot of interesting emails after that."

According to myth, Atlantis disappeared beneath the waves during a volcanic cataclysm, but the most scintillating thing about Kelley's Lost City is that it is not volcanic. Its vents were formed by the chemical reaction between mantle rock and seawater; they differ radically from volcanic vents. Black smokers boil with metal sulfides; Lost City's vents are warm (below two hundred degrees Fahrenheit) and metal-free. The fluids from black smokers are as acidic as vinegar; Lost City's emissions are as alkaline as Drano. Black smokers are fairly fragile, vulnerable to being smothered by lava, and because of that, relatively short-lived. Lost City's sturdy vents are at least 150,000 years old. After twenty-three years of studying hydrothermal systems, scientists had thought they were getting a handle on them; that even unique vents shared similar characteristics. Lost City upended all that.

When the *Atlantis* returned to Woods Hole, the press was waiting. "It was insane," Kelley recalled. As word spread and samples were scrutinized in labs and research papers were written, among the many reasons people were agog over Lost City was that its chemistry made it a front-runner in the search for life's origins. This type of vent system is a factory for hydrocarbons—the organic molecules that serve as cellular life's building blocks—and Kelley and her team were able to prove that Lost City produces them from nonbiological sources. It's an ideal petri dish for life to break out—on earth, and potentially elsewhere in the cosmos. "It opened up a new way of thinking," Kelley said.

In the years that followed, Kelley, Früh-Green, Karson, and others returned to the site with reams of equipment. They mapped the field—which is still growing—and found more than thirty pinnacles. They investigated its peculiar life-forms. James Cameron dropped by with four submersibles to film his IMAX documentary, *Aliens of the Deep*. The novelist Clive Cussler set one of his thrillers there. Marine geologists and biologists, geochemists, astrobiologists: they all got on ships and made the pilgrimage. In 2016, UNESCO designated Lost City a World Heritage Site of "outstanding universal value."

To date, Lost City stands alone, but Kelley believes there are others like it—"The combination of mantle rocks, heat, and faulting along mid-ocean ridges should provide rich opportunities to find more Lost Cities"—and that if we cared to explore more than a piddling fraction of the seabed, we would discover them. In the deep past when life was getting its toehold, the seabed was a bonanza of mantle-like rocks: vents like this would have been common. One way to view Lost City is as a kind of aquatic time machine that allows scientists to peer into the ocean as it was four billion years ago.

"It's just so . . . stark's not even the right word," Kelley mused. "But there's something about it. You know, like *how could this exist?*" She lowered her voice, as if divulging a secret: "It's spiritual, and I don't say that very often."

*

El Guapo, the Handsome One, filled the screens with its shaggy splendor, its rock-star hairdo of tubeworms and limpets, its black dragon breath. The fifty-five-foot black smoker was in excellent form. "It's very *Game of Thrones,*" someone observed from the van's gallery, which was full, with people crowded along the back wall. "It's booming," Kelley agreed. "Yeah, it's definitely booming." She turned to Verhein, in the pilot's seat. "Can you zoom in on the top?"

If you do not like worms, you should not go to El Guapo. It's a writhing worm-a-palooza. There were tubeworms with feathery oxblood plumes and violet-white tubes kinked like antique plumbing, and palm worms that resembled tiny, burgundy-colored palm trees, and hardy sulfide worms that secrete their own protective metal armor.

"Those little red guys that look like pill bugs are scale worms," marine biologist Mike Vardaro said, standing behind the hot seat and narrating into a microphone. He was webcasting our tour of Axial's most formidable vents. "They're predatory. They crawl around the vent and nip bites out of the other worms. So as the scale worms approach, the tubeworms will duck inside their tubes."

"Fine job," Kelley quipped. "Soon you'll be seeing him on NBC."

Vardaro grinned. "So, one of our viewers has emailed to ask: 'What's the most significant thing you've learned while studying this volcano?'"

"Well, we've seen a lava flow thick enough to cover most of the

Seattle Space Needle in an eruption with thirty thousand explosions and eight thousand earthquakes," Kelley replied offhandedly, her attention on the screens. "So we're learning how active it can be."

Jason's eyes panned across El Guapo. From my data-logging seat I could see it on a dozen camera feeds. Dark fluid billowed from its top, perfectly resembling smoke. When the cameras zoomed in, it was easy to imagine the vent as an Impressionist painting, stippled with verdigris, fuchsia, sienna, umber, carbon gray, jet black. Flossy filaments of white bacteria streamed along its sides. As Jason moved closer, its thrusters stirred up whorls of orange and yellow microbial floc that looked like crushed Cheetos. "There are snails and other mollusks," Vardaro continued in the background. "Sometimes at its base you'll see a lot of pycnogonids—sea spiders." "Lots of weird things here," Verhein cut in cheerfully.

"This is a nice oxidized area," Kelley said, pointing to a screen. "When the dust settles let's get some 4K footage." She turned to a student named Rachel Scott, who was logging video. Scott nodded, engaging the ultra-high-definition 4K camera.

Gonzalez, standing behind Scott, offered to take over. "You need to get some sleep," she urged. Scott shook her head: "Sleep is for the weak."

"The water here is superheated," Vardaro informed viewers. "Three hundred degrees Celsius or so. And then a millimeter to either side of the chimney, it's two degrees Celsius." A snaggle of tubeworms appeared in a close-up, luxuriating in their hydrogen sulfide Jacuzzi. "Oh, perfect, perfect," Kelley said. "When you see worms with those nice red heads, they're very happy."

Ever since tubeworms were spotted in the Galápagos, they'd been the mascots for deep-sea vents and chemosynthesis. The red in their plumes is hemoglobin-rich blood (adapted to transport hydrogen sulfide as well as oxygen). They have no eyes, no mouth, no gut, no anus—which would seem like a raw deal for any animal. But the worms' biology is suited for their environment. They feed by absorbing the bacteria through their skin, and housing them in a special organ called a trophosome. As the vent fluid washes over them, they breathe it in through their plumes—which act as gills—and then the bacteria take over, ingesting the chemicals, converting them to energy, and sharing the meal with their hosts. For tubeworms, life is a dinner party.

I sat back and took it all in. For all its searing heat and biting worms and bluster, El Guapo was hypnotic to watch. You couldn't look at the black smoker and think about a work deadline or that you needed to make a dentist appointment or how your pants felt tight. You could only be immersed in its presence; it admitted nothing mundane. Human thoughts, beliefs, involvement: none of these were required in this alternate world. "It's just going all the time," Kelley said. "It doesn't care that we're here at all."

In the van, the disco ball spun, David Bowie sang about the space odyssey of Major Tom, and everyone was absorbed by the counter-reality on the screens. Without a doubt, this was the closest I could get to the abyss without being in the water. But I needed to be in the water—and I was about to meet a group of people who felt the same way. Before I left Oregon, I would have two phone numbers jotted on a piece of paper—an unexpected gift from a person I planned to visit after I disembarked from the *Atlantis*—and those contacts would open up the deep to me in ways that I could never have predicted. The first number belonged to a man whose company had recently built the world's first commercial full-ocean-depth submersible, meaning there was nowhere on the planet it couldn't dive to safely and repeatedly. The second phone number was for the man who owned it, and was now preparing to take his new sub and its support ship out to sea.

CHAPTER 4

What Happens in Hades . . .

I could tell myself that this was just another dive.
It wasn't, of course, and I knew it.
—JACQUES PICCARD

DORA, OREGON
NUKU'ALOFA, TONGA

left the *Atlantis* when it returned to Newport, but I didn't leave Oregon. I had another appointment in the state. In truth, it was more like a pilgrimage, or a leg of *The Amazing Race*. I drove east from Newport, then south past Eugene to Roseburg, before turning west on the Coos Bay Wagon Road and following the East Fork Coquille River through the mountains of the southern Oregon Coast Range. The route demanded four-wheel-drive, full concentration, and no fear of thundering logging trucks. It took me deep into a land of giants, noble Sitka spruces and red cedars and Douglas firs, draped with tentacles of moss, standing near scars of clear-cut earth so raw they hurt to look at.

After a while the roller-coaster road flattened out and opened into a lush valley, ringed by foothills. Elk grazed in emerald fields under cement-colored skies. I'd arrived in Dora, Oregon: population 150. It was an unexpected spot to find the man classified as U.S. Navy Deep Submersible Pilot No. 1: Captain Don Walsh.

From the beginning, Walsh had topped my deep-sea call list. The word *legend* gets overused, but in this case you need it. To miss out on talking to Walsh would be ocean book malpractice. I had contacted him through a mutual friend, and he'd responded immediately. "I would be happy to meet with you," he emailed. "I'm not sure I can offer much that you don't already know. But I am willing to try." He'd attached a six-page professional biography that read like a bucket list for the ages, except that Walsh had already done everything on it.

Where to start? After becoming the first men to the bottom of the ocean, Walsh and Piccard were feted at the White House by President Eisenhower. Walsh was awarded the Legion of Merit medal; Piccard was given the Navy's Distinguished Public Service Award. Piccard returned to Switzerland. The *Trieste* remained in the Navy's tool kit, but its fragility was a concern and it was confined to shallower depths. Before long it was phased out completely. (There was a new submersible on the Navy's drawing board—ideas being pitched for what would eventually become the *Alvin*.)

As a submarine commander, Korean War and Vietnam War veteran, and now the undersea equivalent of an Apollo 11 pilot, Walsh had brilliant prospects. He went on to cram as much novelty, service, learning, and adventure into his career as a person possibly could. Over the years, he would make sixty polar expeditions, earn three graduate degrees (including a PhD in oceanography), fly biplanes, seaplanes, and gliders, and hover in a Russian *Mir* submersible above the bridges of the *Titanic* and the *Bismarck* in their abyssal graves. Walsh was deputy director of all eleven of the U.S. Navy's research laboratories. He did ocean-related duty in the Pentagon, at the State Department, in an advisory group at NASA. He was the founding director of the Institute for Marine and Coastal Studies at the University of Southern California. In Walsh's bio there is a section devoted to "Presidential Appointments" and an index of honors that stretches for pages, and at a certain point the largeness of his life just takes over and all the words converge into a monosyllable: *Wow.*

I pulled up in front of Walsh's red cedar ranch house, set in a cathedral of forest. The front door opened and Walsh came out on the porch to greet me. He was a compact man with sharp blue eyes and a full head of curly snow-white hair, wearing a rumpled green polo shirt, khakis, and slip-on boat shoes. I must have looked shell-shocked from the drive, because he launched into the story of how he and his wife, Joan, had come to live in this secluded valley. Thirty years ago, he explained, they'd escaped "the parking lot known as Southern California" after Joan had scouted for a home that melded her love of old-growth trees with his love of the ocean. Walsh, working on a ship off Panama, learned that her search had ended upon receiving a fax that said: "Congratulations, you're the owner of a ninety-acre ranch in southwest Oregon."

"Barn, shop, all fenced in, half mile of river with three kinds of salmon, hawks patrolling the pasture, trees—" Walsh enumerated before cutting himself off: "Anyway, I do go on. I'm enthusiastic about the place."

I followed him into the house and up a flight of stairs to his office, a lofty room with wraparound windows and vaulted wooden ceilings. It reminded me of an eagle's nest, if the eagle were a librarian: the room contained eight thousand books. "They're kind of like your children, you know?" Walsh said with a chuckle. "I find it hard to part with them."

Originally I'd hoped to talk for a couple of hours; Walsh was eighty-seven years old when I visited, and I wasn't sure how much bandwidth I could expect. But it quickly became clear that he had the energy and acuity of someone half his age. During the previous year, he told me, he'd traveled to fifteen countries: "I think the worst sin is boredom." He was a natural raconteur with a droll sense of humor and a recall of arcane detail. It was fun to listen to him talk. And of course, he wasn't short on material.

We sat in a corner of the office, drinking tea and debriefing. Every few minutes Walsh would come out with some zinger like "Arthur C. Clarke was my dive buddy," or "There were times when our missions had us totally submerged for two months," so it took a while to get around to what he referred to, simply, as "the deep dive."

I asked him about an article in which he'd recalled his first impressions of the *Trieste*—and they weren't good ones. The bathyscaphe had struck him as "this collection of odd bits of metal" that looked like "an explosion in a boiler factory." At the time, he remembers thinking, "I'd never get into that thing."

"So why did you?"

Walsh smiled slyly. "Well, as Scott Carpenter told me . . . ," he began, chortling at the memory. "He said, 'You know, you get up there and you're sitting in this thing, and the fuse has been lit—and you suddenly realize that it was built by the low bidder.'" Walsh channeled the Mercury 7 astronaut's second thoughts: "'Hmmmm, I don't think I'll go today. Could you put that ladder back up here?'" He leaned back in his chair and laughed. "Sometimes you get propelled by circumstance."

That made for an excellent story, but it wasn't the whole story. Yes, circumstance put him in command of the *Trieste;* others might have been chosen. But luck favors the bold—and Walsh was bold. He wouldn't have been anywhere near the *Trieste* if he hadn't volunteered for deep-sea duty in the first place. ("I mean, I couldn't *spell* bathyscaphe—is that one word or two?") His fellow submarine officers weren't exactly lining up for the opportunity: later, Walsh would learn that he'd been the only volunteer. "I define exploration as curiosity acted upon," he said, stressing the power of that one little verb: *to act*. Nobody ends up orbiting the moon, or touring the hadal zone, or doing anything truly extraordinary by accident.

Walsh and a tight-knit team had labored in Guam for six months to make sure the deep dive was anything but a flier. They made a series of test dives, putting the *Trieste* through its paces. "We were listening to the squeaks and groans and what was going to fail and how do we fix it, and was that a normal noise or an abnormal noise," Walsh recounted. "So we knew its moods."

By the time he and Piccard closed the hatch—next stop, Challenger Deep—they felt confident, or close to it, despite the hairy launch conditions. "You don't have time to be scared," Walsh said. "We never let fear in the door, and I never felt it. We just got on with the job." And as a seasoned submariner: "Being underwater in claustrophobic conditions was a way of life for me."

That quiet professionalism was a hallmark of the submarine force. "It truly is silent service," Walsh observed. "We don't talk about what we do." On team *Trieste* there was, as he put it, no "personal glorification in the form of heroic pajamas festooned with flags and patches." No flamboyance. No preening. No fuss. "I wanted everything to be low-key and low visibility."

There was one emblem of the deep dive, a small but iconic object. Traditionally, submarine officers wear a gold pin that depicts two dolphins—the masters of diving and resurfacing—flanking the bow of a sub. Walsh designed a special version of the insignia for the Navy's deep submersible pilots, incorporating the *Trieste*. When I asked if I could see it, he opened a display case, extracted a jewelry box, and passed it to me. The pin felt warm in my hand. It was a rich matte gold, delicately sculpted in bas-relief. *You can keep the diamonds*, I thought. *It's all about the dolphins.*

Walsh put the pin away. I glanced out the window and noticed the daylight was draining fast. We'd been talking for at least six hours, and though Walsh was still going strong, I wasn't keen on tackling the Wagon Road after dark. As we walked to my car, I asked him what he'd heard about an expedition called the Five Deeps. Recently, I'd read that a Texas businessman and explorer named Victor Vescovo had contracted with a private company, Triton Submarines, to build a two-person sub that could dive to (and return from) the deepest places in the ocean—including the Challenger Deep. I wasn't sure if it was true.

"Oh, Victor!" Walsh said brightly. Not only was Walsh familiar with the expedition, he was part of it: he planned to accompany the group to the Mariana Trench. He told me that he knew the expedition's players, that it was a deep-sea all-star team.

Walsh gave me contact information for Vescovo and Triton's president, Patrick Lahey, and offered to introduce me via email. He encouraged me to reach out to them soon, because the expedition was under way. I said that I certainly would, and thanked him again, steeling myself for the drive out. It had started to drizzle and Walsh turned

"A conglomeration of monsters may be found":
an inset from the *Carta Marina* by Olaus Magnus, 1539.

William Beebe (*left*) and Otis Barton (*right*) with the *Bathysphere*, Bermuda, 1934 (*above*). The bathyscaphe *Trieste*, 1959 (*below*).

"Was it sheer madness to dive seven miles into the sea under such conditions?": Don Walsh (*right*) and Jacques Piccard inside the pressure hull of the *Trieste*.

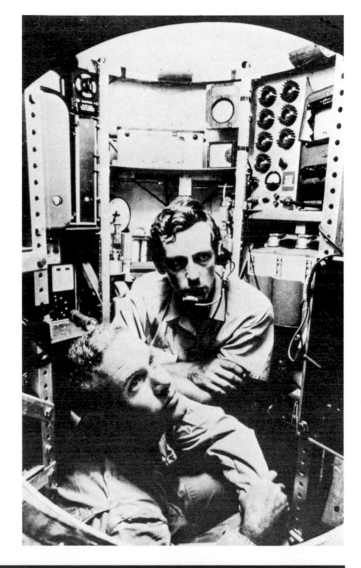

"A yawning crevasse in the bowels of the Pacific": the bathymetry of the Mariana Trench's Challenger Deep, the ocean's absolute nadir (*below*).

Mariana Trench: Challenger Deep
Exploration History

● *Trieste*, 1960 (1 dive)
● *Deepsea Challenger*, 2012 (1 dive)
● *Limiting Factor* dives, 2019 (4 dives total)

-10350
-10425
-10500
-10575
-10650
-10725
-10800
-10875

Depth in Meters

"Western Pool"
(Site of *Trieste* dive in 1960)

"Eastern Pool"
(Site of *Deepsea Challenger* dive in 2012,
Limiting Factor Dives #1, #2, #3 in 2019)

"Central Pool"
(*Limiting Factor* Dive #4 in 2019)

Map prepared by the
Five Deeps Expedition (2019)

Sonar Mapping by the *DSSV Pressure Drop* employing a
Kongsberg EM124 Multibeam Sonar (April 26 – May 4, 2019)
Note: All depths above 10,300 meters shown in red

"It's a spiritual thing": Terry Kerby (*left*) in his element.

Below: The *Pisces* investigates the wreck of the Japanese midget submarine that provoked the first shots in the Pacific War on December 7, 1941. It was sunk by the USS *Ward*, an American destroyer patrolling the entrance to Pearl Harbor.

"Pele's handiwork": the iron-based ecosystem (*above*) and pillow lava (*below*) on Lōʻihi volcano, with the *Pisces IV* sub in the background.

"A knack for finding remarkable vents": Deborah Kelley (*left*).

The summit of Inferno, a black smoker hydrothermal vent on Axial Volcano (*below*).

"Weirdfish": also known as *Genioliparis ferox*. Photographed swimming 9,518 feet down at Slope Base, on the Regional Cabled Array (*left*).

Inside the Jason control van (*left*).

The ROV Jason being recovered at night; a rattail, one of the deep's regular customers; and a spider crab on Axial's sheet lava (*below*).

Sully, a black smoker hydrothermal vent on the Juan de Fuca Ridge, resplendent with tubeworms (*below*).

Lost City's IMAX
Tower, a dramatic
outgrowth on the
two-hundred-foot-tall
Poseidon edifice (*left*);
carbonate flanges trap
clear vent fluid, forming
upside-down reflecting
pools (*below*).

to go back inside the house, leaving me with a thought: "Maybe there will be an opportunity for you to join the ship."

*

I decided to call Lahey first. Walsh had filled in some details, but I wanted to know more before reaching out to Vescovo. I was aware of Triton's reputation for building cutting-edge manned submersibles. If you've ever seen a sub that looks like something George Jetson would drive—an aquatic spaceship in which the pilot and passengers sit inside a clear acrylic bubble—you've seen the company's level of innovation. Its subs have revolutionized the way people experience the deep. To fly through the twilight zone in one of Triton's transparent pressure hulls, with its 340-degree panoramic view, is to view the ocean as one big psychedelic aquarium.

So if you are someone who would like to buy a submersible, and you would like it to exceed the performance of other, less revolutionary submersibles, and you're not the type to balk at large numbers on an invoice—Triton is the company to call. In September 2014, Victor Vescovo had done exactly that. Unlike other clients who'd chosen Triton's acrylic-sphere models—subs that could dive to a few thousand feet and were lightweight enough to launch from the average mogul's yacht—Vescovo wasn't interested in ordering off the menu. He had something different in mind: a submersible that could dive to thirty-six thousand feet, giving him free run of the hadal zone. At a time when robots were trundling across Mars and artificial intelligence had learned how to smell, that probably seemed to him like a straightforward request.

Far from it. A safe and reliable hadal passenger sub—was that an oxymoron? There are some stubborn laws of physics to contend with, basic questions of how to keep a person alive in a bubble that's being stomped on by sixteen thousand pounds of pressure per square inch. (For reference, that's 292 fully fueled 747s stacked on top of you.) The sub would need to be strong enough to withstand implosion, but agile enough to handle in rugged terrain; brawny enough to trust with human cargo, but compact enough to launch from a midsize ship. Every wire, bolt, circuit board, and battery on it, every last capacitor, O-ring, and gasket, would need to be fail-safe under gargantuan pressures, in freezing temperatures, in corrosive salt water, and they would need to stay that way over time.

For a raft of technical, financial, and even psychological reasons, such a vehicle had never existed before. To date, no nation had deemed it economically feasible, or otherwise worth the hassle, to build a modern full-ocean-depth sub that could dive repeatedly with crew (although China claimed it was trying). No institutions had gone for it, no military agencies, no tech-bro billionaires. Few people were clamoring to visit Hades—not even scientists, who thought it was more prudent to send robots. But there wasn't a surplus of those, either.

The engineering challenges of sending *anything* on regular round trips below thirty thousand feet were so complex that, so far, only four robots had managed to work in the trenches—and not without trouble. Kaiko, a forty-million-dollar Japanese robot, broke away from its tether and was lost at sea in 2003. Six years later, the Woods Hole Oceanographic Institution debuted Nereus, a hybrid autonomous and remotely operated vehicle, the most sophisticated deep-sea robot ever built. There wasn't much that Nereus couldn't do—except survive in the hadal zone, apparently. In 2014, it imploded in the Kermadec Trench, directly south of the Tonga Trench. Scientists watched in despair from their research ship as bits of the robot floated to the surface.

During that time of hadal frustration, there was one exceptional event. On March 26, 2012, the film director and ocean explorer James Cameron became the third person in history to touch down in the Challenger Deep—and the first to do it solo, in a custom-built one-man sub—fifty-two years after Walsh and Piccard had descended in the *Trieste*. When you consider that during that half century two hundred people flew to the International Space Station, and thousands stood atop Mount Everest, it's boggling that it took humanity so long to make a second jaunt to the bottom of the world.

Cameron's dive was a reminder that the great depths were still down there—and still unexplored. "In the space of one day, I've gone to another planet and come back," he said upon surfacing. (And Walsh was there, too, standing on deck to congratulate Cameron as he climbed out of the hatch.) The event felt so monumental to me that I can remember crying as I sat in my office, glued to *National Geographic*'s website so I could follow the dive's progress in real time. To anyone who cared about the ocean's deep abiding mysteries, it mattered. A lot.

But Cameron's submersible, the *Deepsea Challenger*—a neon-green, rocket-shaped vehicle—never dived again. During its two hours and thirty-eight minutes on the trench floor, eleven of the sub's twelve thrusters failed, among other mechanical problems. Though the *Deepsea Challenger* completed its mission safely, it had plainly suffered in the hadal zone.

Seven years had passed since then, an eternity in technological time. Everything had advanced: batteries, materials, electronics, software. (Even attitudes: Maybe we shouldn't just ignore a big chunk of our own planet?) Triton had accepted Vescovo's commission, and more important: it had completed the task. In late 2018, the company successfully debuted a full-ocean-depth submersible that resembled nothing that had come before it.

The two-person sub was shaped like a padded briefcase, with streamlined elliptical edges. Its pressure hull—a five-foot-diameter titanium sphere—was positioned at the bottom of the craft. Three acrylic viewports gazed out of the sphere, giving it the appearance of a chubby alien's face. And on December 19, 2018, Vescovo had piloted his new sub on its first hadal dive, plunging 27,480 feet to the bottom of the Puerto Rico Trench.

That alone was a champagne-popping feat, but Vescovo was just getting started. His goal was to dive to the deepest points in all five of the world's ocean basins—thus the expedition's name, the Five Deeps.

His reason for pursuing this goal was simple: because nobody had done it before. Along with the sub, Vescovo had bought and refitted a 224-foot ship, hired an experienced support team, and invited top scientists along. Moving on from Puerto Rico, the Five Deeps team had continued its run of hadal dives in the Southern Ocean's South Sandwich Trench—celebrating with a stop on South Georgia Island to do a shot of whiskey at Ernest Shackleton's grave. If, as Walsh had suggested, I might join this crew, I would be observing deep-sea history in the making.

*

When I reached him, Lahey was in his office at Triton's headquarters in Sebastian, Florida, but he was about to fly to Indonesia for Vescovo's next dive, into the Indian Ocean's Java Trench. He gave me a rundown of the itinerary. After Java, the Five Deeps would sail to the Mariana Trench (where Walsh would attend), followed by the Tonga Trench. Then the ship would return to the Atlantic, drop by the wreck of the *Titanic* (because why not?) and continue north to the Arctic Ocean for some final dives at a site called the Molloy Hole. When I said that it seemed like the expedition was going well, Lahey hesitated, and then replied somewhat cryptically: "We've learned some hard lessons." But he sounded upbeat.

Lahey was fifty-six years old. He was friendly and talkative and didn't seem to mind answering dozens of questions as I extracted his autobiography. I liked him instantly because he is as obsessed with the ocean as I am, even though, like me, he grew up in Southern Ontario. Lahey was raised in Ottawa, but when he was seven years old, fate intervened and his family moved to Barbados for three years. "I'd never seen the ocean before I went there," he told me. "I fell head over heels in love with it."

Salt water or bust: anything else was uninspiring. Upon return to frozen Ottawa, Lahey had fixated on the aquatic to a degree that confounded his parents. While other thirteen-year-olds read comic books, Lahey studied for scuba certification. While his friends watched

Hockey Night in Canada, he prepared for dives into icy lakes. When the time came, it wasn't a hard choice to scrap university in favor of commercial diving school. "My dad was gutted," Lahey recalled. "He was like, *'Are you kidding me?* You're a smart kid—are you just going to throw it all away?'" His father's professional visions for his son did not involve undersea welding and jackhammering, followed by hours of decompression in a hyperbaric chamber. Predictably, Lahey had a different take: "Imagine sitting in an office doing something you loathe just so you can bring home a paycheck."

At age eighteen he began his career as a hard-hat diver, doing construction work on deep-sea oil rigs and pipelines—a profession in which even a minor bungle can cost a person his life. Lahey had just hit his twenties when he got his first chance to pilot a submersible, a one-man craft called the *Mantis 11*. Clamping the sub's manipulator arm onto a guide wire, he descended to fourteen hundred feet to inspect the blowout preventer on an oil well. Despite the prosaic mission, Lahey, like William Beebe before him, was awed by the clarity and color of the water, altered by its radiant blues: "From my very first dive I was hooked."

The 1980s were a golden age of manned subs, a time before robots assumed the undersea labor, so Lahey's piloting skills were in high demand. He worked on oil platforms in the North Sea—"That was the Super Bowl of diving"—and in the Gulf of Mexico. When the space shuttle *Challenger* blew into pieces in 1986, Lahey was dispatched to find those pieces on the Atlantic seafloor. In the northern Mariana Islands, he took tourists on flybys of sunken World War II wrecks. In South Korea, he worked on the construction of a four-lane undersea tunnel.

He hopped from job to job, sub to sub—and back then, there were all kinds of wacky subs. There were subs shaped like flying saucers and footballs and crabs, with names like the *Deep Jeep,* the *Snooper,* the *Ben Franklin,* and the *Guppy.* For a moment, it was possible to glimpse a future in which submersibles were an everyday recreational vehicle, with people jetting around underwater like they'd drive down the highway. That moment didn't last, but Lahey—who had no intention of leaving the deep to the robots—envisioned another possibility. If the subs were smaller, sleeker, easier to operate and maintain, and desirable enough to become status symbols, then a promising market did exist: yacht owners. They were already at sea; money was no object. Why *not* have a sub as beautifully designed as an iPhone parked next to the cigarette boat and the helicopter?

In 2007, Lahey had cofounded Triton Submarines to create a new generation of eye-catching, user-friendly subs, with acrylic spheres, decent headroom, and plush leather seats. Their maximum depths didn't extend beyond the twilight zone, but that was deep enough for a taste of the sublime. The subs were a hit with yacht owners—and with filmmakers and scientists, who reveled in the immersive visual experience.

When the hedge fund magnate Ray Dalio loaned his Triton sub to a group of marine biologists in 2012, they captured the first footage of a giant squid hunting off Japan. Previous encounters with the animal had meant ogling a dull mauve corpse: before that dive, nobody had ever witnessed a giant squid in action. The scientists were stunned, because far from being blandly colored, the massive creature looked metallic, as though it had been dipped in silver and bronze. It moved with the fluidity of water itself, its long tentacles studded with suckers, its hubcap eye gazing directly at the camera.

How many more epiphanies were down there? A good rule of thumb is the deeper you go, the stranger things get—and now Lahey had mobilized his decades of experience to build a sub that could explore the deepest reaches of the underworld. As far back as 2011, Triton's website had included a mockup of the 36000/3, a sub that would take three people to thirty-six thousand feet, or full ocean depth. But this dream machine existed only in pixels: no vehicle was headed to Hades unless someone with a burning desire to dive below twenty thousand feet stepped up to write a series of seven-figure checks. And extremely wealthy people who want to be sealed inside a metal ball and sent plummeting for miles into the ocean's sepulchral blackness are not found on every street corner.

But it took only one: enter Victor Vescovo. On the phone, Lahey described him as "a friggin' unicorn" and "a bit of a Vulcan," which sounded like an interesting mix. I was excited to speak to Vescovo, but he was en route to Indonesia, so at Lahey's suggestion I emailed him to ask if I could accompany the expedition on a dive. Sometimes weeks can go by before you hear back on a query, but as I would learn, that's not how Vescovo operates. His reply was prompt. "I'd be delighted to have you onboard for a spell," he wrote. "My suggestion would be Tonga."

*

Air New Zealand Flight 270 landed with a bounce on the palm-fringed runway, the pilot braking hard and wheeling the plane around in a brisk U-turn. We didn't have to taxi far. Fua'amotu Airport was a matchbox building on a wee island in a tiny country: the Kingdom of Tonga. Viewed from the sky, this archipelago of 170 islands is a string of commas and periods, a semicolon or two—punctuation marks in the Pacific Ocean's epic novel. Tonga's main island, Tongatapu, occupies a mere hundred square miles.

But if Tonga lacks in terrestrial heft, its surrounding waters are magisterial. To sail 180 miles south from the capital city of Nuku'alofa, on Tongatapu, is to find yourself floating atop thirty-five thousand feet of unquiet ocean above a seabed laceration known as the Horizon Deep. It's the deepest point in the 850-mile-long Tonga Trench, and the world's second-deepest spot, period—just a whisper shy of the Mariana Trench's Challenger Deep. You won't read about it in tourist brochures, but Tonga's ultradeep realm is one of the wonders of the underworld.

In a contest of extremes, the Tonga Trench and the Mariana Trench are well matched. They're the inverted summits of Hades—a regal pair of forbidding destinations, as implacable as interstellar space. Like all hadal trenches they were created by subduction: as one tectonic plate dives beneath another, that collision bends the down-going plate, forming a deep, V-shaped trench. There are approximately twenty-seven hadal trenches in the ocean, twenty-three of which are located on the Ring of Fire, the belt of subduction zones around the Pacific margin. Only four of these trenches plunge below ten thousand meters (32,800 feet)—the Mariana, Tonga, Kermadec, and Philippine—and though they're well hidden from us, these titans are among the earth's most dramatic features.

The Mariana Trench has starred in undersea horror movies, but the Tonga Trench is scarier, and that's before you factor in the eight pounds of plutonium in its depths, jettisoned during the aborted Apollo 13 mission. It's steeper, more severe, more seismically volatile—*busier*. At the Tonga Trench's north end, the Pacific plate is subducting beneath the Australian plate at the startling rate of nine inches per year. Nowhere else is a tectonic plate being gobbled with such relish, seamounts and volcanoes ingested like dinner rolls. It's a buffet of geological havoc.

Every so often the Tonga Trench gets indigestion and belches out an earthquake from way down in the mantle: a majority of the world's deepest quakes originate there, rumbling hundreds of miles beneath the seabed. In 2009, a slab of the Pacific plate cracked as it was being subducted, and the Tonga Trench roared. A magnitude 8.1 earthquake triggered two magnitude 7.8 earthquakes, and all three earthquakes shook simultaneously, generating tsunami waves that ravaged Tonga and Samoa.

As an encore, one of Tonga's seafloor volcanoes burped up a new island, more than two miles long and a half mile wide, now known as Hunga Tonga-Hunga Ha'apai. (In January 2022, this same volcano would be historical in its fury, unleashing an eruption that blasted steam and ash thirty-six miles up into the mesosphere, created a two-hundred-and-ninety-foot tsunami at its epicenter, and sent shock waves around the globe.) In 2019, another Tongan island called Lateiki disappeared into the depths during a submarine eruption, only to pop up again in a slightly different location.

A set of stairs was towed to the plane, and a flight attendant unsealed the door. I hauled my bag from the overhead compartment and stepped into the heat and sunshine and jasmine-scented air. In a gallery facing the airfield, Tongan families awaited relatives returning from Auckland, the nearest metropolitan city. The returnees waved in delight, juggling boxes and overstuffed shopping bags; a ukulele band serenaded their arrival.

I walked across the tarmac to the terminal, swept along with the crowd. After a brief inspection—"How about you, lady? Do you have any guns? Knives? Drugs?"—I was shunted into the baggage claim, a sweltering room packed with people. A ceiling fan ticked uselessly overhead. I looked around to find the seven men I was traveling with, a group I'd met for the first time at the boarding lounge in Auckland.

It was Lahey and his team of experts from Triton, fresh off a dazzling high. In May 2019, three weeks before I arrived in Tonga, Vescovo had become the fourth person ever to touch down in the Challenger Deep. He and Lahey had gone on to pilot four more dives in the Mariana Trench—for a total of five plunges below thirty-five thousand feet. And now the expedition was on to its next Pacific quest: the first manned exploration of the Challenger Deep's twisted sister, the Horizon Deep.

*

In the customs area I saw two Triton technicians, Frank Lombardo and Tim Macdonald, being pulled aside for extra scrutiny. Their bags were clanking with machine parts, thruster motors, pipes, hoses, electronic components sprouting wires and dials—kit that'll earn you a hard second look in an airport.

Macdonald, a thirty-year-old Aussie, was the youngest member of the Triton tribe. He had curly auburn hair, an athletic build, and the sunny vibe of someone whom life had deposited exactly where he wanted to be. One of Macdonald's passions was big-wave surfing; a little turbulence on land didn't rattle him. Lombardo, too, looked unperturbed as his suitcases were torn apart. He was a tall, tough Floridian with salt-and-pepper whiskers, a "no bullshit, I've seen it all" southern drawl, and decades of hazardous commercial diving work on his résumé.

Outside the terminal I caught up with Lahey. He was standing on the sidewalk, surrounded by luggage. "We're always humping equipment for the ship or the scientists or the sub," he explained, laughing. "We're mules, basically." Lahey had a lively sense of irreverence: watching him with his team, it was obvious that he was as much a ringleader as a boss. Even his hair—silver-white and spiked in a brush cut—was energetic. "We've wanted to build a sub like this for a long time," he told me, the excitement clear in his voice. "There's only so much doodling you can do on the back of a napkin. Victor gave us that opportunity."

Soon we were joined by senior Triton tech Kelvin Magee, a brick of a guy with a wide smile. Magee, a fifty-three-year-old from British Columbia, had a reputation for MacGyveresque problem solving. He was a submersible pilot, all-around ocean veteran, and someone you'd want in your camp for the apocalypse: his nickname is "the Kelvinator." "We're missing a bag," he informed Lahey, adding: "It's got ten thousand dollars' worth of titanium parts in it." Then Macdonald emerged from the terminal.

"Hey, Tim!" Lahey called. "Did you get the old cavity search, man? As soon as they see that Australian passport . . ."

"Yeah, you know it," Macdonald replied with a grin.

Lahey turned to me. "Tim is actually a very talented mechanical engineer. He's highly educated, although he doesn't look like it."

Lombardo and a few others were missing in action, presumably lost in customs. I was cotton-headed and lousy with jet lag, so I decided to leave for my hotel. Vescovo's ship, the *Pressure Drop*, was still offshore, mapping the Tonga Trench; it was due to arrive in the harbor the next morning. (The expedition was arranged to allow people to fly home between dives, with a transit crew sailing the ship to its next destination.) I told the Triton crew that I would see them on board, and then I grabbed a seat on a departing shuttle van.

*

"Welcome to the *Pressure Drop*," said Rob McCallum, standing on deck in khaki shorts, river shoes, and a polo shirt with the Five

Deeps crest, a black shield bearing the Latin phrase IN PROFUNDO: COGNITIO (In the Deep: Knowledge). I walked up the ship's gangway and shook his hand. Don Walsh had prepped me about McCallum, rating him as "the key to all of this." Given that McCallum was the expedition leader, it seemed like a fair assessment.

With its global scope, sky-high risk factors, and weather-dictated timetable, the Five Deeps was a Gordian knot of logistics, but McCallum was steadily untangling it. "We're taking a group of people who've never worked together, a ship that's been refitted but never tested, and a prototype sub, and we're going off to do something that's never been done before, and wasn't really thought possible," he told me. "Other than that, piece of cake."

The fifty-four-year-old Kiwi was a former New Zealand national park ranger who grew up in Papua New Guinea; he'd spent his career solving gnarly problems in remote locations. His company, EYOS Expeditions, runs about fifty ambitious journeys each year for clients who don't like to hear the words *no* or *can't* or *impossible*. McCallum had led James Cameron's expedition to the Challenger Deep, which made him a logical choice to lead Vescovo's, but more than that, McCallum is in a class of one. He is *the* guy you want to handle your exotic life-threatening adventure.

I set my duffel bag down and took in the scene. The *Pressure Drop* was a working ship; you couldn't mistake it for a yacht. It is a U.S. Navy *Stalwart*-class vessel, one of eighteen originally built to eavesdrop on Russian submarines during the Cold War. (Back then its name was the *Indomitable*.) Lahey had found it dry-docked in a Seattle shipyard and recommended it as a support vessel. The ship needed a herculean overhaul, but its spying background meant that it was quiet, which is ideal for subsea communications. In the deep, Vescovo speaks through an acoustic modem: it takes seven seconds for his voice to travel from the trench floor to the ship (and vice versa). The last thing anyone needs is for urgent hadal messages to be garbled by engine noise.

The *Pressure Drop*'s decks sported an assortment of cranes, three smaller boats, a satellite dish, and an outdoor lounge above the bridge, called the Sky Bar. A white structure stretched across a mission deck at the stern: the sub's hangar. Shrouded by walls, sheltered from the elements, the vehicle had a mystique. I was dying to see it, and Lahey had promised to give me a tour later.

In front of the hangar, three instruments known as landers were lined up like Smart Cars in tight parking spots. The landers were boxy and bulky and unexciting to look at, but they were as advanced as the sub, and integral to the dives. "They're the pickup trucks of the science operations," Lahey had explained. Equipped with cameras, baited traps, probes, sensors, and tracking and communications gear, the landers are sent down in advance to measure water temperature and salinity—information that's needed to ballast the sub correctly. On the bottom, they lure animals hoping for a meal, film the diners, collect specimens, take water samples, and serve as navigational beacons. (There is no GPS underwater.) When their work is done, they're recalled with an acoustic code that causes them to drop their weights and rise to the surface.

The landers' odd names were emblazoned down their sides: Skaff, Flere, and Closp. The submersible itself was called the *Limiting Factor*. Vescovo, a science-fiction fan, had named the ship, sub, landers, and support boats after the precocious superintelligent machines in the late author Iain M. Banks's Culture novels.

McCallum explained that Vescovo would arrive the following morning and then we'd leave, timing our departure to skirt a major storm. It would take us twenty hours to sail to the trench, with four days planned at the dive site, a contingency day for weather, and a day to return to Nuku'alofa, making it a weeklong trip for forty-three passengers and crew. McCallum asked a deck officer to show me to my cabin, and left me with instructions: "Right, so at four you'll head up to the bridge for a safety brief. Please join us for drinks in the Sky Bar at five. And if you need me I'll be down in my little office, or running around like a madman."

*

Vescovo boarded at noon the next day, coming into McCallum's office with a gust of energy so intense I wondered if it might knock the skull-and-crossbones flag off the wall. A dozen or so people were milling around in the adjacent dry lab, a spacious room that served as a mission control area and central gathering spot. Vescovo greeted everyone with enthusiasm. His body language said: *Let's get this adventure started.*

Vescovo had a striking presence. He didn't look like anyone I'd met before. He is a lean, kinetic man, six feet tall, with ice-blue eyes, blond hair that he wears in a longish ponytail, and a close-cropped silvery beard. His aquiline features give the impression of a raptor—but a polite one. When Vescovo smiles, he uses all his teeth, and you can see flashes of a ten-year-old kid, amped on curiosity. Yet there's also something ageless about him, as though he might've teleported from another era, or possibly another planet. ("I think he looks like a Bond villain," McCallum would say at one point. "Or a White Walker," Vescovo had countered, casting himself in *Game of Thrones*.)

"So let's see, I'm at the bottom of the deep, and I'm in my own titanium submarine," he said, laughing, when I asked him how things were going. "I can do whatever I want. I'm looking around, I'm seeing sea creatures—oh yeah, I'm having a blast!"

One glance at Vescovo's biography tells you that his idea of a blast is the high-octane sort. He has completed the Explorer's Grand Slam, summiting the highest peak on every continent, and skiing to the North and South Poles. (On Mount Everest, he survived what he calls a "minor avalanche" in the Khumbu Icefall.) He's conversant in seven languages. His proficiency in Arabic came in handy during his twenty years as a U.S. Navy Reserve intelligence officer with top-secret clearance, especially right after 9/11. He pilots his own Euro-copter 120 helicopter and Embraer Phenom jet, which he uses to fly rescue dogs to new homes. For relaxation, he studies military history.

According to Vescovo, the roots of his go-big-or-go-home philosophy can be traced back to age three, when he snuck into the family sedan, released the emergency brake, and sent the car hurtling down the driveway into a tree. The resulting crash cracked his skull in three places, shattered his jaw, broke his leg and some ribs, and provided an early whiff of mortality: "I realized that every day is precious, and you may not get another one—best make full use of them."

No one could say that he hadn't taken his own advice. At Stanford, he'd completed a four-year political science and economics degree in three years; at MIT, he finished his master's in defense and arms-control studies in ten months. He was recruited by the Navy while taking his MBA at Harvard, then served on active duty in the Persian Gulf, at NATO's Allied Joint Force Command in Italy, and at the Joint Intelligence Center Pacific at Pearl Harbor.

If you're thinking that sounds like enough to keep anyone busy, think again. Simultaneously, Vescovo was building his business career. He worked in management consulting, on Wall Street, at a successful dot-com start-up. Now, at age fifty-one, he was a founding partner of a private equity firm—a job that funded his adrenaline habit and pursuits like the Five Deeps, with its price tag of fifty million dollars (and counting).

Vescovo walked over to view the bathymetric maps of the Tonga Trench that were displayed across screens in the dry lab, with the Horizon Deep identified by a yellow dot at 10,820 meters (35,499 feet). The maps had been made with the expedition's multibeam sonar array, welded to the ship's hull. In the beginning, Vescovo hadn't known that he needed a sonar system. He'd figured the coordinates of the world's deepest spots were known. (Spoiler alert: they weren't.) In some areas like the Mariana Trench, the depths had been charted—with room for improvement. But in the Indian Ocean, no one had identified the deepest spot; and in the Southern Ocean, the bathymetry of the South Sandwich Trench was largely unknown. Before Vescovo could dive to the Five Deeps, first he had to find them.

This wasn't a simple proposition. It wasn't something you could determine by scrolling through Google Earth. In world atlases the trenches are visible as dark clefts in the seafloor, and they appear in approximately the right locations. But most of those data come from satellite altimetry, and they're fuzzy at best.

Even if you have a ship and a state-of-the-art sonar system, it still isn't easy to make precise depth measurements through six or seven miles of water. As the sonar passes over an area it shoots sound waves downward, and then measures the speed of their return to create a three-dimensional profile of the terrain below. But if you're after the *exact* depth, here's the tricky part: sound waves travel at different speeds depending on water temperature and salinity, which vary because the ocean isn't homogeneous. It's more like a layered cocktail that keeps shifting and moving. (These layers, known as thermoclines, aren't necessarily subtle. On their descent in the *Trieste,* Walsh and Piccard had smacked into colder layers, gotten stuck, and had to adjust the bathyscaphe's ballast until they were heavy enough to break through.) The only way to account for these fluctuations is to painstakingly sample the water column all the way to the bottom. With a full-ocean-depth sub and landers on board, the Five Deeps team was in a unique position to do this, to get complete sound velocity profiles for any stretch of ocean, no matter how deep—to measure Hades in situ.

*

The question of where to dive, which came up often, was answered by an impressive group of scientists who'd become involved with the expedition. The chief scientist, Alan Jamieson—at the time, based at England's Newcastle University—was a well-known hadal biologist who had written a definitive book on the hadal zone. After reading it, Lahey had reached out to gauge Jamieson's interest in joining the expedition.

"I dragged Al into this, and I think at first there was some question as to whether I was just a raving lunatic," Lahey had recounted the previous night over beers at the Sky Bar, giving me the backstory on Jamieson, who'd gone into Nuku'alofa for some land time after weeks on the ship. As Lahey was talking, Jamieson had shown up, sat down with us, and continued the anecdote.

"I got a call from some crazy guy in Florida, banging on about some other crazy guy," Jamieson recalled. "But when somebody phones you up and says, 'Do you want to go around the world on a privately funded hunt with a naval intelligence guy with his own multi-, multi-, multimillion-dollar submersible?' you automatically have to say yes."

Jamieson was tall, in his early forties, with blue eyes, brown hair, and a tendency to glower even when cracking a joke. He wore a black T-shirt, black cargo shorts, black sunglasses, and black steel-toed boots. "He's a good one. You'll want to spend time with him. Wicked sense of humor," Don Walsh had advised me, and then added: "He's a Scotsman. You can hardly understand what he's saying."

Sure enough, Jamieson is from Longniddry, a village in East Lothian, Scotland. When I asked how he'd come to study the ocean's most inaccessible region, he replied, "Oh, I just got drunk in a bar one night." What actually happened was that while studying industrial design at university, he became intrigued by deep-sea landers. They were essential tools for marine scientists, less of a production and more available than ROVs. But the landers were expensive, unwieldy, and limited in function, like experiments someone had cobbled together. Jamieson thought they could be smarter, smaller, less costly—better in general. As a final-year project, he designed one.

His lander worked; its success opened doors. After graduation he was hired by the University of Aberdeen's Oceanlab, a marine science and engineering group. Within months he found himself on a ship in the North Atlantic, building increasingly elaborate underwater gear. As his work progressed, Jamieson realized that many of the baited landers were failing to attract deep-sea fish: "Not because of the technology, but because we didn't understand fish behavior." In order to lure wily creatures thousands of feet down, he needed to think like they did.

When he published a paper titled "Behavioral Responses to Structures on the Seafloor by the Deep-Sea Fish *Coryphaenoides armatus*: Implications for the Use of Baited Landers," it became evident that Jamieson, who "never had aspirations to be a marine biologist," had a serious talent for deep-sea biology. At his boss's urging, he wrote a dissertation and got his PhD. Before long, Jamieson noticed another shortcoming within his field: almost nobody was working at full ocean depth. Scientific ambitions seemed to stop at the abyssal plain. "So I started thinking, *Why doesn't anyone ever go into the trenches?*"

Expense? Difficulty? Remoteness? Those didn't seem like insurmountable problems, so Jamieson aimed deeper. He built innovative, low-cost hadal landers—tripod-shaped platforms with cameras, sensors, and two types of baited traps—and rustled up opportunities to use them. For years, Jamieson cadged ship time from a group of German brain surgeons whose funding to study the weird eyeballs of deep-sea fish dovetailed nicely with his arsenal of gear. While the brain surgeons trawled for fish with binocular eyes, or eyes that could swivel in any direction, or eyes with superpowered retinas, he dropped landers onto the seafloor. He made it a priority to join any research cruise that would give him access to the hadal zone.

Back then, in the early aughts, no one had documented fish below twenty thousand feet with any measure of certainty. Specimens had been caught in nets, but because the nets didn't close, there was no way to guarantee the animals had come from the very bottom. Landers could solve that problem by documenting the scene with video, and in 2007, Jamieson scored. One of his landers filmed a snailfish at 22,600 feet, an angelic-looking, translucent pink creature with a wide head, fluttery pectoral fins, and a body that tapers to a point. It was the first confirmed record of a hadal fish: the first glimpse of its behavior, its movements, how it fed, what it ate. All of which were startling.

The trenches were thought to be food deserts, the organic particles that rained from surface waters having mostly been consumed on the way down. Any creatures that managed to adapt to the pressure, therefore, must be starvelings creeping around to save energy, floating skeletons, virtual ghosts. But Jamieson's snailfish was a fat little predator that swam vigorously, like a shallower fish. Its prey was amphipods—insect-like crustaceans that swarm the lander's bait—and it was eating them like cocktail peanuts. (In the hadal zone, there is no shortage of amphipods.)

It was an auspicious start to what would become a downward quest among hadal biologists: Who could find the deepest fish? The deepest—anything? Who could take the census of a trench? What animals flourished below twenty thousand feet, and how did they manage it? How deep was too deep for a species?

*

This ultradeep focus meant that Jamieson was always heading off into twitchy subduction zones, and he'd experienced their tantrums firsthand. When the 2011 megaquake struck Japan, he was loading his landers onto a ship in Tokyo's port, preparing to sail to the area of the Japan Trench that would later be identified as the epicenter. Nine months later, he endured another major earthquake in New Zealand—so strong it shook him awake in his hotel room—caused by lurchings in the Kermadec Trench.

If investigating Hades demanded more from him—more risk, more time at sea, more effort to crack its secrets—he was willing to pay that price. He preferred adventures to desk time, and the pace of discovery was worth it. He and his wife, Rachel, had three sons under age ten, so family life beckoned, but Rachel Jamieson was also a marine scientist: the two met on a research cruise in the Indian Ocean. She understood the calling, though it's likely she wasn't thrilled when Jamieson missed the birth of their second son while hunting for giant amphipods off New Zealand. ("Kid didn't arrive on time. He was too late. It was his fault," Jamieson joked.) When the Five Deeps invitation came along, he signed on without hesitation, though it meant months away from home. He made one agreement: "I categorically promised my wife that I wouldn't get into the sub."

It hadn't seemed like an option anyway. Vescovo was going for solo depth records; diving alone was his personal grail. Jamieson was concentrating on the landers. Along with Vescovo's three full-ocean-depth bruisers, he'd brought along two landers of his own. The chance to deploy five landers in multiple trenches, back to back to back, was reason enough to be on the ship.

But after his own record-setting plunge into the Java Trench, Vescovo had turned to Jamieson and asked, "If we put you in the sub, where would you want to go?" It was a tantalizing question. The Java Trench had turned out to be far more rambunctious than expected. Jamieson was struck by the bathymetric maps, which showed evidence of enormous landslides along the trench's walls—huge collapses from earthquakes. These represented major geological events that had surely caused past tsunamis, and no one had even known they were there. "All right, if we're gonna dive, let's do something really mental," Jamieson recalls thinking to himself. "Let's not ease into this. Let's find the most ridiculous structure at the deepest point, and just go for it."

Two days later Jamieson squeezed through the *Limiting Factor*'s hatch, off to become the first hadal scientist to visit the realm that he studied. Lahey was the pilot. Their destination, a vertical pitch at twenty-three thousand feet, would be among the most extreme terrain ever explored in a manned submersible. Jamieson, though not a claustrophobe, wasn't relaxed in the five-foot sphere: "The hatch leaked the whole way down, which was unnerving."

But the dive itself was a revelation. They found themselves in a cave near the bottom, its walls and ceiling covered with sponges that resembled bats. As Lahey maneuvered the sub from under the overhang and ascended a slope, Jamieson stared in wonder at a rainbow of chemosynthetic bacterial mats—canary yellow, tangerine, twilight blue—woven into the stark rock. There were crystal-white anemones and purple holothurians. He watched, enthralled, as a snailfish ambled by. (Not only was Jamieson face-to-face with his elusive study subject, it was the first-ever encounter between a human and a live hadal snailfish.) Video cameras and ROVs weren't doing the deep justice, he realized. They conveyed none of its brilliance. None of its magnitude.

It was a turning point. Frustrated by the expedition's early months, Jamieson had been planning to drop out. He'd felt that science was being treated as an afterthought, amid a rash of delays and problems. Both of his landers had been lost in the Southern Ocean. He'd had the flu.

Now Jamieson was a wide-eyed convert, a kid unleashed in a toy store. He called home from the ship's satellite phone. "Remember I said I wouldn't dive in the sub?" he asked Rachel. "I knew you would," she replied, happy to be hearing about it after he was safely back. He confirmed that he would stay on the ship as it sailed into the Pacific, but assured her that he wouldn't be seeing any of its ten-thousand-meter trenches in person: "I won't do the stupid deep ones. I'm not on the list. I'm not going on the Apollo missions." Recounting the conversation, Jamieson had paused for effect before delivering the kicker: "So then ten days later we get to the Mariana Trench."

First, Vescovo had made two solo dives to the Challenger Deep. Then Lahey dived with German engineer Jonathan Struwe from DNV-GL, the marine classification society that officially certified the *Limiting Factor* for unlimited dives to full ocean depth. On a fourth dive to the Challenger Deep, Lahey was joined by Triton's principal engineer, John Ramsay, the sub's designer. Then Jamieson climbed into the passenger seat again, diving 35,151 feet with Vescovo to the Sirena Deep, the Mariana Trench's second-deepest spot.

Once more: magic. The Sirena Deep was an undulating carpet of fine gold sediment, punctuated by sharp rocks. Vescovo and Jamieson flew up and down rolling hills and little cliffs. They saw lemon-yellow sulfur mounds and crinoids that looked like daffodils. The trench floor was pocked with burrows and dimples and holes—the condominiums of countless creatures. Indigo- and ochre-colored bacterial mats furred the rocks like moss. The Sirena Deep sang in its own key of mystery, and they were the only two people on earth who knew the tune.

*

Now, as the ship prepared to leave for the Horizon Deep, Jamieson was examining the Tonga Trench maps with Vescovo, marine geologist Heather Stewart from the British Geological Survey, and the Five Deeps' lead ocean mapping scientist, Cassie Bongiovanni. The trench looked savage, with steep walls, jagged outcroppings, and what appeared to be a seamount rearing up in its center. In fact, Stewart explained, it was a giant ridge—a fold in the earth's crust, buckled by subduction.

"Tonga's just cool," Jamieson observed. "Look at the bottom—it's stunning. There are really, really interesting things going on."

"This one's crazy," Vescovo agreed, looking pleased.

Bongiovanni nodded. "This is a very dynamic area. I mean it's *really* aggressive."

"The sheer size of some of these fault escarpments and massive cliffs . . . ," Stewart said, her voice hushed. For her, the maps weren't theoretical: she was slated to accompany Vescovo on his second dive. She was visibly jazzed and quite nervous, not only because she would be the first woman in history to dive below twenty-three thousand feet, but also because of the intensity of this particular trench. "Nobody's ever done a transect of one of these walls at depth," she said quietly, as if to herself. *"At all."*

It was beginning to sound like a refrain. Every dive into the hadal zone was a dive to somewhere we had never been (with the exception, ironically, of the Challenger Deep). Every deep dive was a first. Every trench was more complicated than we knew—and no two trenches were the same. Earlier, Jamieson had quipped that he categorized them into two groups: "There are trenches that like us, and trenches that don't like us." The Java Trench had been generous. The Mariana Trench had been gracious. Would the Tonga Trench welcome visitors—or not? In forty-eight hours, we would find out.

CHAPTER 5

. . . Stays in Hades

Let everything happen to you: beauty and terror.
Just keep going.
—RAINER MARIA RILKE

THE HORIZON DEEP, 23.3° S, 174.7° W
SOUTHWEST PACIFIC OCEAN

As we steamed away from Nukuʻalofa, Rob McCallum gathered us for an all-hands meeting. It was a preliminary run-through of the dive plans, and a chance to hash out questions and concerns. The marine forecast was a worry. The winds and the waves were tame now, but starting tonight conditions would deteriorate. Then they'd improve briefly before becoming even worse. Submersibles can't be launched in heavy seas. On the Beaufort scale of one to twelve—from glassy flat to howling hurricane—anything above five is a non-starter. But if the weather gods were kind, that sliver of (relative) calm between storms would offer just enough time to complete both dives.

The dry lab was standing room only, with every chair taken and people leaning against the walls. Sitting next to McCallum was the ship's captain, Stuart Buckle. There was none of the leathery, bewhiskered old salt to Buckle; he was a good-natured, redheaded thirty-eight-year-old from a hamlet in the Scottish Highlands, who'd gone to sea as a teenager and cut his teeth in the North Sea oil fields. For

Vescovo, signing Buckle was a coup because he was the only captain alive who'd supported a manned dive (Cameron's) to the Challenger Deep. It's not easy to position a ship so that a submersible launched from its deck lands in the deepest divot in the Mariana Trench. To get a sense of the vertical distance and the relative scale of the target area, imagine an airline pilot trying to drop a car from thirty-five thousand feet into a specific parking lot. Buckle's condition for taking the job was that he be allowed to handpick his entire crew.

McCallum let the room settle. "Right, so we're on our way to the Horizon Deep, in the second-deepest trench on earth. We're seeking to identify and dive to the deepest point."

Vescovo, standing nearby, added: "First manned dive ever."

"Dive One is Victor solo to the Horizon Deep," McCallum continued. "Dive Two is Victor and Heather to a twenty-five-hundred-foot wall at ten thousand meters. Because the weather window is tight, we'll be doing back-to-back dives with no maintenance day in between." He paused to let that sink in. Each dive lasted twelve hours. The post-dive maintenance took even longer, and it wasn't like it was optional. Executing back-to-back ten-thousand-meter dives was a stress marathon, and it hadn't been attempted before.

McCallum went on, his voice turning sober. "The weather is what's shaping this week. There's a big low moving across the north of New Zealand—it's a whopper of a system. So we will see swell all week, with a wave height of about five meters and some pretty good wind, twenty to twenty-five knots."

"We're going to get our asses kicked," Jamieson offered.

"It'll be an ass-kicking," Lahey agreed.

"Tomorrow we'll have a full briefing for the dives, and discuss timing and where to put the landers," McCallum advised. "But I want to say one thing now. We're coming off the massive high of the Challenger Deep performance, and maybe you think after that it's all downhill. But this is serious stuff. We're going to send two humans down 10.8 kilometers in boisterous weather. It requires our A game. This is still as difficult as you can possibly get in submersible diving."

He scanned the room over the top of his reading glasses. "Okay, so we have some new faces on board."

The lineup of guests was eclectic. Along with me, it included a British artist, a Tongan geologist, a Japanese ocean-mapping specialist, and a Canadian deep-sea icon. Throughout the dives, Vescovo had invited eminent ocean explorers like Walsh to join the ship. On this leg, that post was occupied by Dr. Joe MacInnis, a soulful eighty-two-year-old physician from Toronto, Ontario.

MacInnis was a celebrated aquanaut, author, and speaker, and a pioneer in the field of diving medicine: the physiology and psychology of how people cope (or not) with exposure to depth. In his career he'd overseen one daring dive after another—and made many himself—as he probed the limits of human ability. At depths where breathing compressed air would cause nitrogen narcosis or oxygen toxicity, MacInnis had experimented with breathing neon and argon, among other things you shouldn't try at home. It was his job to prevent saturation divers—who flood their bodies with inert gases so they can work hundreds of feet down for extended periods of time—from dying in gruesome ways, such as having their lungs explode or their flesh sucked into their helmets.

MacInnis had been the first person to dive under ice at the North Pole (where he planted a Canadian flag), and he discovered the world's northernmost known shipwreck, the HMS *Breadalbane,* preserved like a time capsule on the seafloor of the Northwest Passage. (The three-masted barque was trapped by pack ice in 1853, while searching for Sir John Franklin's lost *Erebus* and *Terror* expedition.) Like Jacques Cousteau and other aquatic visionaries in the mid-twentieth century, MacInnis was intrigued by the prospect of people living undersea in high-tech habitats. "It is a chance to find harmony with the major portion of the planet," he wrote in his 1974 book *Underwater Man.*

In 1969, as an experiment, he'd built a research station called Sublimnos and installed it in Lake Huron, forty feet below the surface. It had windows and a domed skylight and was roomy enough for four people. Compressed air, hot water, and twelve volts of power were piped in from shore. On one night in July that year, MacInnis would later tell me, he sat in Sublimnos and gazed up at the moon through the water, knowing that his friend Neil Armstrong was striding across it at that very moment.

Sublimnos had captivated Lahey in his youth, along with another Ontario teenager, James Cameron, who wrote to MacInnis for advice on building his own subsurface lab. These days, MacInnis still advised Cameron on his undersea endeavors. "Joe's famous in the deep submergence world," McCallum told the group. "But he's also a doctor, so he's here if you don't feel well. If you've got a nasty rash, please show it to Joe."

Across the table from MacInnis was another legendary aquanaut, the former French naval commander Paul-Henri Nargeolet. Early on, Vescovo had hired Nargeolet, who was seventy-three, as a technical advisor. Nargeolet was a dashing character and a habitué of the abyss; he'd helmed France's six-thousand-meter submersible, the *Nautile*. He was also an undersea demolitions expert who had cleared thousands of World War II mines from the seafloor—an inventory that included many booby-trapped bombs, courtesy of Hitler's troops. "There were tons and tons of them," he'd explained in his mellifluous French accent. "It was not nothing, you know? It was not a little stuff."

Nargeolet had retrieved all manner of urgent items from the deep—historical artifacts, military weapons, downed helicopters and airplanes, black boxes and bodies—but he was best known for his expertise on the *Titanic*. He piloted the first manned dive to its wreck site in 1987, and made more than thirty subsequent visits. (On one of those dives he'd taken MacInnis, who produced an IMAX film about the doomed ocean liner—and talked up the *Titanic*'s spooky beauty to his friend James Cameron, who proceeded to make a little *Titanic* movie of his own.)

As the meeting wrapped up, Buckle had a final note. "For those of you who haven't been on board, we are quite a lively ship. We do roll around a lot."

McCallum nodded. He pointed to a tray outside his office that was set up like a minibar of seasickness medications. "Please help yourself. I'll chuck a Stugeron if I know it's going to be a real beating."

"You're not a wuss if you take them, but you're an idiot if you don't," Buckle warned.

"Are there any landmasses out there?" someone asked from the back of the room.

Buckle shook his head. "Nothing. There's nowhere to hide."

*

After the lifeboat drill I met up with Jamieson in the wet lab, a narrow room lined with metal countertops and sinks. He had two extra-cold freezers stocked with hadal animals, and a collection of formaldehyde-pickled deep-sea fish that I was eager to see. By this point in the expedition Jamieson had racked up sixty-three lander dives, filmed ten terabytes of video, and captured a trove of beasts that would all be donating their DNA to science. "We're looking at the population of the trench from the shallow end all the way to the deep end," he said, pulling on a pair of thick rubber gloves.

The chemical locker was outside on deck. Jamieson unlatched its door and removed a plastic cask that was clamped tightly shut. "Don't inhale," he said, prying off the lid. Dunking his hands into the formaldehyde solution, he removed a cod-size fish. "This is the shallowest of the grenadiers, a classic rattail."

I leaned forward to look. The rattail had a big pointy head, a sad droopy mouth, a sturdy plum-gray body, and a long tail that pinched to a point. Its eyes bulged from decompression, giving it a shocked appearance. "They're very curious, rattails," Jamieson noted. "They'll explore the landers with their barbels, like dogs who just *know* there's a sausage in there somewhere." He pointed to the barbel, a thick whisker under the fish's chin. "That's where its taste buds are."

This specimen was caught at eighty-two hundred feet, Jamieson said, but other rattail species have been found as deep as twenty-five thousand, nine hundred feet. "They're one of the commonest fish in the sea. They never get on TV because they're not charismatic enough." He turned the rattail over, checking its condition. "Its skin is amazing—really tough," he said, returning it to its liquid crypt and wiping his gloves on a towel. "These ones preserve all right. It's the snailfish that don't. They actually melt."

Hadal snailfish have no swim bladders or air cavities in their bodies; their innards are encased in a buoyant, transparent gel. Their bones are demineralized, giving them a floppy skeleton, and they lack a fully closed skull. "They need the pressure to maintain their body form," Jamieson said. "When you pick them up they're very, very delicate. It's like handling a water-filled condom. They slip around in your hand."

He reached into the locker for a Tupperware container. Inside, there was a baggie filled with clear peachy goo, and what appeared to be lumps of aspic: the remains of a snailfish. "It's not much to look at now," Jamieson said. "But they're absolutely beautiful in life." He pointed to some darker lumps. "That's its liver. Stomach. Intestines. And those are its eyes, the two black bits."

By now, Jamieson and other hadal biologists had determined how deep a snailfish could go: around twenty-seven thousand, three hundred feet. They'd also figured out the reason why. Hadal snailfish are saturated with trimethylamine-N-oxide (TMAO), an organic molecule that acts like scaffolding in their cells, allowing them to equalize at greater depths than other fish. But TMAO has its limits: with enough pressure even a snailfish's cells will collapse. Until evolution changes that equation, among the myriad life-forms at the bottom of a thirty-thousand-foot trench, you won't find a single fish.

The more I learned about snailfish, the more they charmed me. Of all the contenders for top predator in the harshest environment on earth—and the winner was a pink gummy bear? "They're great because they're the deepest fish in the world and they're not even a deep-sea fish," Jamieson explained. "They're a shallow-water fish that's so audacious it's overtaken all the deep-sea fish." Even now, hundreds of snailfish species live closer to the surface. But twenty million years ago, some of them began to venture downward. They moved from sunning themselves in tide pools and estuaries to hunting amphipods in hadal trenches at evolutionary warp speed. Now they dominate a realm where they can snack all day long with few competitors. Even better, nothing eats them.

There is one trade-off. A hadal snailfish's life span is relatively short, about six to twelve years. (Some abyssal fish can live to be eighty.) But that, too, makes evolutionary sense. Trenches are such unstable areas that their residents can't count on longevity. After the 2011 earthquake in Japan, scientists dropped cameras into the Japan Trench and found a seafloor Pompeii. "They've adapted to spawn quickly," Jamieson said. "It's just to do with sustaining a population in a ridiculously dangerous place."

He put the snailfish back in the box, snapped it shut, and returned it to the locker. "Have you come across anything surprising?" I asked. "Like something you really didn't expect?" He paused for a moment, thinking. "It's a lot of the usual suspects," he said, and then brightened. "Oh, in the Java Trench—I don't know if you saw the video with the big sort of transparent thing with the dog head and the tentacles?"

"*What?*"

"Here, I'll show you."

*

We walked back to Jamieson's office, a nook off the dry lab just big enough for a desk and two chairs. The tiny room was awash in science papers, coffee mugs, cameras, books, and deep-sea memorabilia. He shoved some papers off a chair, sat down at his computer, and pulled up a highlight reel of lander footage. A patch of seafloor from the South Sandwich Trench appeared on-screen. "It's the only subzero hadal zone," Jamieson noted. "A beautifully complicated trench with all sorts of different habitats in it." In the foreground, rock shards were strewn across a bed of tawny ooze. "So this is a pyroclastic flow from a volcano at six thousand meters. Lots of little volcanic nuggets that were ejected. And look—there's a new species of snailfish."

The snailfish swam by slowly. You could see right through its ribbony tail, as though it were a hologram. The lander's lights reflected off the fish, giving it a pearly gleam. It was popsicle size and tadpole-shaped, with most of its body mass front-loaded in its head. Inside that noggin, Jamieson told me, hadal snailfish have two mouths. With mouth number one, they suck in amphipods: "But if you put an amphipod in your mouth, the first thing it's going to do is eat its way out of your head." To solve that problem, the snailfish's second internal mouth consists of two grinding plates.

The video played on, switching locations to the Java Trench. Two metal bars reached out from the lander, holding a dead mackerel above a silty expanse that was spotlighted like a stage set. The mackerel's skin glinted silver, indicating its fresh arrival on the seafloor. Already the bait was swarming with amphipods, the trench's master scavengers.

To watch amphipods devour a carcass is to know for a fact that you don't want to be buried at sea. Still, you have to admire their alacrity: from a mile away, a morsel of dead flesh draws them like a dinner bell. Amphipods are not dainty eaters: they can ingest three times their body weight. "When the fish skeleton comes back it's completely clean," Jamieson said. "You can put it under a microscope and you won't find anything. It's just gone."

Unfortunately for the amphipods, there's no such thing as a free meal. Soon a snailfish appeared in the frame, lurking tactically. Then a round brown head jutted in from the side, attached to a long thick body. It was a Hindenburg of a creature with beady pinhole eyes and a huge pouty mouth. Jamieson identified it as a cusk eel, and in particular a species known unpoetically as the robust assfish. Like the snailfish, the assfish wasn't interested in the bait, but in the crustaceans that came for the bait. It hung there, motionless, as though it hoped nobody would notice it. Then a big red prawn paddled by and the assfish struck, its cavernous mouth snapping open and sucking in the prawn and a dozen amphipods like a vacuum cleaner.

The next visitation was less thuggish. "Dumbo octopus at six thousand meters—who would've thought?" Jamieson said, as an ethereal white and pink-tinged Pokémon character floated past the camera. Dumbos are rare, primitive animals that propel themselves with the help of two flappy fins perched on their heads like cartoon elephant ears. They have a bell-shaped mantle, black button eyes, and webbed arms they unfurl like an umbrella to fly through the water. The previous depth record for an octopus had been 5,145 meters (16,880 feet), so finding one three thousand feet below that was remarkable. (Jamieson would later film another Dumbo at nearly seven thousand meters, or 22,966 feet.)

The Dumbo sailed off, the video skipped forward, and a new time sequence appeared. "Watch," Jamieson prompted, as a spectral figure materialized from the darkness, gliding toward the bait. We stared in silence as it approached the camera. It was an apparition, a phantom, a psilocybin vision—but, as Jamieson had said, it also resembled a gelatinous dog's head trailing a white tendril. The head was luminous, and as crystalline as a bubble. It glimmered in pale shades of violet and topaz, with twinkles of aquamarine and white, and there were glowing orbs suspended inside it, like the electrodes of a cyborg brain dreamed up by Ridley Scott. When it arrived at the lander the creature made a ninety-degree turn, showed off its canine profile, and then exited the frame stage right. "That's the weirdest thing we've found so far," Jamieson declared, adding that it was a stalked ascidian, or sea squirt—albeit a species that nobody had ever laid eyes on before.

The video ended with an improbable shot of the snailfish, the robust assfish, the Dumbo octopus, and another hapless red prawn assembled around the mackerel, as though the trench's megafauna had been summoned for a class photo. Jamieson clicked the file shut, pushed his chair back, and smiled dryly. "As you can see, there's a lot going on," he said. "The idea of the deep as this barren, lifeless place is just horseshit."

*

McCallum's weather prediction proved accurate. During the night I was jolted awake when the ship's motion changed, from churning ahead to surging and rolling. After a light breakfast of biscotti, Coke Zero, and Dramamine, I pinballed down hallways and braced myself as I climbed up ladderlike stairwells and made my way across the deck to the hangar, where Lahey and his team were running the sub through its pre-dive safety check. The sky was crowded with big-bellied clouds and the ocean looked muscular. Later in the day we would arrive at the trench. If conditions permitted a launch, Vescovo would make his solo descent at eight o'clock the next morning.

The hangar was open to the stern, and the *Limiting Factor* filled most of the space inside it. The eleven-ton sub was thirteen feet tall, nine feet wide, and fifteen feet long. It was white, with sleek lines, its complexities hidden from casual view. Underneath its smooth skin there was a labyrinth of circuitry, systems upon systems with fail-safe redundancies, all designed for one purpose: to keep its passengers alive under the weight of eleven hundred atmospheres.

I stood for a moment watching the men work. Lahey was talking on a radio to Macdonald, who was inside the sphere testing the control panels, while Magee and Lombardo were adjusting the manipulator arm, speaking what seemed like their own private language. "I want to redo the deploy," Magee said, disappearing behind the sub. "HPU?" Lombardo asked. "Stowed it?"

Lahey finished up and walked to the front of the hangar. "So you've come to see the beast," he said, greeting me. I looked up at the sub, taking in its distinctive appearance. It resembled the *Pisces* or the *Alvin* about as much as a Porsche resembles a UPS van. The *Limiting Factor* was built for speed—specifically vertical speed, since up and down are its main directions of travel. Imagine a piece of toast being slotted into a miles-deep toaster: that's how the sub descends through the water column.

Earlier, Vescovo told me that he'd initially asked Triton for a no-frills sub, like an updated version of William Beebe's bathysphere. It's a memory that he laughs at now, but at the time he was worried about the sub becoming "some rocket science project" that spiraled in complexity and cost—a moon shot that could get bogged down in its overambition. "You just have to bolt me into a steel ball and make it go up and down," he remembers saying to Lahey. "That works."

"He didn't care if it had viewports," Lahey recalled now, with a snort. "I said, 'I'm not doing that. I'll build you a sub that takes two people, with viewports. It needs a hydraulic arm so it's a legitimate tool for science. And it has to be certified—not a one-hit wonder that makes a single dive and ends up in a museum somewhere, but a sub that can do thousands of dives to hadal depths and will change our relationship to the deep ocean. After all, it's not just about getting down there. It's about what you can do when you get down there.'"

In front of us was the proof that he'd pulled it off, though the road to Hades had been as infernally hard as anyone could've imagined. "When I say that we had to develop every part of the submersible, that's not an exaggeration," Lahey said. "Because there was nothing, quite literally nothing, that we could buy off the shelf that was ready to go to full ocean depth."

Leveraging his decades in the undersea business, Lahey had scoured the globe for vendors that were capable enough, and intrepid enough, to adapt their products for the hadal zone. In Spain he found a company willing to tackle the batteries, an especially thorny problem. A British engineering group agreed to take a swing at making ultradeep syntactic foam. The sub's ten-inch-thick acrylic viewports came from Germany, its sonar system from Canada, and its undersea modem from Australia, to list just a few of the components.

Even the sub's most basic element—a hollow metal sphere—was a wild production that spanned three continents. It began on computers in the UK, where Triton's principal design engineer, John Ramsay, and principal electrical engineer, Tom Blades, work from their hometown of Devon, England. Their design plans were relayed to Florida and Texas, as Ramsay, Lahey, and Vescovo deliberated on whether to use steel or titanium or a nickel-chromium mix to fabricate the sphere. They settled on Grade 5 titanium, with dashes of aluminum and vanadium; they hired Trent Mackenzie, an Australian metallurgist, to advise the forge—ATI Metals in Wisconsin—on how to optimize the alloy's ductility and strength. The sphere would be made of two identical hemispheres molded from four-inch-thick titanium slabs, a fiery process that Lahey described as "almost primal."

There was one key requirement: the sphere had to be perfect. Not sort of perfect or close to perfect but *absolutely* perfect. Its two halves had to match exactly. They were machined in Los Angeles and Barcelona—with the viewports and the hatch cut out as precisely as jewels—and then bolted together. "Most pressure hulls are welded," Lahey explained. "But when you weld something you introduce discontinuities into the material, which can create areas of high stress." Any weakness, no matter how tiny, would be a vulnerability in the trenches. The slightest manufacturing misstep could lead to, as MacInnis had colorfully put it, the sub's occupants being "turned into pink hash."

The next stop on the sphere's odyssey was St. Petersburg, Russia. It was shipped to the Krylov State Research Center, a Soviet-era marine engineering facility. Hauling an irreplaceable four-ton Fabergé egg halfway around the world wasn't anyone's idea of a good time, and Russia isn't a soothing place to do business, but the trip was essential. Before the sub could be assembled, every piece of it had to survive being squeezed at pressures equivalent to forty-three thousand feet deep—20 percent greater than full ocean depth. The Krylov Center was the only site in the world with a high-powered pressure chamber big enough to accommodate the sphere. (Even then, it was a tight fit.)

Anxiously, Lahey watched as the technicians lowered the sphere into the chamber and flooded it with water to buffer the shock waves from a potential implosion. (When a sphere implodes, it's akin to a bomb going off.) Then they torqued up the pressure. The sphere remained in this crucible for two days, simulating repeated dives to a trench deeper than any on earth. Any warps, any cracks, any trace of metal fatigue would've doomed the project. But there were none. The sphere passed its torture test, the Krylov Center echoed with applause, and Lahey, pacing in the control room, could finally breathe—for a few minutes, anyway.

*

When the sphere was safely back in Florida, the work of building the sub could begin—installing its miles of wiring and thousands of parts—and yet even after it was assembled it would be far from finished. There is a large gap between crafting a completely novel submersible and feeling confident enough to drop it into the Mariana Trench. Typically a new Triton sub undergoes months of sea trials before its owner ever sets foot in it—especially when that owner intends to take it to twenty-seven thousand feet on his first solo dive.

But the Five Deeps schedule didn't allow for that. The polar weather imposed strict deadlines due to the encroaching ice. The Arctic dive had to be made in August, the Antarctic dive in February. Missing that slim window at either pole would set the expedition back a full year, a crippling delay. Dozens of people had been hired, quit other jobs, and were ready to go to sea. McCallum and his team had spent months cajoling permits from heel-dragging bureaucrats around the globe, and none of those permits would be valid if the dive dates changed drastically. Vescovo had cleared his calendar. Lahey couldn't have been under more pressure if he'd climbed into the chamber in Russia.

There were all-nighters, tensions, glitches. On Triton's shop floor, the problems came like buckshot. Seals didn't seal. Cables didn't fit. The batteries were in limbo, because their lithium polymer chemistry made them too volatile to ship by air. When a semi-operational, version 1.0 of the sub was finally dunked in the water, so many systems malfunctioned that it had to be returned to Triton, disassembled, and reassembled. By the time the rebuilt sub was loaded onto the *Pressure Drop*, there was time for only a bare minimum of test dives.

Vescovo had trained in a simulator sphere that Triton installed in his garage, but he obviously needed some experience in the real vehicle, in the ocean, before he met his first trench. And he would get it—along with a crash course in all the ways that things can go wrong in a submersible.

His test dives with Lahey were action-packed with trouble. Alarms blared, warning lights flashed. Dives were aborted due to water gushing through the hatch. One time, at seventeen thousand feet, smoke eddied into the capsule. They initiated emergency procedures, but they were two and a half hours from the surface. (Luckily, it wasn't a fire: the insulation on a wire had melted.) They lost a $500,000 lander when it failed to resurface. They lost the $350,000 manipulator arm when a bolt snapped, sending it tumbling into the depths. Launches and recoveries were rocky, with the potential for peril, because the A-frame crane wasn't quite long enough: the sub had to be deposited in the water, and plucked from it, uncomfortably close to the *Pressure Drop*'s stern. More than once, the submersible was damaged when it collided with the ship.

Each night the Triton squad worked to solve problems that were impossible to anticipate before the *Limiting Factor* was tested in its element. Troubleshooting is standard business in sea trials, but the degree of difficulty was far higher on this sub, the client was watching over their shoulders, and they had to work with a beat-the-clock urgency that was guaranteed to fray nerves. Thinking back on that time, Lahey grimaced. "We were fixing the plane while we were flying it."

On the day before Vescovo's inaugural solo into the Puerto Rico Trench, any bookie observing the scene would've set long odds on the likelihood of success. The sub was balky. Morale was shot. Tempers had flared—and the hatch still leaked. Vescovo was fed up by the delays, the aborted dives, the yard sale of expensive parts that lay scattered in the Atlantic. "We'd had so many gut punches," Lahey said, shaking his head. "So many setbacks."

But after a thirty-six-hour binge of rewiring and repair, recognized by all as a Hail Mary, something unexpected happened: everything went right. Vescovo plunged 5.2 miles to the trench floor without hearing any nerve-jangling alarms, touched down on a field of soft brown ooze marked by fallen clumps of sargassum seaweed, tooled around wide-eyed for an hour, and then, not wanting to tempt fate, dropped his ballast weights and headed uneventfully to the surface. He climbed out of the hatch with an index finger raised triumphantly. Completed: the first of the Five Deeps.

"The mood on the ship completely reversed," Lahey told me. "Suddenly, anything was possible."

Now, six months and four trenches later, those early battles were in the rearview mirror—only the scars remained. The only problems anyone fretted about now were the ones that might arise on the next dive. And the difficulties of exploring the deep, well, those were ever-present. "You've got to pick yourself up and dust yourself off," Lahey concluded. "That's the nature of this type of endeavor." As he spoke, a frigate bird soared overhead, its wings outstretched to surf the air currents, its forked tail feathers riffling in the wind.

Lombardo came out on the stern for a smoke, followed by Magee, who emerged from the tool bay with a towel to wipe a slick of oil off the deck. They didn't need to hear about the *Limiting Factor*'s rough start—they'd lived it.

"Yeah, we got the shit kicked out of us," Magee acknowledged with a half shrug. Lombardo reached over the railing to tap the ash off his cigarette. "Well, the sub's done twenty-five dives. That's really nothing."

"It's still a prototype," Magee agreed.

Lahey nodded. "That's right. So Victor has actually been roped into a situation where he's test piloting. Which isn't normally the way we would do it."

Magee turned to me and grinned. "You want drama?" he said. "We've got drama."

*

In the morning the wind was blowing twenty knots and the ocean was choppy with waves and the sunrise was a strange dark scarlet, but everything was a go for the launch. I dressed quickly, anxious to get out on deck. In the predawn hours I'd heard the landers being craned over the side, leaving early so they'd be in position when Vescovo arrived on the seafloor.

Breakfast was not a leisurely affair. People ducked into the mess, grabbed a bite, slugged down coffee, and hightailed it out. The T-shirts, cargo shorts, and baseball caps were gone, replaced by flame-proof coveralls, hard hats, and adrenaline. I ran into McCallum in the hallway; he was off to pilot one of the support boats. On a whiteboard outside his office he'd listed the day's hazards: "Extreme depth. Tropical sun. Night recovery." And then at the bottom, highlighted in red: "COMPLACENCY." "This is the time when karma will kick us if we're not careful," he'd emphasized at the final dive meeting.

I went outside and stood by the railing, clear of the heavy machinery. The previous afternoon McCallum, Magee, and Lombardo had given the first-timers on the ship, including me, a stern orientation about how to stay out of the way when the cranes and winches cranked up. Apparently it was all too easy to lose an eye or crush a foot or get your arm torn off—or worse. During the sea trials, Lahey had barely dodged decapitation when a cable snapped under tension.

"Launching the sub is *very* stressful," Magee had informed us. "I'm telling you, my stress level's through the roof when we launch this thing."

Lombardo seconded him, warning in a gruff tone: "If something's gonna go wrong, that's where it goes wrong. You have eleven tons of titanium and syntactic foam swinging around in the air, so it's serious."

"It was a disaster until we figured it out," Magee added, chuckling darkly. In the Southern Ocean, I knew, that learning curve had resulted in him collapsing on deck from a stress-induced migraine and losing all feeling on his left side during a difficult launch. And this was a guy who'd once broken his back, boarded a plane, and sat stoically through a seven-hour flight.

At eight o'clock, as the countdown began and the ship's crew lowered McCallum's support boat into the pitching seas and the Triton crew manned their stations, I felt my own adrenaline rising. The ship was a buzzing hive of nerves, but one person seemed utterly calm: Vescovo. I'd watched him go through his pilot's checklist, load his gear into the sub, eat a bowl of oatmeal—all with a quiet focus and no indication that he was off to do anything out of the ordinary. He might have been a commuter getting ready to hop on the subway.

This nonchalance wasn't the result of denial. Rather, it was a display of self-discipline. Vescovo knew where he was going, and he knew he was going there in a sub that was still a work in progress. As both Don Walsh and Terry Kerby had pointed out, no worst-case scenario has ever been helped by a burst of hysteria. Debilitating jitters must stay outside the hatch. A veteran practitioner of life-or-death activities, Vescovo was aware that runaway emotions were not his friend. "Patrick and I always say, 'Let's have a boring dive tomorrow,'" he told me, outlining his philosophy. "If there are heroics on an expedition, someone screwed up."

The boring dive was about to start, so Vescovo stood on deck as Magee, conductor of the industrial orchestra, directed everyone into position. Up on the bridge, Buckle maneuvered the ship to create a lee, sheltering the launch area from wind. Out on the water, the two support boats rocked in the swells. Winches whined as the hangar rolled back, and the *Limiting Factor* was moved along steel tracks and then lifted by the crane and placed inches above the ocean, pinned against a bumper across the stern. Tim Macdonald, wearing a wet suit, safety vest, helmet, and neoprene booties, climbed atop the sub. He was the "swimmer," a risky job that involves grappling with the sub's hooks, lines, and straps during launch and recovery, and securing the craft so Vescovo can enter and exit. "I'm the most expendable," he'd joked. "That's why they put me in."

With the *Limiting Factor* suspended over the water, Macdonald helped Vescovo into the hatch and then closed it. Carefully, still attached to the crane by multiple lines, the sub was lowered into the sea. Simultaneously, Buckle nudged the *Pressure Drop* forward to open some distance between the vessels. "When these two floating things are right next to each other, that's the most dangerous time," he'd explained to me earlier. "The waves can just pick them up and bash them together."

Macdonald, now riding the sub like a bronco, worked fast to unhook the lines. It helped that he was a surfer: from my vantage point, I could see a set rolling in. "Having Tim as the swimmer has made a big difference in the speed that we can get the sub disconnected and away from the ship," Buckle told me. "The first two swimmers we had were good guys, but they were much older, not as flexible. One didn't even have all his fingers."

The sub seesawed on the surface, and not in a gentle way. The ocean tossed it up and down, batting it like a punching bag. ("First submersible I've ever had to put a seat belt in," Lahey noted.) Macdonald finished up, leaped into the water, and was picked up by McCallum. The *Limiting Factor* was free to descend. We watched as it bobbed in the waves, a white star in a blue universe. Then it was gone.

*

I found Lahey in the dry lab, huddled over a laptop alongside Triton's principal electrical engineer, Tom Blades, a tall, soft-spoken Brit in his thirties. The two would stay glued to this spot for the next ten hours, monitoring the sub's movements using software Blades had developed. Currently Vescovo was descending at a meter per second, the speed of an average elevator. Every fifteen minutes he would transmit his depth and heading via acoustic modem, his voice coming in faintly but audibly amid eerie squeals of static. In the hadal zone the sub's downward velocity would slow, as the water pressure gripped it in a tightening vise.

Life on the ship was quiet while the dive was under way. People who'd been up all night went off to their cabins to sleep. I felt restless and found it hard to do anything but stare at the screens that charted Vescovo's progress. It is an odd thing to go about your day knowing someone is miles below you in a world that nobody has ever seen. I could only imagine what it would be like to roam around Hades, the only warm-blooded witness to its silence and darkness and wonder. It seemed to me that an experience so existentially big and phantasmagorically cool would change a person forever. It would alter your reality: you couldn't ever perceive the ocean or the earth or your life or life itself in the same way you had before. In an indelible way, you would know your exquisite but insignificant place in the cosmos. Maybe that's why the solitude was Vescovo's favorite part. He loved to dive alone.

He'd taken some heat for that, for leaving an empty seat on such scientifically significant dives. Deep-sea cognoscenti who were not on the expedition carped about it on Twitter. "IMO [the Five Deeps is] more of an ego project than research (in the vein of 19th century 'gentleman explorers')," read one typical complaint, tweeted by a marine geologist from California. "Think of how many actual grants could have been funded with the colossal volume of $ that's gone into this thing." Vescovo was accused of being selfish, a showboat, an avatar of white privilege and machismo—though I suspect none of his critics would have turned down the chance to dive with him.

In part, I understood the sniping. Who knew how long it would be before anyone got another opportunity to visit these sites? Decades? Centuries? Never? On the other hand, it hardly seems unreasonable that a guy who spent eight figures of his own money on a submersible would have strong opinions about what he wanted to do with it. Who wouldn't love to be Captain Nemo for a day? I would have sold my lesser internal organs to trade places with Vescovo in the Tonga Trench.

"Any experience is made richer when you share it with somebody else," I heard Lahey say one time—explaining why every Triton sub has at least two seats—and I nodded absently, but in my mind I was thinking, *No*. Some of us aren't built that way: most writers and artists, for instance, and plenty of extreme athletes. Spiritual seekers, committed readers, introverts of all stripes. Yes, having company might be pleasant or fun, but for a certain type of person that comfort is less interesting than the wilds of the psyche's inner space—a place that doesn't admit you with a plus-one.

On Vescovo's 24,388-foot descent in Antarctica his communications systems had cut out at 10,500 feet, severing his connection to the surface, and that new, heightened level of aloneness had unnerved even him. But only for a moment. The sub's performance wasn't affected by the outage, so he'd opted to continue the dive. "I've got a perfectly good vehicle, I'm just going to stay down," he recalls thinking. On the ship, however, there was no such resolution: for all anyone knew, there was a dire reason for his sudden radio silence. (People who were on board when Vescovo was MIA described Lahey as rattled, repeating the word *fuck* like a mantra, his silver hair turning whiter by the minute.)

Aware that everyone would be freaking out, Vescovo had confined his bottom time to one hour instead of three—but the point is, in that position, as arguably the most isolated man on the planet, he was content. For him, that solitude was soul food. Steady doses of it were what he needed, not merely what he wanted, and he'd arranged his life accordingly: "I don't think it's a surprise to anyone that I've never been married, never had kids."

The morning passed, people worked, lunch was served—and still Vescovo was falling, falling, falling into the Tonga Trench. Joe MacInnis sat with Lahey and Blades, emanating calm support. Heather Stewart, who was training for a marathon, pounded out miles on the treadmill in the ship's closet-size gym. Jamieson pored over lander videos in his office. P. H. Nargeolet circuited the main deck, holding an egg that had survived a round trip to the Challenger Deep—placed in the sub's outer hatch compartment, subject to the full force of the water pressure—with the title "World's Deepest Egg" inscribed on it in black magic marker.

At his computer, McCallum wrangled with logistics. Over the next three months, the expedition team would hopscotch from Tonga to Puerto Rico to Newfoundland to Norway. The ship would follow to meet them, refueling in Samoa, threading through the Panama Canal, veering left when it hit the Atlantic, and meandering north to the Arctic Ocean. The Five Deeps finale would be in London, with a presentation at the Royal Geographical Society, the Victorian-era institution that had supported the explorations of Ernest Shackleton, Roald Amundsen, and Robert Falcon Scott.

Just before one o'clock, I watched on the screens as Vescovo passed through 10,700 meters, or 35,100 feet, and shortly thereafter touched down in the Horizon Deep. "Life support good," he relayed, his voice thin and faraway, as though it were coming through a tin can on a very long string. "At bottom. Repeat: at bottom." "Roger, L.F.," Lahey responded, enunciating every syllable loudly and slowly.

"Understand, depth one-zero-eight-one-seven meters. Life support good. On the bottom. Congratulations."

A cheer went up; everyone's shoulders relaxed. But we also knew the dive wasn't over. "I don't get excited until he's back on deck and the sub's back on deck," Lahey had confided earlier. "Just because, I don't know—call it superstition. Luck can run out." Every safety precaution had been taken, but a trip to Hades is never 100 percent jeopardy-free. There was still time for trouble to arrive in the Tonga Trench. And fifteen minutes later, it did.

*

If you could have projected yourself into the Horizon Deep at that moment, you would have seen Vescovo landing on a blanket of pale gold ooze, as smooth as new-fallen snow. You would have sensed an unsettling ghost-town vibe, as though all the creatures, even the amphipods, had fled. You would have seen the *Limiting Factor* set off across the trench floor, its thrusters stirring up sediment so fine it swirled like cigarette smoke in the water.

When he arrived on the bottom, Vescovo used his sonar system to ping Skaff, the lander stationed at the trench's deepest point. His instruments told him that Skaff was about a thousand feet away. During the forty-five minutes it took to locate the lander, he found himself on edge, wary of his austere surroundings. Even as he moved forward, he felt as though he were running in place. In his twenties Vescovo had lived briefly in Saudi Arabia, and the terrain outside his viewports reminded him of the Rub' al Khali, a pitiless stretch of desert known as the Empty Quarter. Adding to his uneasiness, he heard a weird noise coming from the back of the sub. From the outside it would've appeared the dive was going exactly as planned. But if you were inside the pressure sphere, you would have seen what Vescovo saw: that many of his systems were failing.

It started with an alarm from the battery bank on the starboard side. The *Limiting Factor* is powered by six 275-pound external batteries, and three of them were tripping. On a control panel behind his head, circuit breakers began to click off. Then two of the sub's ten thrusters quit. Vescovo's front control panels flashed with red warning lights and haywire voltage readings, as his internal power supplies drained. Now all of the batteries were losing their charge. When he tried to run the manipulator arm, it was dead.

Vescovo had an inkling of the problem, a guess that would later be confirmed: There was an electrical fire in Battery One—a cold, high-pressure fire with no flames and no combustion. It was a special kind of deep-sea conflagration, a furious surge of current that scorched a junction box made of burly carbonate plastic, and melted all the circuitry within it. Seawater had penetrated an oil-filled compartment that housed parts of the sub's electrical system, shorting out fuses that, in turn, flared off searing heat. One by one on the starboard side, the electrical functions shut down.

At 35,500 feet from the surface, this wasn't the best news. It also wasn't the worst: the sub's systems are distributed and compartmentalized, so the portside was unaffected. And Vescovo had run out of power before—in the Challenger Deep, no less. He knew that as long as his life support systems were fine, he could drop his weights and ascend safely, even without power. He switched off everything he could to conserve juice and contain the damage—and continued to explore the Tonga Trench for another ninety minutes. When he was down to the last whimper of battery life, he decided it was time to leave.

*

Five hours later, around six o'clock, I stood at the stern in a hard hat and coveralls, ready to watch the recovery. Nargeolet and MacInnis stood beside me, identically attired. Magee and Lombardo waited at their posts, scanning the water for the submersible's lights. We knew that Vescovo was coming up slightly early, but we didn't exactly know why; his terse dispatches didn't allow for a lengthy explanation. Whatever the cause for the abbreviated dive, it didn't seem to be a five-alarm emergency. We were more curious than worried.

The sun had slunk below the horizon but there was still some light in the sky. Steely clouds massed above the darkening ocean, softened by a wash of apricot dusk. The wind and waves were less chaotic, which was helpful. Macdonald sat in a Zodiac with several crew, prepared to go for another high-stakes swim. As the twilight dimmed, their features faded, Macdonald's headlamp and the reflective stripes on the men's coveralls offering the only evidence of their presence in the slate-black sea.

"There it is!" someone yelled, as the *Limiting Factor* popped up three hundred yards off the stern, LED lights blazing. The Zodiac raced over and Macdonald jumped in and swam to the sub, hooking it to the towlines trailing from the ship. He climbed aboard and crouched by the hatch as the winches reeled him in, and he noticed with alarm that the sub reeked of burning electronics. Twenty yards from the *Pressure Drop*, Macdonald opened the hatch's top compartment and pulled out a harness of thick straps that attached to the crane's monster-size hook. When the hook swung by overhead, he caught it, and in one deft motion slammed the eye of the harness into it. The *Limiting Factor* was home.

Once the sub was lifted from the water and braced against the bumper, Macdonald unsealed the inner hatch and Vescovo exited, looking unusually pale and wrung out. He waved, smiling weakly. He stepped on deck, shook hands with Lahey, and reported, "It was hard," before beelining to his cabin. No battery power had meant no heater: Vescovo was deep frozen.

The post-dive maintenance began immediately, and before long, Lahey and the electrical engineer Blades were holding the burnt-out junction box, puzzling over it in the dry lab. The box smelled noxious, like a fire in a plastic factory, and looked like it had been through a war. "It's a difficult one," Blades said, frowning. "There's so much damage we can't see what the original cause was. There's a seal that could've leaked slightly, but that seal's long gone now. We just need to rebuild it." "Tom's unflappable," Lahey said. "Unlike me."

Blades corrected him: "I'm screaming inside."

One thing was certain: no one would be diving the next day. "We have to confirm that the water contamination didn't migrate into the motor controllers," Lahey explained, adding that the repairs would take at least forty-eight hours, and some were finicky battery surgeries that would be safer to do on land. This was crushing news for Stewart. Her dive would be rescheduled, but the Tonga Trench—a geologist's dream—wouldn't be the site. If we waited out here, the New Zealand storm would be upon us. The weather window had closed.

*

Vescovo was sitting alone in the ship's mess, bundled in a heavy sweater, drinking tea and eating spaghetti Bolognese, trying to restore his body temperature. A movie poster for a Cold War thriller called *Ice Station Zero* hung on the wall behind him, which seemed appropriate. I asked if I could join him, and sat down. I wasn't sure if he'd want to relive the dive when he was still recovering from it, but he seemed happy to talk—and he wasn't a fan of this latest trench.

"It was *hostile*," he said emphatically. "I would use that word. Dead. Cold. It wasn't welcoming. It didn't want me there."

"So it was unlike the others?"

"Oh yeah," Vescovo said. "In the Mariana Trench I saw sea anemones. I saw a holothurian flapping away within ten minutes of arriving on the bottom. Bacterial mats in orange, red, and yellow. That place was *alive*. The Java Trench—it had tons of life. And the Southern Ocean was a darn grocery store." He shook his head. "Not here. This was the most alien."

He reached into a pocket for his iPhone. "Do you want to see the Horizon Deep?"

Did I want to be the second human to set eyes on it? That would be a yes. I had to restrain myself from snatching the phone out of his hand. Vescovo tapped on a video he'd taken through his viewport, and turned the screen to face me. It glowed a vivid cerulean blue, as the sub's lights revealed a place that had never known light. The water was piercingly clear. In the background, the thrusters emitted a high-pitched keening, like whale song from a distant planet. No rocks were visible, no pebbles, no lava—nothing to break the baby-soft skin of sediment. The bottom was flat with subtle undulations, but when a boulder did appear, it was the size of a truck. Vescovo was right: the Horizon Deep did not invite lingering. There was a surreal, hypnotic quality to its stillness, and it came across like an unspoken threat. I asked him to replay the video.

While I watched it for a third or fourth time, MacInnis walked in and pulled up a chair. He leaned in to see the footage. All week, for a podcast, he'd been interviewing people on the ship about their relationship with the ocean, coaxing them to express it from the heart. I knew that he recognized the Tonga Trench's dark grandeur, and that Vescovo would, too, once he thawed. (Later, after viewing hours of video from the dive, Stewart and Jamieson would conclude that the Horizon Deep's uncanny smoothness was the result of a "recent mass transport deposit," otherwise known as a fresh submarine landslide.)

"So Joe," Vescovo said, nodding to MacInnis, "I was just telling Susan, this is the most inhospitable trench I've seen yet. And it was bloody cold—like sitting in a freezer for ten hours."

MacInnis, who had spent time beneath Arctic ice in an undersea habitat called Sub-Igloo, commiserated. Then he asked Vescovo about the battery fire. "I got some indications that I was having a 'thermal event,' as we euphemistically call it," Vescovo replied casually, as though describing a minor incident like a flat tire. "I could see the battery power being eaten away, so I just went into power conservation mode. When you have electrical issues, the best thing to do is de-energize the system." He took a sip of his tea, warming his hands on the mug. "It would be an abort condition if you're hypersensitive, but I'm not." The sub was operational, he said, if not perfectly so. "It was like trying to fly a really misbehaving helicopter." *If its engine was burning,* I thought.

"And here you are, eleven kilometers beneath the surface of the ocean," MacInnis said, flinching at the idea. "That's a long way from home."

Vescovo sighed. "This stuff always seems to happen when I'm on the very bottom. It never happens on the descent or the ascent." His eyes lit up, as if an exciting thought had occurred to him: "Deepest sub fire ever!"

There was a question I'd wanted to ask, so I wedged it in now. I wondered what had prompted Vescovo to put so much on the line—time, energy, money, risk—to explore the deepest of the deeps. It wasn't, after all, an ordinary quest; the average centimillionaire is more interested in tax avoidance schemes and golf. If all he wanted was accolades, there were easier, showier, and less expensive options. "Really, it's the adventure," Vescovo replied. "If you're not going to live with some degree of adventure, I think it's a life half-lived." The mission "to map and explore the deepest places on this planet" was a long-neglected one that he figured he could pull off, so he simply decided to do it. Admittedly, there was "a bit of ego involved"—"I very much like being the first"—but the more compelling reward was the experience itself, the opportunity to visit a truly unknown realm.

"Did you have a love of the ocean before you started the expedition?"

Vescovo hesitated. "I had a *like* of the ocean. That's a better way to describe it. I have a deeper appreciation for it now. And just seeing what I'm seeing—how mysterious it is. I'm growing attached to it."

MacInnis smiled. "She's cast her spell on you."

"Yeah, she has, hasn't she?" Vescovo said, laughing. "Damn her."

*

That night I lay in my bunk contemplating the Pacific through a porthole: the rise and fall of the swell, the waves rushing by with their whitecaps, the ship's wake lit silver by the moon. By now the *Pressure Drop*'s rolling motion felt almost soothing, but I didn't fall asleep quickly. There was too much to think about.

"The surface of things is not where attention should rest," the psychedelic sage Terence McKenna had counseled, and that is certainly true of the ocean. But how many people know its depths? For now, precious few, which is why even a peek at the Tonga Trench was beguiling. You could look at Vescovo's video and choose to find it scary or foreboding—in one article I'd read, a writer referred to the Mariana Trench as a "terrifying wormhole," as if that were an accepted fact—or you could be fascinated by the hadal zone, the immemorial beauty and fury and truth of it, and see it for what it was: proof that even in our own world, there are worlds yet to discover.

Why have we ignored so much of the deep for so long? It's as if we live in a mansion filled with treasures and artworks and fabulous animals, but haven't bothered to look in most of the rooms. It's a failure of curiosity, to say the least, a hobbling myopia that leaves us oddly unacquainted with our own home. For a species long on creativity and imagination, we've uncharacteristically limited ourselves in range, our attention fixed outward and upward as if those were the only dimensions that counted. Maybe that's because seven miles down, we aren't in a position to call the shots. In the deep, humanity can't even pretend to be in charge. Of course we're not in charge of space, either, but exploring upward gives us the illusion of expansion, as though we're conquering territory, extending our ever-acquisitive reach. In this mindset, to go inward, into the abyss, is to be stuck with what we already have.

The *Limiting Factor* doesn't conquer—it submits. It allows access to the hadal zone, but only on Hades's overwhelming terms. No matter who is piloting the sub, the ocean is always in the driver's seat. Personally I found this thrilling, and so did everyone else on this ship, and many others, but in general, consciously or subconsciously, we've long held the mistaken belief that the deep wasn't worth the ceding of control, the trouble, and acted as if the earth's largest domain weren't the cornerstone of everything above. That attitude was changing, but only slowly, and only because it had to if we hoped to survive.

"Ninety-nine percent of us on this manned spacecraft we call Planet Earth, we're not going to Mars or the moon," Don Walsh had mused to me in Oregon. "Somehow we have to understand the place where we live and how it works and how it doesn't work and what effect we're having on it."

The *Pressure Drop*, with its peerless submersible and landers and sonar, was a floating front-row seat to the unveiling of the deep.

Olaus Magnus, Edward Forbes, Charles Wyville Thomson, William Beebe—what would they have given to see it? As Lahey had said, his voice revving with excitement: "We're turning on the light in these wide swaths of ocean. And you'll see, the bathymetry's *incredible*. You'll see pinnacles, you'll see seamounts, you'll see the walls, the shapes . . ." His sentence had faded at the impossibility of listing every revelation, but where his words ended, my imagination continued. When we returned to Nuku'alofa, I would leave the ship as planned, but I intended to stick close to this group. After a personal half century of longing to know what was down there, this was my chance to see.

CHAPTER 6

"This Is the Mother of All Shipwrecks"

This was also how he learned that four nautical leagues
to the north . . . a Spanish galleon had been lying under water
since the eighteenth century with its cargo of more than five
hundred billion pesos in pure gold and precious stones.
The story astounded him, but he did not think of it again until
a few months later, when his love awakened in him an
overwhelming desire to salvage the sunken treasure so that
Fermina Daza could bathe in showers of gold.
—GABRIEL GARCÍA MÁRQUEZ, *Love in the Time of Cholera*

CARTAGENA DE INDIAS, COLOMBIA, AVENTURA, FLORIDA

Thirty miles outside Cartagena Harbor, two thousand feet down in the Caribbean Sea, a robot called REMUS 6000 flew in a grid pattern above the seabed, hunting for a Spanish galleon that had been lost for 307 years. Twelve feet long and shaped like a finned cigar, REMUS had a titanium spine and a nervous system of artificial intelligence, and it carried sensors, cameras, and four types of sonar, including one that could see like an X-ray beneath the sediment. Even if it were buried, no bit of wreckage would pass undetected beneath the robot's acoustic eyes.

But was the galleon even here? Was it *anywhere* in the area? For decades, people had been searching for it without success. This

November 2015 expedition was only the latest in a series, and it seemed on track to be as frustrating as the rest. The ninety-square-mile search zone was divided into six sectors, and already five of them had come up empty, the sonar images revealing nothing but rock and sand. REMUS completed its circuit, rose to the surface, and signaled its location. It was recovered with a grappling hook and craned aboard the ARC *Malpelo*, a Colombian Navy ship, where a team of Woods Hole engineers awaited it. Like its cousin Jason, the robot traveled with an entourage.

REMUS swims freely, without a fiber-optic tether, so no one knew yet what it had spied in the abyss: that information was recorded on the robot's internal computer. Its data were swiftly transferred onto a hard drive, and handed off to a coast guard vessel that would make a fast run to shore. A Colombian naval officer then delivered the hard drive to an apartment near the harbor, where two men worked at a long table covered with computers and bathymetric maps.

One of these men, marine archaeologist Roger Dooley, had made an audacious bet. He'd staked his career and his reputation on the belief that he had turned up a long-forgotten clue to the galleon's final resting place. It was Dooley who had pinpointed the search area, Dooley who raised millions to conduct the high-tech hunt, Dooley who convinced the president of Colombia to support the search effort, Dooley who labored in archives and studied archaic documents and sweated every detail, whose lifetime of obsession had led him to this moment. He projected confidence, but the deep seafloor wasn't offering any guarantees.

With him was Garry Kozak, a Canadian sonar expert. To the untrained eye, raw multibeam, side-scan, and sub-bottom profiles of the seafloor are the visual equivalent of white noise; discerning a signal within reams of acoustic data is both an art and a science. Kozak was a wizard at finding needles in ocean-size haystacks. He was part of a team, funded by the late Microsoft cofounder Paul Allen, that had located dozens of important World War II wrecks in the deep, including the USS *Indianapolis*—the ill-fated ship that was torpedoed by a Japanese submarine and sank eighteen thousand feet in twelve minutes. (Nine hundred American sailors were left adrift, fending off sharks for four days before rescuers arrived. Only 316 men survived.) But Kozak was capable of spotting objects much smaller than a battleship. More than once, he had found the lone bodies of drowning victims.

For weeks, throughout the search for the galleon, Kozak had scrutinized REMUS's data and slumped back in disappointment. But the previous day he'd identified what sonar techs refer to as an anomaly—a cluster of bright, raised bumps that stood out like braille on the seabed. "We've got something here that is not geological," he told Dooley. "See the scatter pattern?" Dooley had radioed the *Malpelo*, asking that REMUS return to take close-up photographs, which would determine whether the anomaly was a sunken galleon's debris field, or something as prosaic as a heap of rusted-out oil drums. Those photos were now on the hard drive that had been rushed ashore.

Dooley leaned forward as Kozak plugged the drive into his computer and pulled up the first image. The screen filled with an overhead shot of a rubbly disturbance on the seafloor—interesting, but not definitive. Then came the second image. On the left side of the frame a rattail hovered, casting a shadow under REMUS's bright glare. To the right, clearly visible and jutting through the sediment, were three stately bronze cannons. Cups that would later be identified as K'ang-hsi-period Chinese porcelain were strewn across the bottom amid a spray of gold coins. "Oh my *God*," Dooley said, clutching a hand to his forehead, overcome by emotion. Kozak stared at the screen, stunned.

There it was, the Spanish galleon *San José*, the majestic flagship of the Tierra Firme fleet, one of the mightiest vessels of its age. It had sailed from Portobelo, Panama, on May 28, 1708, loaded to the gills with gold, silver, emeralds, and other treasures that Spain's Bourbon king, Philip V, was desperate to haul back from the New World to fund the ongoing War of the Spanish Succession—which pitted Spain and France against England, the Dutch Republic, and virtually every other country in Europe. Times were turbulent for the Spanish monarchy. They would soon get even worse.

The *San José*'s last voyage was never going to be a joy ride. The

Caribbean was crawling with pirates, and a squadron of British warships prowled off Colombia's coast, led by the hulking, seventy-gun HMS *Expedition*. The English commander, Charles Wager, was well aware that the galleon was a floating Brinks truck. In the late afternoon on June 8, 1708, before the *San José* and the sixteen other ships in its armada could reach safety in Cartagena Harbor, the British advanced and a battle ensued. As night cloaked the water, the ships exchanged cannon fire until suddenly in the smoke and chaos and darkness, the *San José*'s bow-front gunpowder magazine exploded. Within minutes, the galleon vanished.

The *San José*'s captain general, Don José Fernández de Santillán, went down with his ship, as did some six hundred officers, noblemen, bureaucrats, merchants, soldiers, and sailors; three drummers, a standard-bearer, a priest or two, and some goats and chickens. Fourteen crewmen survived, clinging to floating mast debris, and were taken prisoner by the British. Wager had won the battle, but he'd lost the treasure to the deep.

Ever since the technology has existed to find the *San José*, both marine archaeologists and treasure hunters—warring tribes that typically agree on nothing but their mutual loathing—have coveted the galleon, because its immense historical value is equaled by its riches. Depending on your perspective, the *San José* is either a priceless piece of cultural heritage, its every last nail worthy of painstaking study (archaeologists), or a big pile of loot that should be vacuumed up yesterday (treasure hunters). *Everybody* wanted this ship. The ocean is large, however, and the British and Spanish records were vague and contradictory about where, exactly, the battle had occurred. Now we knew. But as the champagne flowed in Cartagena, a bitter truth of deep-water archaeology was about to assert itself: finding the wreck is the easy part.

*

The deep's vast archive holds the skeletons of every vessel imaginable: Phoenician galleys, Viking dragon ships, Roman warships, Chinese junks, Portuguese caravels—ships of every dimension and purpose, ships from every war that has ever been fought at sea. UNESCO, the United Nations Educational, Scientific and Cultural Organization, estimates that some three million ships linger on the seabed, all but a tiny fraction of them unknown.

The majority of the wrecks lie in coastal waters, dashed on reefs, broken and splintered, scrubbed to pieces by storms. But the ships that come to rest in the cold abyss can be remarkably well preserved: nestled in sediment and protected from time's scouring, the lashings of waves and currents, the appetites of wood-boring shipworms, the fishermen's nets, the grasp of humanity. Deep shipwrecks are like stowaways from the past. They hide in the purgatory between absence and presence; gone but still here, lost for eternity until, against the odds, they are found.

From REMUS's photos, it was evident the *San José* wreck was in superb condition. The ship had landed upright, with its hull sunk into the silt. It was sealed in mud, where mollusks couldn't feast on it. The bow was gone, but Dooley, who'd studied the galleon's architecture, realized immediately that the remaining structure was sound and its cargo intact. To excavate it would be to open a portal directly into the seventeenth century.

What would it take to do that? To carefully and rigorously dissect a shipwreck that lies beneath a half mile of ocean? To raise its stories along with its treasure; to wring knowledge from every inch of the site? In four words: expertise, patience, robots, and money. Especially money. Only a handful of abyssal wrecks have been thoroughly examined, because the costs quickly soar into the millions. And if there's any group that doesn't have that kind of cash, it's marine archaeologists.

Fishermen and sponge divers had been pulling antiquities out of the sea for ages, but marine archaeology as a science didn't begin until the late 1950s, when scuba became widely available. Even then, few people were interested in working underwater. The first half of the twentieth century had made land archaeology prestigious—with discoveries like Machu Picchu, the caves at Lascaux, and King Tut's tomb—but digging into seafloor ooze didn't have the same cachet. Marine archaeology was considered a sideshow. From the start, funding was scarce to nonexistent.

In 1960, a group of archaeology graduate students began to investigate shipwrecks off Turkey's south coast. They cobbled together equipment and lived rough on the beach, too broke even to buy tents. When they successfully recovered the remains of a Bronze Age merchant vessel at a depth of a hundred feet—the first ancient ship to be excavated on the seabed—and their findings revealed that thirteenth-century BCE maritime trade was far more sophisticated than anyone had guessed, the archaeological world scoffed. The meticulous, time-consuming work of surveying a site, mapping and measuring it, recording the position and elevation of every artifact, extracting fragile items without damage—none of this, land archaeologists declared, could be done properly by someone bobbing around in tanks, fins, and a rubber mask.

"It was a real uphill battle," one of the students, George Bass from the University of Pennsylvania, recalled fifty years later in an interview. "I mean, we were really sneered at. And we just lived on beans and rice and tomatoes, and some days we ran out of food altogether. We had nothing to eat—zero."

Somehow that hardscrabble aquatic lifestyle agreed with Bass, who would ultimately become known as the father of marine archaeology. He and his group kept bootstrapping along in the Aegean, which was littered with intriguing wrecks. They pushed their scuba depths, risking the bends daily. They used a decrepit barge as a diving platform, and slept in the smelly hold of a fishing boat, and endured a plague of biting flies. In those early days, Bass wrote, "We were always wet and cold, always tired, and, more often than most would admit, a little frightened." But they also jury-rigged new tools, devised new methods for undersea excavation, and made find after stunning find. Bass viewed the seafloor as the world's greatest museum, since "virtually everything ever made by humans, from exquisite pieces of jewelry to massive building blocks for Egyptian pyramids, has been carried at one time or another by watercraft."

A seventh-century CE Byzantine ship yielded a trove of pottery, and a glimpse of life during the empire's Heraclian Dynasty. One archaeology student spent years piecing together fragments of its wooden hull to reveal its unusually sleek design. Others examined its contents and learned that its captain was a priest, sailing on a Church-sponsored mission to supply wine to Byzantine troops during a holy war with Persia. At another wreck site, where a merchant ship sank around 1025 CE, the scuba-diving archaeologists pried three tons of multicolored glass shards from the seabed with dental picks and tweezers (dodging a territorial octopus that liked to snatch the pieces from their hands), then spent two decades reconstructing them into the world's largest collection of medieval Islamic glass artifacts. See? Patience.

*

In the sixties, Bass and his semistarved crew couldn't have predicted that in a matter of thirty years they would be hailed for making one of the twentieth century's top archaeological discoveries—a royal ship dating thirty-three hundred years back to the reign of the Egyptian pharaoh Akhenaten and his queen, Nefertiti. Now known as the Uluburun wreck, it held splendors from all corners of the ancient world: Mycenaean cups, Canaanite lamps, Mesopotamian stone seals, Cypriot copper, African ebony, Asian ivory, Egyptian gold. There were cosmetics containers sculpted in the shape of ducks, and a trombone carved from a hippopotamus tooth, and a gold scarab etched with Nefertiti's name. Over eleven years, making 22,500 dives to the borderline-crazy depth of two hundred feet, they would recover fifteen thousand objects and a portion of the ship's Lebanese cedar hull. They would bridge cryptic knowledge gaps and solve age-old mysteries and add chapters to the history books, and the naysayers would end up looking silly. That was the future. But it would take a lot more scrounging, scrimping, and struggling to get there.

When Turkish sponge divers dragging nets at three hundred feet raised two classical Greek statues, Bass yearned to investigate their source. Any ship that was carrying these masterpieces had to be significant. In sixties-era scuba gear, though—forget it. You'd end up becoming part of the wreck. For depths beyond two hundred feet, another tool was needed. "I was determined to use a submarine," Bass recalled in his memoir, *Archaeology Beneath the Sea*. "But how?"

Luck stepped in, and the Electric Boat division of General Dynamics agreed to provide a two-man sub, underwriting the cost. The company was eager to create a market for privately owned submersibles, and having one of its models off hunting relics from the dawn of civilization made for great advertising copy. In 1964, the sub was transported to Turkey. Bass was convinced that along with searching for deeper wrecks, it could be used to lower excavation costs at shallower sites, surveying a ship in hours—work that took divers weeks to complete. He was right, but the benefit was short-lived: the archaeologists had to give up the sub because they couldn't afford its liability insurance.

Still, it was only a matter of time before technology displaced sponge divers as the key to finding lost ships. The breakthrough came in 1985 when an expedition led by the oceanographer Bob Ballard, working with the towed camera system Argo, located the most famous abyssal wreck of all, the RMS *Titanic*. The ship lay thirteen thousand feet down in the Atlantic gloom—its humongous hull snapped like a pencil—but the cameras were able to capture it in crisp detail. "All around the *Titanic*, we saw thousands and thousands of objects that had drifted to the bottom or somehow spilled out intact," Ballard would later write, "including a porcelain teacup sitting upright, a silver serving tray, many bottles of wine, and—saddest of all—pairs of empty shoes with old-fashioned buttons, splayed out on the seafloor at an angle and distance suggesting that their owners had worn them all the way to the bottom."

Ballard was evangelistic about using robots to explore the deep. He envisioned them as "unmanned tethered eyeballs" that could rove anywhere, find any wreck. Testing an early version of the ROV Jason, Ballard began to search the seafloor beneath historical trade routes, where crossing the open waters always carried the risk of meeting unexpected storms. In 1989, he found a site in the Strait of Sicily where several Roman ships lay at twenty-five hundred feet. After mapping the area, Jason gently recovered clay amphoras and other artifacts with a set of tongs attached to its manipulator arm. It was a leap forward—and proof that with the right tools, an excavation could be done with precision even at great depths. Though of course, it wouldn't be cheap.

In 2005, Norwegian archaeologists got the opportunity to try it, funded by an improbable backer: Norsk Hydro, an oil company that wanted to install a 725-mile-long pipeline across the seabed. When completed, the pipe would run from Norway's deep-water Ormen Lange gas field all the way to northeast England, sluicing 20 percent of the UK's natural gas. It was the kind of grand-scale project that oil companies love—an eleven-billion-dollar industrial complex sunk three thousand feet down in the Norwegian Sea, enabling them to make even more billions. There was one pesky problem: a historical shipwreck lay right in the middle of the proposed pipeline route.

Norway has cultural heritage laws with teeth, so Norsk Hydro couldn't do what it surely wanted to do, which was bulldoze the ship into oblivion. And it couldn't solve the problem by rerouting the pipeline, because there were historical shipwrecks everywhere else, too. Also, the bathymetry was complicated. The seafloor around the Ormen Lange field is crenellated by scars from the Storegga Slide, an enormous submarine landslide that occurred eighty-two hundred years ago. (Geologists argue about what caused the slide, but they do know that a slab of Norway's continental shelf collapsed, creating a mega-tsunami that walloped northern Europe.) There was no way around it. The interfering shipwreck would have to be surgically removed.

The project's leader, Fredrik Søreide from the Norwegian University of Science and Technology, wrote about the Ormen Lange wreck in his book *Ships from the Depths*. I have to admit that I found it particularly delightful to read about the world's first robotic deep-sea archaeological excavation—the pioneering methods, the cutting-edge equipment, the customized software, the ROV built specifically for the job (complete with soft-touch manipulator arm to pick up frag-ile items, and lasers to take precise measurements), the second ROV used for high-definition videography—knowing that an oil company was being forced to foot the ten-million-dollar bill. And the downed ship? Well, let's just say it wasn't a Spanish treasure galleon.

The archaeologists identified it as a late-eighteenth-century mer-chant trader of unknown name that had been ferrying a boatload of booze to Russia. At the site, more than a thousand bottles lay scat-tered across the seafloor, as though flung by a drunken, angry giant. "The vessel seems to have been carrying a profitable cargo of spirits," Søreide noted optimistically, "which may have been accompanied by a load of grain, salt, or a similar perishable cargo that did not survive." Besides the bottles, objects recovered included earthenware casks, Russian coins, and stone plates.

Regardless of its humble nature, the Ormen Lange ship left a legacy: a best-practices playbook for anyone hoping to scientifically dissect a deep wreck. And, Dooley had vowed, that's how it would be done on the *San José*. No corner would be cut, no expense spared. An army of archaeologists, robotics engineers, marine biologists, histo-rians, and conservators would be assembled. There would be auton-omous robots and remote-controlled robots and maybe a manned submersible. He estimated that the galleon's excavation would cost fifty million dollars.

*

There is something innately haunting about a sunken ship. Doom-struck and slumbering underwater, it's out of place and out of time—an emblem of human progress trapped somewhere it shouldn't be. Once strong and seaworthy, now it's the corpse of a fallen crea-ture, its ribs ghosted with silt, its iron oxidized into filaments of rust that drip down its sides like candle wax. And in that veiled decay, it's hard not to sense your own mortality.

This beyond-the-grave quality is what makes the *San José* so irre-sistibly alluring. Recovering it is like transmuting legend into reality,

death into life—less of an excavation than a resurrection. The ship's
tale of valor, tragedy, and lost treasure lights up the dopamine recep-
tors of every person who hears it. Days spent at a desk, in commuter
traffic, scrolling on your phone, hustling to get to the dry cleaner—
they lack mystery, they lack awe. The *San José* provides both. But
the galleon was so well cached in its underworld that only the most
devoted seeker would ever find it. That was Dooley, and befittingly,
he was not at all easy for me to find.

I first learned about the *San José* in December 2015, when its discovery hit the news. As always I was scanning for stories about the
deep, and headlines like "Diving Robot Finds $22 Billion in Sunken
Gold" and "After Centuries in the Sea, Colombia Says the 'Holy
Grail' of Treasure Ships Is Finally Found" were guaranteed to catch
my eye. I read everything, but the clips were frustratingly short and
repetitive—light on facts and heavy on hype, especially about the
value of the treasure. The dollar amount ranged from one billion to
thirty billion, with most media settling on seventeen billion, with no
explanation of how that number was determined. The reports credited a marine archaeology team, but no names were given. Woods
Hole was involved, but only as a contractor. The Colombian Navy
was guarding the wreck, the coordinates of which were a state secret,
for fear of looters showing up.

It was a valid concern. Illegal salvaging had become ruthlessly
efficient, even in deep water. You might think it's impossible (or at
least off-puttingly hard) for a small band of criminals to dismantle a
sunken World War II battleship, raise the pieces hundreds of feet from
the seabed, and cart the whole thing off to sell for scrap metal. You
would be wrong. Between 2006 and 2016, in the Java Sea alone, three
British warships, three Dutch warships, an Australian warship, and
an American submarine had been plucked from the bottom. And this
metal piracy was happening throughout the world. If thieves would
go to that kind of trouble for copper and steel, imagine what they
would do for gold.

Certainly it made sense to keep quiet about the galleon's location,
and other specifics. But restless curiosity is my default setting, so the

bare-bones news stories were unsatisfying. I had a ton of questions and nowhere to direct them, and after a while I sort of forgot about the *San José*. Then in 2019, while reporting on the lost Flight MH370, I was referred to Garry Kozak, who had served as an advisor on that search. Prepping for the interview, I looked at his website and saw that he'd also "provided AUV mission planning and sonar data analysis for the *San José* search project."

When I got Kozak on the phone I launched right in. It turned out that he had participated in more than one quest for the galleon. "I first got involved in looking for it in the early 1980s," he told me, and then chuckled. "Funny story about that. One night in Cartagena Harbor we were boarded by pirates."

"Really?"

"Really. Right there. Boarded by pirates. Luckily the cabins were locked, but they took everything on deck, the outboard engines, anchors—just anything they could get. So that was exciting."

Kozak was an engaging storyteller with forty years of undersea experience. He had participated in scores of aquatic hunts, including numerous pursuits of the Loch Ness monster (one of which resulted in a photo of what looked like a large plesiosaur fin). I kept steering the conversation back to the *San José*. Kozak chronicled the process of finding the ship—"There isn't one instrument that's perfect for everything. So you use side-scan sonar for surface objects, sub-bottom profilers and magnetometers for buried objects"—but was hesitant about revealing the identity of the man who'd hired him. When I pressed for a name, Kozak sighed, and offered: "I'll touch base with him. If he gives the okay I'll send you his contact details." He was silent for a moment. "When he told me he was going to look for the *San José*, I sort of snickered and thought, 'There's no way he'll pull this off.' But Roger's a pretty amazing guy. And this had been his dream."

*

"What I'm gonna tell you right now, it's a little confidential, right? Nobody knows—not the president of Colombia, not my wife.

Nobody knows the whole story. I really have not told the story to anybody. Just so you understand." Roger Dooley, pausing for breath, directed me toward a white leather couch in his living room. Through the floor-to-ceiling windows of his high-rise condo I could see boats cruising along Florida's Intracoastal Waterway, the azure-blue Atlantic, and the towers of Miami Beach, all baking under a blazing sun.

If the setting was an ode to languid recreation, Dooley himself was not. I knew from hours of talking to him on the phone that he was excitable. He didn't look anywhere near his age, which was seventy-four. "I'm like a one-man orchestra," he told me. He is tall, with pale wavy hair and pale blue eyes and a pale rosy complexion and a silver beard and mustache framing an expressive, angular face. His most distinctive characteristic is his voice. Dooley was born in Newark, New Jersey, spent much of his childhood in Brooklyn, New York, and then moved to Havana, Cuba, at age thirteen. The resulting accent is an unplaceable patois of silky Spanish consonants and flat New York vowels, with gravelly undertones of whiskey drinking and cigarette smoking—though I don't know if he does either of those things—all whirled together in a verbal blender. Also, he talks at the speed of a runaway train.

As I sat, scribbling as fast as I could in my notebook, Dooley strode around the room pointing at paintings of ships that lined the walls, describing them at about a million miles an hour. He was starting at the beginning, with a history lesson. "People call all kinds of ships galleons," he said. "But the galleon was a specific ship in a specific period. They were built from the early 1580s to 1700—that's *all*."

When the *San José* was christened in 1698, therefore, it represented the end of an era. In the sixteenth and seventeenth centuries, Spain had dominated the Atlantic and grabbed Cuba, Florida, the West Indies, Mexico, Central America, and most of South America; the conquistadores, led by Hernán Cortés and Francisco Pizarro, plundered the Aztec and Incan civilizations with a single-minded rapaciousness. The galleon was the ship that fit their imperial ambitions, sturdy as a vault, bristling with guns, large enough to carry hundreds of men and tons of silver and gold. But no empire lasts forever, and by 1700—a time of rising British naval power and a new generation of faster, more agile warships—Spain's maritime supremacy was being challenged.

Not only was the *San José* one of the last galleons, it was an exceptionally brawny and beautiful one. It was 140 feet long, with a 40-foot beam and a 117-foot keel, two gun decks, three towering masts, and sixty-two bronze cannons, their handles cast in the shape of leaping dolphins. According to Dooley, the *San José* had been elaborately decorated with gilded carvings and paintings of saints. He beckoned me to his desk, in an alcove at the end of the room. Above it hung four posters that he'd created to depict the *San José* for the non-history-book-reading, un-galleon-wise public. The posters were graphic extravaganzas, dense with illustrations, diagrams, and captions. "I want people to know what it really looks like," he said in an exasperated tone. "I'm tired of all these museums and books and publications—I mean, they have it all wrong."

Art historians, military historians, naval architects; specialists in seventeenth-century shipbuilding, numismatics (the study of coins), Hispanic ceramics, Dutch glass, religious iconography, cannons, pistols, anchors: Dooley had enlisted a troop of experts to help add to his own stockpile of facts. He had amassed so much research material that he'd had to rent a warehouse. His favorite place in the world was the General Archive of the Indies in Seville, Spain, with its dizzying collection of eighty million documents from the Spanish colonial empire. "I could spend a month in the archives, no problem!" he told me, adding emphatically: "I cannot live without this."

Locating a deep shipwreck requires serious detective work: combing through records, interpreting old wonky maps, overlaying modern oceanography. The best shipwreck hunters also deploy a touch of intuition. Choosing a search zone means sticking a pin into a map of the ocean and declaring that *this* spot—this one inscrutable scrap of seafloor—is worth spending a fortune to examine because an even smaller object might be there, buried under mounds of silt. Unsurprisingly, the success rate is low. I'd read that only five sunken galleons had ever been found, but that seemed to me like a very small number, so I asked Dooley if it was true.

"Yeah," he replied. "And I found two of them."

I stared at him. "You found another galleon *besides* the *San José*?"

Dooley nodded. "This all started in 1984."

*

You need to know a bit of background on Dooley's life to understand how he came to unearth multiple Spanish galleons, because the stories are inseparable: they fit together like a key in a padlock. If Dooley had remained in Brooklyn, it's unlikely that he would have turned up any galleons at all. But when his Cuban mother divorced his Irish American father and remarried a Cuban hotelier, Dooley, like the Tierra Firme fleet itself, was routed through Cuba.

Nineteen fifty-seven was a hell of a time for an American kid to land in Havana. The Castro brothers, Che Guevara: guerrilla war was raging. Fidel Castro frequented the Hilton, where Dooley's stepfather worked as a night manager, and the two men became friends. Dooley recalls eating dinner with Castro, who liked to cook in the hotel kitchen. "Then came the revolution—I was part of that." Dooley was fourteen years old; he carried a gun and belonged to a militia. During the Bay of Pigs invasion in 1962, he was stationed at the Air Force base in the city's center. He was told to dig a hole, climb in, and await further instructions. Aware that he was sitting in the bull's-eye center of a bombing target, Dooley considered sneaking off to a bar, but didn't.

He turned his attention to diving and spearfishing, which he far preferred to revolutionary war. In 1968, he enrolled at Cuba's Academy of Sciences to study oceanography, but happened upon a book about marine archaeology that made everything else feel irrelevant. Though it probably didn't seem like it, Dooley was well positioned. Limping, busted, communist Cuba: good luck getting anything at the grocery store, but historical shipwrecks? There were plenty of those. For Spanish ships, Havana was the gateway to the colonies. They *all* stopped there. And Cuba's treacherous reefs and pop-up hurricanes meant that many of them sank there.

"In all of Cuba there were only two archaeologists," Dooley said. "*Land* archaeologists." The sea was wide open. After learning the basics of the science and getting a master's degree in archaeology, he embarked on a cross-country trip to interview fishermen, asking if they'd come across artifacts or snagged their nets on anchors. He trawled the national archives for accounts of wrecks. "The information was incredible," he recalled. "I had a list of targets. But in Cuba there was no money. If I could get two books in a year, that was a thing."

By 1984, Dooley was working for a government-sponsored dive company that shared his interest in finding lost ships. After a lengthy search, he located what he considered to be Cuba's most significant wreck: the *Nuestra Señora de Mercedes,* a Spanish galleon that hit a reef in 1698. The *Mercedes* was a vice flagship from the same armada as the *San José,* but in a slightly earlier period. It, too, had been carrying treasure—though much of its cargo was salvaged right after it sank, because it lay in only thirty feet of water. Since then, storms had smeared the wreckage over a larger area. At the site, Dooley's archaeological survey revealed a large anchor and two cannons. Finally, here was a chance to do a real excavation. When his boss ordered him to pillage the site instead, Dooley quit on the spot.

But the *Mercedes* was more than a failed opportunity. While researching it, Dooley had managed to travel to the archives in Seville. It was there, while studying correspondence from the governor of Havana to the king of Spain, that he stumbled upon a misfiled packet of letters about the *San José.* Written by the governor of Cartagena, the letters discussed the battle and its aftermath, the disappearance of the galleon and its gold. Dooley was riveted. He began a file that would expand like an accordion over the next thirty years.

Time went by. Dooley made an ocean documentary, wrote a book about Caribbean reef fish, found lesser wrecks, made his way home to America. The *San José* was never far from his mind. But once again—how? To even *think* about a deep-water search was expensive. And there was another wrinkle. An American treasure-hunting outfit called Sea Search Armada was suing the Colombian government over the terms of an agreement to salvage the *San José*. The treasure hunters claimed to have located the ship in 1981, offering dubious evidence: scraps of waterlogged wood and a set of coordinates that proved to be nowhere near correct. Colombia had responded by passing a law that entitled marine salvors to only 5 percent of a vessel's value, taxed at a rate of 45 percent—a fee that wouldn't cover their costs—revoking the company's search permit, and declining to issue any others. Setting aside the fact that Sea Search Armada hadn't actually found the galleon, the lawsuit was a cautionary tale: valuable shipwrecks tend to spawn nasty fights.

There had been a terrible scene in 2007, when a company called Odyssey Marine Exploration raked up seventeen tons of silver and gold coins from a wreck thirty-six hundred feet down, off Portugal, squirreled the loot into boxes, and flew it to Florida in a Gulfstream GV and a chartered Boeing 757. Odyssey was a publicly traded treasure hunting operation so it trumpeted its $500 million find in the press, sending its stock price soaring. When asked about the treasure's provenance, however, the company was strangely coy. It wasn't sure where the coins had come from. Maybe someone had tossed them overboard? When it turned out they'd been scooped from the wreck of a Spanish frigate sunk by the British in 1804, Spain battled Odyssey all the way to the U.S. Supreme Court, and won. (Notably, there are many laws that take precedence over "finders, keepers.") In 2012, the treasure hunters were forced to pack up the coins and fly them back across the Atlantic. "This is not money, it is our history," a Spanish official huffed.

Problem is, in the deep you need money to get at the history. As awful as Odyssey's behavior was—and there was nothing archaeological about how it treated the site—there would've been no coins to argue about if the company hadn't spent millions to find them. So if the goal is to become better acquainted with our submerged cultural heritage, too often nobody wins. Marine archaeologists have the expertise. Governments have the rights. Commercial enterprises have the robots and the money but won't do anything for free. In a society that's figured out how to split the atom, couldn't there be a way to balance those interests?

If we fail to investigate deep wrecks because we're too busy suing each other, there's far more at stake than gold coins. Unlike the Library of Alexandria, the seafloor hasn't burned. Entombed in the silt are antiquities with the power to recast history. In 1900, for instance, a Greek sponge diver came upon a wreck site two hundred feet down, near Crete, and saw a tangle of bodies inside it—not dead people, but marble statues. When the two-thousand-year-old ship was probed further, divers found a collection of artworks worthy of the Louvre. But one object was more sublime: a hunk of bronze that turned out to be a mind-bendingly complex astronomical calculator with hundreds of intricate gears—an analog computer that predicted the motions of celestial bodies. More than a century later, researchers are still studying and x-raying and pondering the device, now known as the Antikythera mechanism; they suspect that it was designed by the great mathematician Archimedes. In a recent paper about the mechanism, the authors concluded: "It challenges all our preconceptions about the technological capabilities of the ancient Greeks."

No, it's not easy to unearth the deep's historical holdings. Yes, we should do it anyway—and find an economic and cultural model that works. In 2013, a door cracked open when Colombian president Juan Manuel Santos changed the country's law: whoever found the *San José* could now claim up to 50 percent of its "nonpatrimonial" treasure. A gold chalice, a pearl necklace, the altarpiece of the ship's chapel: one-of-a-kind objects like those could never be sold. But a bag of uncut emeralds or millions of near-identical coins? Those were more like commodities, Santos reasoned, and some could be parted with. If the right partner came along, Colombia was willing to make a deal.

Hearing the news, Dooley mobilized. He gathered his research, cast around unsuccessfully for a Latin American investor, then found the ideal person in Britain: a businessman who was willing to bankroll a search for the *San José*. The investor, who remains anonymous, agreed to back Dooley. He supported the idea of doing a meticulous archaeological excavation, and was also willing to build a state-of-the-art conservation lab and a *San José* museum in Cartagena. (If the ship were found he would recoup his costs, and, one assumes, profit from the sale of the nonpatrimonial treasure.) Funding in hand, Dooley sought permission from the Colombian government. He decided to go right to the top.

Telling the story now, Dooley was wild-eyed, acting out his desperation to get an audience with President Santos. (As you might guess, cold-calling a head of state does not yield instant results.) It took months of persistence but finally at a reception in Manhattan, Dooley got the chance to make his pitch. He presented Santos with a gift: a previously unknown eighteenth-century map that contained a tantalizing clue to the *San José*'s location—and it was miles away from where everybody else had been looking. "I know I can find the ship," he told the president.

*

"You see the *Bajo del Almirante*, the Shoal of the Admiral, 1729, right here," Dooley said, gesturing to a copy of that same map on his living room wall. The map was drawn in sepia ink on ivory paper, and it gave off an air of antiquity, even as a replica. It depicts the waters around Cartagena, including an islet (or shoal) that had eluded other cartographers: "Twenty years later that name was not on *any* maps of Colombia. It disappeared." That little inscription, written in a fountain pen's barely legible scratching, was a big hint. "I believe that this name was given to this shoal because that's where Admiral Wager fought," Dooley explained. "And it was drawn not long after the battle, when the memories were still fresh."

He stepped over to his dining room table and unfurled a modern nautical chart. "Let me show you—I have never showed this to anybody." I examined the chart, trying to determine its orientation to Cartagena, but the scale was too large; I couldn't see the coast. Dooley jabbed a finger on some bathymetric lines. "The people from

Sea Search Armada, they say they found the wreck right here. There's *nothing* there." He traced his finger to a different part of the chart. "Where was it? It was right here."

He paused, running a hand through his hair. "Why did I know the wreck was there? *Ahhhhh*, it's a very complicated story. It's a lot of complicated, this shipwreck. This is what happened." Then Dooley was off and running, speed-talking about British logbooks, Spanish logbooks, survivors' accounts, wind direction, ship positions, battle strategy, timing, the locations of various islands—channeling the logic of a seventeenth-century navigator. "You know, they don't say, 'The ship sank in the middle of the ocean.' They tried to name the closest place." He pointed to the antique map. "And when you measure from there to where the wreck is, that's the closest place. *Bajo del Almirante.*"

So far, so good: Dooley had won his bet. "We owe you a debt of gratitude forever," Santos announced when the *San José* was found. In 2016, Dooley returned to the wreck with a bigger ship, more robots, and sixty researchers, spent more of the investor's millions to study the site, took 104,000 photographs, compiled a two-thousand-page report, and made an excavation plan: "We had everything ready to go."

Instead, everything stopped. Before the work could begin, Colombia elected a new president, Iván Duque, who immediately squelched the idea of selling off "nonpatrimonial" treasure to fund the project. *All* of the *San José*, his administration declared, was patrimonial: "Not a single splinter, vase, coin, or stone—nothing that is in the wreck area—can be marketed." But if Colombia wanted to keep every last doubloon (and cover the costs in perpetuity to insure millions of coins and house them in a vault), the country would have to pay for the excavation itself . . . somehow. The project was stalled. More than four years had passed since REMUS had revealed the *San José*'s location, and there had been no further progress in the deep.

It was painful to watch Dooley grappling with this—his dream, right there, so close and yet so damn far. "It's a *disaster*," he moaned. Part of the problem, he added, was that the gold overshadowed everything else. Billions of dollars on the seafloor: it made people crazy. After Duque's pronouncement, the Colombian media—which had previously celebrated the galleon's discovery—began to pile on. Who was this interloper who thought he could waltz into the nation's territorial waters with high-powered robots? A few critics were almost hysterical: Dooley was accused of being a "scoundrel," a "neo-pirate," and the progenitor of an "infamous criminal plot."

"I'm an archaeologist!" Dooley said, pacing the room, his voice rising to a shout. "I don't care about the gold! I can tell you hundreds of things that are more important on the ship than gold coins! Unique things that no one in the world knows about! That's the whole point! The *least* important thing on the *San José* is the treasure."

I believed him. The man was not a treasure hunter. "In itself gold is of no greater value than lead or wood to the archaeologist," George Bass had written, and Dooley echoed that philosophy. He seemed most excited about solving arcane historical puzzles like whether the *San José*'s gun carriages had two or four wheels. In photos of the wreck, Dooley had spotted items that looked to the uninitiated eye like carbuncled lumps of nothing, but to him were thrilling riddles. For instance, a box of enema syringes. "So what the hell are those doing on the ship?" he asked rhetorically, and then laughed. "That was the French." Who knew that enema syringes were all the rage in Louis XIV's France—that, in fact, they were fashion accessories? "Some merchant was bringing them back to Europe to sell," Dooley conjectured. "There was a huge black market."

And what about the square bottles of gin used for medicinal purposes ("You know how many of those square bottles are in museums? None!"), and the curved Turkish sabers carried by the *San José*'s soldiers ("That's worth more than any piece of gold!"), and the hundreds of sealed boxes of contraband that might contain absolutely anything. "We're never gonna know until we excavate it," Dooley said glumly.

The seventeen-billion-dollar question, of course, was when that day would come—or whether it would come at all. What frustrated Dooley the most, it seemed, was the sheer illogic of the standoff. What good did it do to declare "every splinter" of the galleon a precious relic and then leave it to molder on the seafloor? (Or worse, the wreck might get looted.) But maybe the tide would change again. Maybe the next Colombian president would green-light the project. Or maybe philanthropists would step up and fund the excavation, the conservation lab, and the museum. Maybe the dream wasn't dead, but one thing was certain: if we were ever to know the *San José*, we would have to pay for the privilege. The deep always demands its toll.

I closed my notebook. The afternoon was fading, and outside the windows the Atlantic had darkened to the color of a bruise. The lovely, mercurial, merciless Atlantic. It had many moods and many secrets, and it would soon tilt my world upside down, although I didn't know that yet. I did know that I was about to get on a plane and cross that ocean, and that, for now, it held at least one magnificent Spanish galleon. "The whole ship is buried there," Dooley said, his voice now hoarse. "The complete cargo! It's a time capsule. *Everything* in it is unique! You could probably build *ten* museums! I mean it's *incredible*! There's no two *San José*s, eh? This is the mother of all shipwrecks."

CHAPTER 7

The End of the Beginning

The ocean is a place of paradoxes.
—RACHEL CARSON

LONDON, ENGLAND

was late for an appointment with a giant squid, so I hurried past the American mastodon and the *Mantellisaurus* and the three-billion-year-old rock, half running through the London Natural History Museum's cavernous central hall. It was nine o'clock on a Sunday morning, the stroller-pushing crowds hadn't descended yet, and my footsteps echoed on granite floors. Overhead, the skeleton of a blue whale, arched in mid-dive, hung from a cathedral ceiling lined with Victorian botanical paintings. This 138-year-old institution with its eighty million specimens, its five-mile warren of underground corridors, its wings and galleries and theaters and labs and storerooms, its hundreds of researchers, its bones and fossils and minerals and plants, its acres of insects and birds and mammals and fish—it all added up to a bit of a miracle. The museum was an ark of DNA in South Kensington, humanity's grandest attempt to catalog the natural world.

Among its collections were some of the rarest deep-sea specimens in existence. They weren't on public exhibit but you could see them with a guide, so I'd arranged for one, and I met him at the end of the hall. My guide, Mark, had a perky manner, a sharp nasal accent, and a talent for mansplaining. Mark was a docent, not a scientist, but he had the door codes and he knew where everything was, which was all that mattered to me.

This wouldn't be an extended visit. My time in London was short, only four days, most of which would be dedicated to the September 2019 finale of the Five Deeps expedition. Later today the *Pressure Drop*—having completed the last, Arctic leg of its forty-seven-thousand-mile circumnavigation—would sail up the Thames and dock at Canary Wharf. There were parties planned, and talks, and tours of the ship, so the sponsors, press, and public could meet the team and see the *Limiting Factor*. Like all deep-sea endeavors, the expedition had unfolded far from the public eye—until now. I couldn't miss the victory lap. And London offered another irresistible opportunity: the chance to see the world's most intact giant squid specimen, which resided in the basement of this museum.

After some compulsory small talk, Mark set off at a fast clip and soon we were behind the scenes in a modern wing called the Darwin Center. It was a maze of labs and facilities, eight sterile stories of steel and glass, and at the moment it was deserted except for a colony of flesh-eating beetles that was working overtime to polish off a carcass. When the museum's scientists need to reduce a specimen to its skeleton, Mark explained, "they'll remove the scales or the fur, and the beetles do the rest." The glass-fronted room was plastered with warning signs and riddled with traps to prevent beetle escape. "These little chaps do a good job," he added with an admiring nod. "When they don't have much work, we'll give them some dog biscuits to tick them over."

We walked on, winding deeper into the building's lower levels until Mark stopped at a heavy door: "We're going into the airlock." The door opened to reveal another door, and beyond that was a long corridor of metal cabinets that reminded me of a server farm. Instead of computers the cabinets were stocked with creatures, carefully filed by species. I noticed the temperature had dropped about twenty degrees. "Now why do we have it so cold down here?" Mark asked in a wheedling tone, as if quizzing a class of third graders. I stared at him until he answered the question himself. "It's primarily because of the evaporation." The specimens are suspended in ethyl alcohol, which prevents deterioration, preserves DNA, and deters microbial pests, but which is also highly flammable and temperature sensitive. If the jars aren't sealed tightly or stored properly they can release noxious fumes, and if too much alcohol evaporates even the most robust animal ends up looking like a piece of jerky.

One wall featured open shelves lined with glass jars, each housing . . . something. "I call this our Little Shop of Horrors," Mark said, giggling. He pointed to a container: "This is a jar of sperm whale eyeballs. Oh, and that one there is a manatee. And some tiger cubs behind there, you can actually see their little stripes. You've got your platypus, your wombat, your zebra fetus. Do you see the labels? Now why do you think they use Latin on the labels? Well, I'll tell you. Because in science—"

Letting him talk, I moved farther down the hall. This was all very interesting, but I wasn't here for the floating eyeballs. Mark sensed my impatience. He strode ahead and opened another door with a flourish. "Now, do you want to see some big things?"

*

I caught a whiff of chemicals when we stepped inside, and Mark flicked on the lights. "This is the Tank Room," he announced as I gaped at the scene in front of me: a yawning industrial space anchored by rows of metal tanks, with exposed ductwork, steel girders, and ventilation hoses running overhead. Around its perimeter, stacked to the ceiling on sturdy shelves, were hundreds of oversize glass casks filled with animals pickled in liquid. It was Dr. Frankenstein's lab if he'd had a fetish for fish. At a glance I saw a row of sharks hanging vertically, and a wolf eel baring its fangs, and an entire Komodo dragon. There were two coelacanths, a prehistoric fish thought to have gone extinct in the Cretaceous period before being found very much alive in the Indian Ocean, and Mark pointed out a dragonfish with carpet-nail teeth: "Have you seen *Alien*? Looks like the chest-buster, doesn't it?"

In one corner, locked behind glass doors, was an irreplaceable collection of "type" specimens—the definitive representatives of a species—and creatures with extra historical clout. The dainty little octopus in one glass jar had shared Charles Darwin's cabin on the voyage of the *Beagle,* living in a saltwater tank while the great biologist marveled at its shape-shifting, ink-squirting, color-changing abilities. Other animals on the shelves had come from the HMS *Challenger'*s dredge. But the main event, bisecting the room, was a forty-foot-long glass tank that contained Archie, short for *Architeuthis dux:* the giant squid.

"Impressive, isn't it?" Mark asked needlessly, because this was undeniable. To walk the length of Archie's tank took me a full thirty seconds, though at twenty-eight feet long she (yes) was only a juvenile. Giant squids have eight anaconda-thick arms and two wildly elongated hunting tentacles, their tips studded with serrated suckers, projecting from a hulking rocket-shaped mantle. (Their heads are wedged between their mantle and their arms, the anatomical equivalent of us having our faces staring out of our hips.) These ten appendages are as stretchy as rubber bands so measurements can be deceptive, but scientists estimate that an adult *Architeuthis* grows to about forty feet and weighs up to a ton.

But bulk alone isn't what earned the giant squid its wicked reputation. What made the Kraken truly sinister in human imagination was our assumption of its cunning and malevolent intent. How could a slithery beast with eyes as big as volleyballs *not* be scouting and scheming and plotting its next attack? Squids are cephalopods, a class of marine mollusks that also includes octopuses, cuttlefish, and nautiluses. Some members of this invertebrate group are famously savvy, with cognitive abilities that surpass many vertebrates (including some mammals). Yet scientists are unsure if *Architeuthis* is a ferocious predator or a shy opportunist, whether it uses its highly evolved brain and eyes to stalk other mammoth creatures—like its nemesis the sperm whale—or to avoid them. In other words, who's hunting whom?

Along with the many sightings of sperm whales cruising along with twenty-foot tentacles trailing from their mouths, there are historical accounts of surface battles in which the giant squid appeared to be winning. In 1875, for instance, George Drevar, captain of a schooner called the *Pauline,* described seeing "a monster sea-serpent coiled twice around a large sperm whale," with the "monster" using its tentacles "as levers, twisting itself and its victim around with great velocity." After fifteen minutes of thrashing, the whale was dragged headfirst into the depths and Drevar felt "a cold shiver . . . on beholding the last agonizing struggle of the poor whale that had seemed as helpless in the coils of the vicious monster as a small bird in the talons of a hawk."

As recently as 2003, a French team sailing in a round-the-world race was surprised when their boat—a 110-foot trimaran—came to a shuddering halt in the middle of the Atlantic. The skipper, Olivier de Kersauson, figured they'd gotten snagged on a fishing net. But when First Mate Didier Ragot investigated, peering underwater through a viewport in the hull, he saw tentacles as wide as his thigh. "The creature seemed to be wrapping itself around the boat, which rocked violently," Ragot recounted. "The floorboards creaked, and the rudder started to bend. Then, just as the stern seemed ready to snap, everything went still." For whatever reason, the squid had released them and retreated into the depths. "I don't know what we would have done if it hadn't let go," de Kersauson admitted later. Unlike the crazed Captain Nemo from *20,000 Leagues Under the Sea,* who fended off marauding giant squids with an ax, "We weren't going to attack it with our penknives." (Ironically, the prize the French sailors were competing for was the Jules Verne Trophy.)

There would seem to be ample evidence of the giant squid's aggression, but the stomach contents of sperm whales tell a different story. At the center of their crown of arms, squids have a parrotlike beak that shreds their prey, and those Cuisinart blades are indigestible. Judging from the number of beaks that scientists have found rattling around in whale bellies, *Architeuthis* may put up a good fight, but sperm whales are the deep's undisputed heavyweight champions.

The corpse in the Tank Room provided no epiphanies about giant squid behavior. British philosopher Alan Watts was correct when he wrote, "No one ever understood the wonder of a bird on the wing by stuffing it and putting it into a glass case." As useful as Archie was to researchers—who had examined her in cellular detail—she was reduced to a tattered mass of tissue suspended in liquid the color of limoncello. Her superhighway of a nervous system was intact (all half-billion neurons of it), as was her donut-shaped brain and her three hearts, but her blue blood with its copper-based hemocyanin (rather than red, iron-based hemoglobin) was gone. Now her skin was forever a pallid beige, rather than flashing with the brilliant colors and iridescence that cephalopods are known for.

Before 2012, when a luminous lure devised by the American marine biologist Edith Widder repeatedly drew the reclusive *Architeuthis* into camera range off the coast of Japan and in the Gulf of Mexico—and the resulting videos were viewed by millions online—few people had ever seen one alive. On occasion, flustered fishermen would find a giant squid writhing in their nets (this is how Archie was caught), but the animals survived only briefly. "When she arrived here she was frozen," Mark explained. "She took four days to thaw out." While preserving Archie, he added, the scientists wore gas masks. A giant squid's body is infused with ammonia—a chemical that's lighter than salt water—which helps it to maintain neutral buoyancy. "Think of the worst toilet stink ever. That's what she smelled like."

Perhaps the most startling thing about *Architeuthis* is that it isn't the biggest squid in the deep. Archie shares her tank with a dismembered chunk of a species known as the colossal squid. Its arms are shorter but its mantle is burlier, its hunting tentacles feature curved hooks like a cat's claws, and overall it's more of a bruiser. Both species are cryptic, living and hunting more than three thousand feet down, but even less is known about the colossal squid—and intriguingly, researchers have found larger beaks from what they suspect is a *super-colossal* colossal squid.

"In no department of zoological science, indeed, are we quite so much in the dark as with regard to deep-sea cephalopods," the science-fiction author H. G. Wells wrote in 1896. A hundred and twenty-three years later that statement was still true, though probably not for long.

Squid biology has advanced steadily since Olaus Magnus weighed in
five centuries ago: "Their forms are horrible, their heads square, all
set with prickles, and surrounded by long sharp horns like the roots
of an upturned tree." Exploring the deep will mean coming to know
giant squids as the paradoxes they are: predators powerful enough
to seize a sailboat, prey that's likely to be slurped down by a sperm
whale, skilled but cautious, curious but retiring, far from diabolical
yet smart enough to avoid us.

I ran my hand along the side of Archie's tank, silently saluting her.
"Do you like calamari?" Mark asked.

"Not anymore."

*

The Admiralty bar in Trafalgar Square is designed to evoke the interior of Admiral Lord Nelson's ship, the HMS *Victory,* three floors of
nautical décor and memorabilia conveying the impression that if it
weren't busy dispensing craft beer, the pub would sail off to the Napo-
leonic Wars. In two showy fountains outside the building, Poseidon's
son Triton cavorts with mermaids, mermen, and dolphins. It was the
perfect venue for the wrap party of an oceanic expedition.

Also appropriately for a group that had spent the last year exploring beneath the surface, the Five Deeps festivities occupied the
floor below street level. If this was a ship, the party was in the hold.
Descending a narrow staircase, I could hear raucous noise and loud
music emanating from below.

The official public event had been held the previous evening, a
cocktail party and presentation at London's Royal Geographical Society, and the weighty history of the place had meant suits and ties and
buttoned-down behavior. Victor Vescovo had orbited through the
crowd accepting congratulations, and in his attire and polish there
were hints of the parallel universe he inhabited as a private equity
banker. The last time I'd seen Vescovo, in Nuku'alofa, he was bare-
foot in jeans, tossing back margaritas at the Sky Bar, his ponytail slipping from its elastic as he belted out karaoke, set to his own lyrics:

I've got the Five Deeps blues
Sitting on a bridge in Tonga
My battery's on fire at ten thousand meters
And I'm just trying to get home

Scanning the society's reception room as the canapés circulated, I got the impression the Five Deeps crew was still adjusting to the novelty of being back on land. Scruffy shipboard beards had been replaced by clean shaves and fresh haircuts. Tim Macdonald had sacrificed his Fu Manchu mustache; he told me that he'd stayed in the Arctic for an extra week, roaming around Svalbard to look for polar bears, before he'd felt ready to rejoin civilization. Socializing with strangers takes practice after living in the hermetic bubble of a ship at sea; some people had brought family members to the event as a buffer.

When the presentation began and the audience filed into a wood-paneled auditorium, I saw Don Walsh being ushered into the first row. Upon introduction, he received a standing ovation. That night he would be among the last to leave the building, beset by reporters, ocean history buffs, and autograph seekers. "I seem to have become a sort of ceremonial object," he said later, sounding amused.

Rob McCallum had acted as the emcee, introducing the team and enumerating the Five Deeps' achievements. The *Limiting Factor* had touched down at thirteen of the ocean's deepest sites, with Vescovo setting numerous records; Alan Jamieson's science team had overseen 103 lander dives, collected more than a hundred thousand biological samples, and identified forty new species (with more to come), documented by five hundred hours of video. The expedition's sonar efforts had resulted in 212,000 square miles of high-resolution maps of the deep seafloor—an area the size of France—60 percent of which had never been charted before. (The maps would be donated to the General Bathymetric Chart of the Oceans, a UN initiative to compile a complete model of the seafloor by 2030.)

In his keynote speech Vescovo had reflected, "It's often asked, 'Is there any true exploration left?' And of course there is. Eighty percent of the ocean is unexplored. When you go below two thousand meters, people say, 'Oh, but there's nothing down there.' Well, I would disagree."

Near the end, during a question-and-answer period, the collective personality of a group that had sailed around the globe with a skull-and-crossbones flag began to assert itself. When one questioner held forth for about ten minutes before asking portentously, ". . . and so do you think there will even *be* an ocean full of life in fifty years?," Jamieson, sitting in the audience beside Walsh, had jumped up, run onto the stage, grabbed the microphone, replied monosyllabically—"Yes"—and run back to his seat.

*

Now Jamieson was hoisting a pint at the Admiralty bar with his fellow Scotsmen, Captain Stuart Buckle and Second Engineer Charlie Ferguson, and all three men were wearing kilts. I greeted them and ordered a pint myself. The bar had a cozy, clubby feel, with thick-planked wood floors and low, arched-brick ceilings. Conversation resounded from tables tucked into berth-like nooks. McCallum walked by in a jazzy Hawaiian shirt, followed by Patrick Lahey, holding a bottle of wine, in search of the buffet.

Across the room, Vescovo was showing people something on his phone. I went over to see if it was video from the sub, but it turned out to be photos of his dogs, three Belgian Schipperkes named Ivan the Terrible, Little Nikolai, and Mishka. They looked like black mini-wolves with alert eyes and streetwise attitudes. "The Russian names fit their personalities," Vescovo explained. "They're chaos beasts."

Before he put his phone away, I asked if he had any footage from his most recent dives, which had included a solo visit to the *Titanic*. I'd heard the wreck described as an entanglement minefield, and pretty much the last place a submersible pilot should go by himself. Sections of the ship had collapsed, weakened by iron-eating bacteria that would eventually consume the whole structure; metal shards jutted from the hull like impalements. Cables, wires, davits, brackets: there were countless ways to get hooked and zero ways to get rescued.

"Yeah," Vescovo said, exhaling sharply. "Rob and Patrick came at me multiple times saying, 'Victor, you don't want to do that. It's too dangerous. You need two sets of eyes in the sub.' And I went and I did it and they were right, and I would never do it again."

On his descent Vescovo had landed a half mile away from the *Titanic*, and realized immediately that he was not in benign waters. Thirteen thousand feet down, the Atlantic abyss was murky, brooding, livid with unusual currents. Shaking off the chill, he adjusted his heading and set off in the direction of the wreck. "I could see the shape of the bow on my sonar," he recalled. "So I'm getting closer and closer, and the seafloor's getting more fractured, and the sonar's telling me I'm ten meters away, and I'm peering out the window and I don't see anything. It's just black, black, black. And then it hit me: *I'm right in front of it*. The blackness I'm looking at is the starboard side of the *Titanic*. So I hit the thrusters to go up, and within two seconds I see a row of portholes. Then another row, and then another . . ." He stepped back for effect. "And I was like, 'Holy crap! This thing is *big*.'"

Nudging the sub forward, right hand clenched on the joystick, Vescovo had toured the site. He saw the gnarled stalactites of rust weeping from the ship, giving it a mottled, almost geological, patina. White sea anemones and corals dotted the wreck like memorial flowers. The bow and the stern lay two thousand feet apart with a debris field between them. It was a piñata of tragedy, busted across the seafloor: twisted pipes, snarls of rope, torn machinery, broken crockery, a pair of pants.

"You can see, the stern just exploded when it hit the bottom," Vescovo said. He thumbed through his photos and pulled up an image of some steel spars—once strong enough to brace a forty-six-thousand-ton ship—bent like Grendel's arthritic claws. "There are these big pieces sticking out everywhere. Stuff like that, you don't want to go anywhere near it. God knows what could snare you." He shook his head, and put his phone back in his pocket. "That was one of the riskiest things I've ever done. Scared the heck out of me. It was worth it, though."

*

Leaving the *Titanic*, the expedition had proceeded north to the Molloy Hole, the deepest spot in the Arctic Ocean. At 5,550 meters (18,209 feet), it was the shallowest of the Five Deeps; unlike the other sites, the Molloy Hole is not part of a hadal trench. It's more like a bowl in the seafloor, at the southern end of the spreading center where the Eurasian and North American tectonic plates are pulling apart.

That might sound like an anticlimax, but the opposite is true: the Arctic deep roils with complexity. It's brutally hard to access, ice-choked and bone-numbingly cold, crammed with unique animals and microbes, and central to the understanding of how climate change will rip the tablecloth out from under our lives. Major ocean currents converge there—the warm, salty waters of the Atlantic meeting the cold, fresh polar waters—creating gyres, vortices, and eddies that rise and sink and circulate heat in a sinuous dance.

To get a sense of how much deep-sea action goes on in the far north, consider the Denmark Strait Cataract, located between Greenland and Iceland. It's the planet's tallest, mightiest waterfall, 11,500 feet high and a hundred miles wide, pumping five million cubic meters of water per second—and it lies two thousand feet beneath the ocean's surface. Also, the Arctic is an undersea obstacle course of volcanic terrain. Peaks, ridges, valleys, faults, cliffs, shelves, vents: if the northern depths were music, they would be a live album by Nine Inch Nails.

Vescovo had made three dives there: by himself, with Heather Stewart, and then a final dip with Jamieson. "It was just *mad*," Jamieson told me. "From start to finish, it was properly rugged canyon diving. You couldn't relax for a second. There were boulders the size of cars coming out of nowhere. We had the sub pitch forward about forty-five degrees because we hit something so hard."

"The Molloy had a different character," Vescovo agreed, recalling the dives now. "It was more like hiking the Alps. And the amount of biological samples we collected was staggering. The lander traps were completely full. It was the biggest haul of the whole expedition." Only in the aftermath of the final dive, he said, did its significance hit him: "Oh my God, we did it. We did the Five Deeps. Everybody's safe."

With the expedition over, I asked, was he experiencing any adrenaline withdrawal? Sipping tequila in London does not exactly have the same thrill quotient as plunging into the uncharted abyss. Extended endorphin highs can drain a person's tank, and they're often chased by post-adventure lows. "Oh, for sure," Vescovo said. "Diving the Mariana Trench three times in a row, I mean, that's not a normal week. In general I love it, but you do need a break. I sort of have to decompress."

Not for long, however. At the Royal Geographical Society, Vescovo had announced a new expedition called the Ring of Fire—a series of hadal dives around the Pacific Rim—and it would begin in less than six months. Initially he'd intended to sell the sub, ship, and landers after the Five Deeps, but no buyer had emerged and he'd become smitten by ocean exploration, so why not head back out to sea?

"We want to do the Philippine Trench," he told me, the plans tumbling out. "It's ten thousand, five hundred meters deep and nobody's been there. The Kermadec Trench, the New Hebrides Trench. The Kuril–Kamchatka Trench if we can get a permit. And we'll go back to the Challenger Deep. We won't just dive it, we'll actually survey all three pools—we'll map the hell out of it. No manned mission has been to the western pool since the *Trieste* in 1960. Then we'll do the northern areas of the Mariana Trench, the Nero Deep, and a few others." He paused for a second, as if letting me catch up. I understood the subtext of what he was saying. This wasn't the finale of anything. It was merely the end of the beginning.

"We're going to be doing so many dives with so much equipment," he concluded.

"With the same team?"

Vescovo nodded. "Everybody wants to come back."

*

The party sailed on: toasts, shots, affectionate insults, laughter. Somehow these people had lived shoulder to shoulder for a year and still enjoyed each other's company. I spent some time talking to John Ramsay, Triton's principal design engineer. I hadn't previously met Ramsay but I'd heard Lahey's assessment of him: "Fucking brilliant. He's the Leonardo da Vinci of submersible design." *The Economist* had seconded that. In an article about personal subs as a must-have accessory for yacht owners, the magazine described Ramsay as "a man who makes Bond's Q look like an amateur."

It was true that Ramsay's ability to conjure beautiful machines was a kind of secret weapon. Deep-sea gear can be brutish and bulky, the engineering steamrolling the aesthetics, but Triton subs stand out for their elegant mix of form and function. James Bond himself would covet one, so it's not surprising that 007's automaker of choice, the British company Aston Martin, had hired Triton to create its first deep-sea vehicle, a sleek two-person sub that handles like a high-performance sports car.

When I walked over to introduce myself, Ramsay was standing by the wall nursing a cider. He was a quiet, dryly funny Brit with rumpled brown hair, just shy of forty. The look on his face said, *I'm here, I'm having a reasonably good time, but I don't really like parties.* Like others whose brains could power a midsize city, Ramsay was comfortable spending time inside his own head. As a boy growing up in southwest England, he'd sketched inventions, products, objects that didn't exist—but in a cooler world, they would.

"I probably shouldn't admit it because I absolutely love the ocean, but I don't really care about submarines," he told me. "I just love designing things, and subs are this unbelievably untapped area for design. It's like being a Victorian engineer, where no one's ever done it before. So you can just do whatever you want, and make changes."

For someone with such a free-ranging imagination, the chance to create the deep-sea equivalent of a lunar module was, to quote Ramsay, "the ultimate." For years, even before Vescovo came along, Ramsay and Lahey had been batting around concepts for a full-ocean-depth three-man sub. Ideally, what would it look like? How would you build it? How could you fundamentally change the way people viewed the deep? The answer, they believed, was glass. It seems counterintuitive, but under pressure, glass can be stronger than titanium: imagine floating in a glass bubble that could take you anywhere in the ocean. In theory, this was possible. In reality, far more research was needed to figure out how to optimize a glass sphere. For Vescovo's sub, titanium was the practical choice. (But glass was definitely on the agenda: "It will be sooner rather than later," Lahey vowed. "I think we're less than ten years away from the first glass pressure hull.")

You have to think that designing any manned submersible comes freighted with terrifying responsibility. It's not like designing a couch or a lawn mower or a toothbrush, where a flaw might result in some slight irritation or mild inconvenience. Ramsay confirmed that for him, the *Limiting Factor*'s test dives were nerve-racking: "That's the bit where my heart is in my mouth." And even after the sub had proved itself, he wasn't entirely carefree when he joined Lahey for a trip to the Challenger Deep.

"Um . . . you can't describe it as a pleasure dive like you would in one of our acrylic models," he said haltingly, when I asked how it felt to accompany his creation into the hadal zone. "It's a very different type of experience. You're in there for twelve hours." It probably didn't help that Ramsay was slightly claustrophobic, and that while they were on the bottom the manipulator arm had gone berserk for no identifiable reason, spinning and clawing spasmodically at his viewport. ("Apparently it's known as 'alien hand syndrome,'" Lahey explained. "Yep, that's a thing." Back on the surface, the fault was diagnosed as "water ingress into the electronics.")

Seven miles of ocean above your head, eight tons of pressure per square inch trying its best to end you, a robotic arm stabbing at your face—and the only thing holding it all back is a thin skin of titanium that was built according to *your* calculations? That's not an ordinary day at the office. That's a tad more intense than giving a PowerPoint presentation or addressing the shareholders. Ramsay's profession was off-limits to dabblers, bunglers, or anyone who had failed math. I mentioned that to him, and he grinned. "I don't think I could find a job that's closer to what I like doing than this."

*

"This is an alcohol-free table," Lahey said jokingly, waving me over with a glass of merlot. He was sitting at a banquette with his wife, Tiziana, an Italian opera singer with a warm demeanor and a cascade of sandy-blond curls. I ducked in and slid across from them. The ceiling was snug and rounded, as though we were hunkered inside a submarine. Noting the similarity, Lahey laughed. "I guess I'm just comfortable in confined spaces."

He reached for the wine bottle to pour me a glass and noticed McCallum passing by: "Hey Rob, join the party, you miserable bastard!" McCallum turned in the direction of the heckling and came over to the table. "Have you been drinking, Rob?" Lahey asked as he sat down. "Because you're staggering a bit."

McCallum looked at me. "He often does this," he said, gesturing to Lahey. I didn't doubt it. The two had worked together, on and off, throughout their careers. There had been other subs and other ships, other expeditions and other clients, and many other drinks. They were members of the same deep-sea clan who, while aligning on most things, reserved the right to argue vigorously and still get along. Refreshingly, neither man was egotistical. It was a trait that made them well suited to their roles, because if the ocean demands anything, it's humility.

Maybe that's why the grand celebration of the Five Deeps was a private affair in a basement pub. It was focused inward, as if the reward wasn't in gaining the world's attention but in the joy of being part of the team itself. "I've never been more proud of a group of people than I was on this expedition," Lahey said, turning serious in an instant when I asked about this. "Because I can assure you, it wasn't easy."

McCallum winced. "In the beginning, my God, it was a mess."

"We had a few things happen," Lahey agreed. "But what's life, if not that? It's a sequence of interesting, terrifying, rewarding experiences, right?"

"I've had some interesting experiences," McCallum replied. "With therapy, some of them are fading."

At that moment a cheer went up in the other room, followed by applause, foot stomping, and whistling. Nobody was ready to leave yet, but soon, if only temporarily, the deepest explorers in history would be going their separate ways. Vescovo was bound for Dallas and his day job, Jamieson for his lab at Newcastle University. Ramsay was off to invent new submersibles at his home office near Cornwall. Walsh, who'd already left London, was en route to Tahiti. In the morning Lahey was flying to Barcelona to test-dive Triton's latest sub, the *DeepView*, a futuristic-looking twenty-four-seat tourist craft with a cylindrical acrylic hull. McCallum was headed to the Monaco Yacht Show, where an "expedition superyacht" his company helped design would be exhibited for sale. It was a polar-class icebreaker with two helipads, a dive center with a decompression chamber, and a hangar with room for multiple subs. "We're doing so much expedition planning it's ridiculous," McCallum said, looking happily overwhelmed.

I sat listening, twirling the stem of my wineglass and fantasizing about life aboard a submersible-carrying superyacht. It was almost painful to think about how much fun it would be to explore any deep-sea place, anytime you felt like it. My yearning must have been obvious, and Lahey must have sensed what was on my mind, because amid the shouting and music and clamor of the party, he leaned across the table and in a quiet voice assured me, "We'll get you on a dive."

CHAPTER 8

You Are Now Entering the Twilight Zone

If one dives and returns to the surface inarticulate with amazement
and with a deep realization of the marvel of what he has seen and
where he has been, then he deserves to go again and again. If he
is unmoved or disappointed, then there remains for him on earth
only a longer or shorter period of waiting for death.

—WILLIAM BEEBE

NEW PROVIDENCE ISLAND, BAHAMAS

Dawn in the Bahamas is a pink-tinted thing, soft and warm and
faultless. Maybe that's less true during hurricane season, but
on my first morning in Nassau I watched a rosé-colored sunrise with
cotton candy clouds. I was idling on a marina dock, smelling the
Atlantic's salt tang and drinking disappointing coffee, waiting for my
6:30 a.m. ride: a Zodiac that would pick me up and run me out to a
ship called the MV *Alucia*. The ship was anchored a half mile offshore,
and I could see it from where I was standing.

From the outside, the *Alucia* was a workhorse—a 184-foot gray-
and-white Clydesdale, tough enough to plow through ice. A crane
protruded from its stern. But that plain exterior was misleading. On
the inside, the *Alucia* was every inch an expedition superyacht, a ref-
uge of blond wood, sectional sofas, and sleek electronics. Its deck
hangar was equipped with Triton subs, jet skis, kayaks, surfboards,

scuba gear—all the accoutrements of aquatic adventure. On the week I boarded it in November 2019, the ship was also equipped with ten submersible pilots.

There is only one person who employs that many submersible pilots and sails them around the globe in such style: the ocean-loving billionaire Ray Dalio. While his peers fixate on space travel, Dalio, a devoted scuba diver, is far more interested in the universe beneath the waves. He refers to the ocean as "our planet's greatest asset," and told a *60 Minutes* reporter that he didn't understand the "resource allocation" of lavishing so much cash on rockets, when the deep sea offers more extraordinary sights, more relevant wisdom, and a superior return on investment. In 2011, he dived with Lahey in a Triton 3300/3— an acrylic-hulled model that descends to a thousand meters—reeled in wonder at the experience, and promptly bought the sub. Then he purchased the *Alucia*—formerly the support ship for France's *Nautile* submersible—and upgraded it dramatically.

Obviously as the founder of the world's largest hedge fund, Dalio could have kept these vehicles for himself. But he and his son, Mark, an accomplished underwater filmmaker, had a more expansive vision: they wanted to share the deep with everyone. They wanted to inspire people to care about it emotionally—because intellectual awareness of the ocean's majesty hasn't stopped us from polluting and plundering it. "The capacity of humanity to save the things they fall in love with, I think, can be great," Dalio explained in an interview.

Reckoning the two most potent forces for engaging hearts and minds are storytelling and science, the Dalios teamed up with James Cameron for the former and Woods Hole Oceanographic Institution for the latter. They launched a nonprofit initiative, known as OceanX, and funded it generously. Its mission was simple: to spread awe. OceanX would be like a smaller, hipper, aquatic NASA, roaming the abyss instead of the stars, using media to take a mass audience along for the ride. "You're not going to see aliens in outer space," Dalio noted. "You're going to see aliens when you go down into the sea."

For marine scientists, an invitation aboard the *Alucia* meant the chance to conduct research in unaccustomed luxury. It was a chance to light out for some pure exploration, the kind of curiosity-based seeking that government agencies no longer fund. And for many, it was something even better: a first-time, in-person journey into the deep. On a typical penny-pinching science cruise, the humans are relegated to the deck while the robots do the diving. Undoubtedly that makes sense when laid out on a spreadsheet, but if you want to win hearts, you need heartbeats at the center of the action. Cameron summed up the OceanX philosophy: "No kid ever dreamed of growing up to be a robot. But they do dream of being explorers."

In the nine years since the Dalios recommissioned it, the *Alucia* had starred in a flurry of TV shows and documentaries, including the BBC series *Blue Planet II*. "Right out the gate Ray was doing these remarkable trips," Lahey told me, referring to the 2012 expedition that captured the first footage of a live giant squid; a subsequent quest to film coelacanths in deep-sea caves in Papua New Guinea; and a survey of the Great Barrier Reef, with David Attenborough narrating from the sub's passenger seat. With the cameras rolling, OceanX pilots flew scientists under Antarctic ice and into submarine volcanoes in the Galápagos. They hovered above the Gulf of Mexico's bewitching brine pools—toxic, ultrasaline *lakes* on the seafloor. For deep-sea cinematography it helped to have two submersibles in the water, so one could serve as a lighting platform and occasionally glide into the frame to provide a human scale: a two-man sub called the *Deep Rover II* joined Dalio's three-man sub, the *Nadir*.

Now there was a third OceanX sub, another Triton 3300/3 called the *Neptune*. It sat poised in the *Alucia*'s hangar, its acrylic sphere gleaming like a soap bubble. It was the first thing I saw when I stepped onto the ship's fantail, climbing out of the Zodiac that had fetched me from shore. The *Neptune* was so sparkling new that it was still being certified to dive to its full depth. I'd arrived on the day of its final inspection, to be conducted by engineers from the submarine class society DNV-GL—the same agency that gave the *Limiting Factor* its stamp of approval. Assuming the *Neptune* passed its test, its inaugural plunge would take place the following morning. Assuming that I didn't die of excitement before then, I was going with it.

*

After I dropped my bag in my cabin, I went up to the ship's bridge for the morning operations meeting. The space was crowded, as was the schedule. Sea trials, maintenance work, emergency drills, pilot training: the *Alucia* had come to the Bahamas for two weeks of deep-sea boot camp. My invitation to join a training dive had been brokered by Lahey, who'd connected me to Ray Dalio's office, which in turn had resulted in a phone call with Buck Taylor, OceanX's senior pilot and submersible team leader. "Ray really wants people to see what's out there," Taylor told me. "And he loves technology, so we're testing the latest to bring the deep ocean back to the world."

As far as jobs go, you couldn't imagine a better one for Taylor. He had made four thousand submersible dives. A former British Royal Navy diver who specialized in undersea bomb removal and piloted rescue subs to save men trapped aboard downed military submarines, he'd had a career heavy on danger—but in person he came across as lighthearted. At age forty-nine, he'd found the equilibrium between serious safety and serious fun. He smiled often. When I met him, his cheeks were shiny with sunburn beneath a five o'clock shadow of ginger-colored stubble.

"This is the biggest team we've ever had on board," Taylor said, introducing the assembled group to me. I scanned the room and noticed a familiar face. Tim Macdonald was here as one of four rookies who were training to fly the OceanX subs with Taylor's squad of five pilots. It seemed like a good fit. Macdonald had maritime experience, an easygoing personality, and the engineering skills that all submersible pilots need. He was nomadic by nature, and so was the *Alucia*. When I spoke to Taylor on the phone, he'd described his crew as "a load of ocean gypsies."

The meeting was brief, an assignment of the day's tasks. There would be two dives this morning, with the *Neptune* run through its final sea trial. The afternoon would be dedicated to repairs, inspections, and a lecture on how one sub could free another sub from entanglement by using a hydraulic saw.

It was eight o'clock. The ship had motored offshore and was now stationed in four thousand feet of water. "We're looking to splash at eight thirty," Taylor said, wrapping things up. He addressed the rookies: "Today I'm going to start making you sweat a little bit." There was some nervous laughter and then everyone headed off to get ready, and Taylor explained his training method to me. "One of the first things we do is put them under pressure to see how they react," he said, smirking slightly. "We want people who can stay calm and work through a problem methodically. If you're behind the joystick going, '*OH MY GOD! OH MY GOD!*' you can make a small problem very big, very quickly."

On the dives, Taylor was looking to weed out the melodramatic, overly cocky, or easily spooked by presenting them with simulated emergencies—because at some point, every pilot will face a real one. "I've had a fire in a sub," he said, ticking the perils off on his fingers. "I've been up to my waist in water in a sub. I've been stuck for fourteen hours in a fishing net. So it does happen."

One time in the Galápagos Taylor got swept up in an internal wave, and the impact was fearsome. "We'd been in the water for two hours," he said, recalling the incident. "And then suddenly out of the corner of my eye, I saw this big green wall coming towards me." Instantly the sub was engulfed in a vortex and flung upward and sideways like a trick yo-yo, careening close to some rocks. Even worse, there were two subs tumbling in the maelstrom, because the other one was right next to him. "We couldn't see each other, and we were just getting *nailed*. We were completely out of control." A collision could have been fatal.

Then as quickly as it had come, the wave left. The visibility cleared, and Taylor asked his shaken passenger, the BBC presenter Liz Bonnin, and the other pilot, Toby Mitchell, if they were okay. For a minute or two they composed themselves—"And then we got hit again." In total, the wave struck them three times, bouncing and tossing and dragging them into deeper and deeper water, before the subs finally broke free of it. Taylor concluded with the moral of the story: "The ocean has a way of reminding you who's in charge."

*

A thousand meters. Thirty-three hundred feet. Two-thirds of a mile: that's how far I would be descending. My dive would take me through the mesopelagic zone—more enchantingly known as the twilight zone—the band of ocean that begins where the sunlight zone stops. It's the Manhattan of the deep, a cosmopolis of creatures that are always on the move. The twilight zone is an eat-or-be-eaten kind of place, lit by the signal flares of bioluminescence: the come-hither lures glowing in front of gaping fanged maws, the pulse and flash of slinky bodies, the flickering of neon-colored photophores that line the bellies of a quadrillion fish.

I'm using the word *quadrillion* literally here, but *quadrillions* is probably more accurate. (By comparison, the Milky Way is thought to contain a mere hundred billion stars.) No one knows exactly how many fish live in the twilight zone—researchers keep revising their estimates upward—but we do know that it contains more fish biomass than all the other regions of the ocean combined. In particular, it teems with a toothy, glittery predator called the bristlemouth. If a lone bristlemouth passed by—they wriggle, rather than swim—you would likely miss it, because it's half the size of a crayon. In fact, aside from marine scientists, few people have ever seen one, or have any idea what it looks like (think barracuda in minnow form), or even know that it exists, which is astonishing because bristlemouths are the most abundant vertebrates on earth. For every human, there are a hundred thousand bristlemouths.

Then there's the lanternfish, another diminutive fish that is present in astronomical numbers. With their googly eyes and extra spangling of photophores—arrayed in patterns unique to each of their family's 250 species—lanternfish are cute. So are hatchetfish, a round, mercury-silver fish with a blade-thin body, wee rudder of a tail, and photophores like the nuclear-blue lights of an airport runway. A runty hatchetfish is no bigger than a poker chip; a large one could fit in the palm of your hand.

Other twilight zone dwellers rank lower on adorability. They appear frightful, even terrifying—straight from nightmare central casting. Some look like disembodied heads; others are all teeth. The viperfish's fangs are so long they burst out of its mouth and start to wrap around its face. To offset this impediment, it can unhinge its jaws and open them to a ninety-degree angle. A dragonfish's teeth are stronger than a great white shark's and reinforced with nanocrystals that don't reflect light, making its pit trap jaws invisible to prey. The fangtooth, the blackdragon, the lancetfish, the frilled shark, the black seadevil, among many others: they've all got mouths full of knives and icepicks. (Fortunately, most of them are small.)

There are also many strange doings with eyes. In the blackout deep it's hard to see and there's nowhere to hide, so it helps to have cleverly adapted vision. The cockeyed squid has one big yellow eye that stares upward and one little blue eye that stares downward; the four-eyed spookfish has two eyes that are split in half and function like mirrors rather than lenses. But in the ocular weirdness Olympics, nothing beats the barreleye fish. Its tubular green eyes are sealed *inside* its transparent head and can rotate directly upward, enabling the barreleye to gaze out of the top of its own noggin.

The twilight zone is unrivaled in its profusion of zany life-forms. See-through octopuses, swimming worms with corkscrew tentacles, an eel with a mouth like a pelican, the three-inch-long firefly squid. The majority of known jellyfish species waft through its waters like wayward UFOs. This gelatinous galaxy also includes the archpredators called siphonophores—glimmering chains of cloned cells that zap prey in a stinging net and can grow up to 150 feet long—and the lovely, carnivorous ctenophores: trippy-looking translucent orbs (also known as comb jellies) that sweep through the water by paddling rows of tiny hairlike cilia. The cilia scatter light, making the orbs ripple with iridescence, and most ctenophores are bioluminescent as well, for an extra kick.

*

By now it won't be surprising to hear that, for a long time, these vibrant mid-waters were presumed to be a desert. Edward Forbes, of course, thought most of the ocean was empty. But even Charles Wyville Thomson of the *Challenger* expedition believed that the deep's fauna either swam near the surface or skulked near the bottom, and that the "intermediate zone" between them could be safely written off. Now we know how wrong that is, yet scientists who've dedicated their careers to investigating the twilight zone still despair of how little we know about it.

Granted, a planet-spanning swath of pitch-black water populated by a gazillion peripatetic creatures—many of which are transparent, too fragile to net, too ephemeral to register on acoustic sensors, or otherwise great at avoiding detection—is not the easiest place to study. What's more, it's always in flux. Each night the twilight zone's residents ascend hundreds of feet closer to the surface. Under cover of darkness these visitors from below are less visible, less vulnerable, as they dine on the sun-nourished phytoplankton (single-celled marine plants) in shallower waters. By dawn they've returned to the deep. This nocturnal commute is the world's largest animal migration—a vertical one that occurs 365 days a year.

For a lanternfish or a krill to swim that far daily is like having to scale a mountain to reach your breakfast table. Not every creature makes the trek; some species pursue a feeding strategy of full-time lurking, ambushing the exhausted migrants as they blunder by. Scientists think that maybe half the twilight zone's population migrates each night—trillions of fish, crustaceans, and others rising and falling in unison, as though the ocean itself were breathing. And in a way, it is. By eating plankton near the surface (inhale) and then swimming back down and excreting it (exhale), the animals are shuttling carbon from the atmosphere into the depths, where it remains sequestered for centuries and, in some cases, millennia. These tiny beasts sink a mighty amount of carbon: an estimated 4.4 billion tons each year, the equivalent of America's total annual emissions.

The twilight zone would be something to see, that's for sure. It's the earth's secret party, its grandest and liveliest spectacle, and in less than twenty-four hours I would be there—a thought that was overwhelming, like Taylor's rogue internal wave surging through my head. I couldn't wait to dive, and yet I also wanted to stay in the liminal space of being about to dive, with that exhilarating rush still ahead of me. The training day passed, the *Neptune* got its official thumbs-up, the subs were craned back on deck, and double- and triple-checked, and cleaned, and readied to dive the next morning. I paced around while the fledgling pilots debriefed with Taylor. They'd handled every faux emergency with admirable calm—but I could've used a Valium.

I was finally able to settle myself in the early evening, when a coral-orange sunset beckoned and we went for a communal swim. People leaped off the *Alucia*'s stern with Soundgarden playing in the background, and the water was so warm that we stayed in for a long time, floating and talking and laughing as the twilight drained from the sky.

*

"My name is Tim, I'll be your pilot today. And this is Buck, he's the trainer."

Tim Macdonald, khaki-clad and wearing a headset, was in the pilot's seat for his debut thousand-meter dive. Taylor and I sat in front of him in the passengers' seats. We were three people in a bubble, wearing socks but not shoes, awaiting our launch. OceanX pilot Lee Frey was the day's surface officer, and I could see him hustling around deck making sure that everything was ready. Earlier, Frey had weighed me and showed me an orientation film, featuring Taylor. "A safety card is located under your cushion," Buck had instructed on-screen in his crisp British accent. "In the unlikely event that we do have a problem . . ." It was hard to imagine a safety card being much help in an undersea emergency, but I'd nodded along: *Yes. Sure. Got it.* Now Macdonald was giving me his pre-dive safety briefing.

"Because nobody had done it before": Victor Vescovo prepares to board his sub in Tonga, on his run of history-making dives.

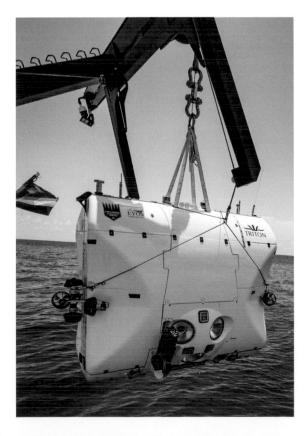

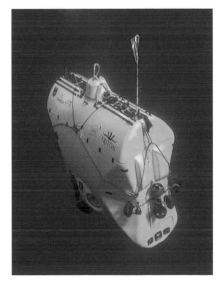

The *Limiting Factor:* a revolutionary deep-sea submersible that can dive repeatedly to the greatest depths (*left* and *above*).

Captain Don Walsh and scientist Patricia Fryer congratulate Vescovo after his first plunge to the Challenger Deep (*above*). Patrick Lahey exits the hatch after the *Limiting Factor*'s full-ocean-depth certification dive. Tim Macdonald, in his role as swimmer, stands behind him (*below*).

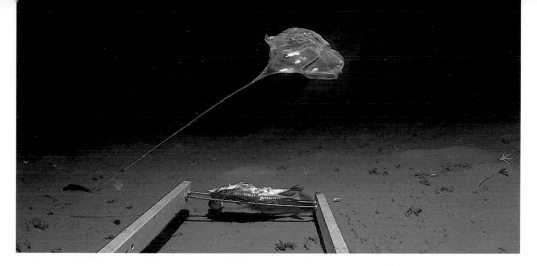

"The weirdest thing we've found so far": a new species of stalked ascidian (*above*) filmed in the Java Trench. Five Deeps' chief scientist Alan Jamieson working on the landers (*below, left*).

A giant isopod (*above*).

Below, left: A gaggle of rattails and two assfish investigate a baited lander.

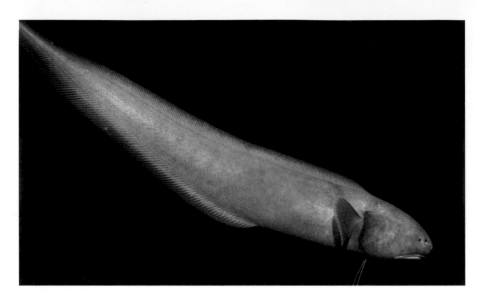

"A species known unpoetically as the robust assfish": amphipods, assfish (*above, below*), and white supergiant amphipod (*below*) at a hadal lander.

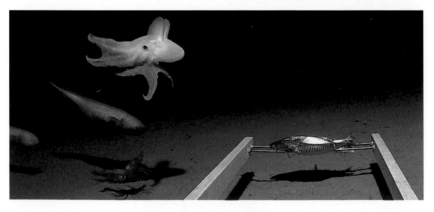

A Dumbo octopus, robust assfish, and red prawn approach a baited lander in the Java Trench (*above*).

Hadal snailfish (*above* and *right*), the world's deepest fish. Alan Jamieson and Tim Macdonald in the *Limiting Factor* (*below*).

Expedition leader Rob McCallum (*above*); Triton team members Frank Lombardo, Kelvin Magee, Steve Chappell (*below*).

From top to bottom: Heather Stewart; Shane Eigler; Cassie Bongiovanni; P. H. Nargeolet.

Captain Stuart Buckle on the *Pressure Drop* (*above*).

The *Limiting Factor* resurfaces from a round trip to Hades (*above*).

The Five Deeps: Victor Vescovo completes his dive to the Arctic Ocean's Molloy Hole (*above*).

The lander Skaff (*above*); glacial eelpouts in the ship's wet lab (*below*).

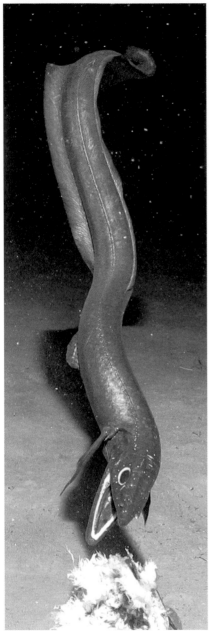

Craggy cliffs on the western side of the Challenger Deep's eastern pool, photographed from the sub's viewport (*above*); stalked crinoids cling to a seafloor rock (*below*).

A cutthroat eel in the Kermadec Trench (*above*).

"We've got a life support system running in here, which controls our atmosphere," Tim continued. "If anything should happen to that, and we have an unstable atmosphere, on your left side you've got a breathing device. It's like a normal scuba regulator, just put it in your mouth and start breathing. And if that becomes totally unusable we've also got rebreathers under here, to avoid smoke inhalation. If, for any reason, Buck and I should become unconscious, you'll need to communicate with the surface."

"Yeah, show me that part," I said.

Macdonald handed me his headset. "You put this on, and push this little button to talk. And you're gonna say, 'X-ray, X-ray, X-ray,' and they'll instruct you on how to come back to the surface. You fill up the main ballast tanks with air, the big yellow ones on the outside, so you'll need to make sure these two valves are closed. And these valves over here, you want to open them up. And then once you've—"

"Very good, but a bit too intense," Taylor cut in. "If your passenger is nervous, that will send them over the edge. Keep it simple. Chat to your guests. Make sure you don't have somebody in the corner crying and dribbling. Just gauge them." Above all, he stressed, make sure that nobody is claustrophobic. "It's much better to find out if someone's having a panic attack here, on deck, than it is when you're in the water—or God forbid, underwater."

Macdonald nodded. "Susan? Are you claustrophobic?"

"No," I replied. Although that wasn't entirely true. During my scuba training in Hawaii, I'd hyperventilated the first time I jumped into the ocean wearing a tank, and had to cling to the side of the dive boat while I gathered myself. The tank felt like a pair of cement boots that would drag me under and pin me to the seafloor forever and never let me breathe again and in that instant, I panicked. I'd gone on to get my license, but that fear still occasionally gripped me. I didn't think it would be like that in the sub, however. We weren't crammed in: we each had our own leather seat. There wasn't enough room to stand up, but the sphere wasn't pressing down on our heads, either. The bubble felt airy and safe.

In front of us, the *Nadir* began to roll forward on steel tracks.

It would launch first, and then wait below the surface so the two subs could descend to thirty-three hundred feet simultaneously. We watched the crane lift it up and deposit it in the water, where the swimmer unhooked its lines. The yellow sub floated for a few seconds, listed to port, and disappeared. Frey walked back from the stern to examine us.

"Surface, *Neptune,* hatch is secure, life support running, safety briefing complete, ready for the water," Macdonald relayed over the VHF radio. Once the sub was submerged the radio channels would become useless, and our communications would rely on an acoustic intercom.

"*Neptune,* surface, safety briefing complete," Frey confirmed. "You are ready to dive."

*

The launch was easy, gentle. Waves sloshed across the acrylic sphere, and the water closed over our heads. It was like being swallowed into a cool blue esophagus. And not just any blue, but the soul-stirring blue that William Beebe had rhapsodized about in his *Bathysphere,* the blue that "admitted no thought of other colors." At two hundred feet down the red objects in the sub had turned black, the yellows were evergreen, the greens were navy. Only the most energetic wavelength of light remained, a pure ultramarine blue. Trying to describe its sensory effect with adjectives, Beebe had admitted defeat: "I felt I was dealing with something too different to be classified in the usual terms."

Personally I would classify it as a Schedule 1 narcotic. It had a dreamlike effect on me as I sat, entranced. The acrylic sphere was an imperceptible boundary; it was as though we were *within* the blue, rather than simply seeing it. "This light . . . ," Macdonald said, equally mind-blown. "You don't get it anywhere else on the planet. You can't re-create it." "Yeah," Taylor agreed. "Welcome to my office."

Three hundred feet, then four hundred feet. We were falling through a blizzard of marine snow, but because our velocity was faster, the particles seemed to be sailing upward. And more surreal: the snowfall was alive. In the sub's lights, preposterously small creatures glittered and darted, jellies flared into sight and then melted away. As Lilliputian as the animals were, their presence provided enormous dimension. Down here, the water was so clear and so still—so absent of the visual cues that signal to the brain: *This is water*—that only by seeing the creatures' movements could you tell that you were traveling through a medium, and not through empty space. Yet at the same time, it was more than a medium. It was a matrix of life.

Passing below six hundred feet, we crossed the border between worlds. We were in the deep ocean; we had entered the twilight zone. Looking up, I could see the last trace of indigo. Looking down, I saw only the dark. At this same threshold, Beebe had reveled in what he called "the blackness of a blue midnight." But strangely, neither he nor I could identify an exact point at which blue was extinguished by black. (Later I would learn the reason why: we'd hit the limits of human vision. In fact, the blue light was still there—and would percolate faintly to three thousand feet—though our eyes could no longer perceive it.)

Suddenly Taylor snapped to attention. "Do you hear that noise?" he asked Macdonald. "That's a new noise to me. It could be a bottle rolling around or whatever, but when you hear something like that, obviously—"

Macdonald flipped on his headlamp to scrutinize the alarm panel, which looked normal, and after a moment or two, Taylor relaxed. "The sub talks to us under pressure," he explained. "The *Nadir* talks a *lot*. Blips and moans as it settles in. Just have a look, be vigilant, make sure you don't have any leaks." Hearing the alertness in Taylor's voice, I realized that I had forgotten this was the *Neptune*'s first certified dive—but clearly, he had not.

I was hunched as far forward as I could get without leaving a nose print on the acrylic, watching the show and trying, fairly unsuccessfully, to identify species. One creature spun by at such close range it almost bounced off the sphere. It was the size and shape of an inverted teacup, topped by a little Mickey Mouse head with wide-set eyes. It looked to me like a vampire squid—a primitive cephalopod with a petite bloodred mantle, earlike fins, two long filamentous tentacles, and webbing between its eight arms. Despite its name the vampire squid is not a squid, but rather a distant cousin of both squids and octopuses. It's a unique species that would be hard to mistake for anything else, but I was startled by the sighting and couldn't quite be sure.

"We're starting to see some bioluminescence now," Taylor observed. He turned to Macdonald: "You can knock the driving lights off."

"Roger that," Macdonald said, and the sub went dark. The sphere was quiet, except for the static crackle of the intercom and the low hum of pumps. To my left, a siphonophore glowed like a nebula; a jellyfish strobed and revealed itself to be shaped like a dahlia. An eel-shaped fish zigzagged by in a mad rush. "That was a scabbardfish," Taylor pointed out. "The teeth on them are unbelievable."

We descended through a rain of fireworks. As the sub traveled, it jostled the animals and microorganisms around it, and they reacted with bursts of light. Most of the twilight zone's inhabitants produce bioluminescence: if it floats or swims down here, odds are it can also glow. On land, only a few creatures light up—fireflies and a handful of other insects, worms, fungi, and bacteria—but in the ocean's mid-waters, bioluminescence is a central fact of life and the key to survival. In the deep, light is a strategy, a tool, a weapon. It's a language—the most common form of communication on earth. Even the bacteria are fluent.

When a creature illuminates itself, it might be trying to attract a mate (by flashing certain patterns) or a meal (by dangling a glowing lure). It might be trying to escape a predator, and turning on its high beams is like flooding a crime scene with a spotlight.

Bioluminescence is handy for both offense and defense. The *Colobonema* jellyfish, for instance, has thirty-two tentacles attached to a transparent bell. When threatened, it ignites the lights on its tentacles, waves them like sparklers, and then ejects them, distracting the attacker with a mass of shimmering spaghetti while the bell makes a surreptitious getaway. (Later, the tentacles grow back.) The stoplight loosejaw dragonfish flicks on crafty red photophores beneath its eyes. Most deep-sea creatures are highly attuned to blue light and can't see red light at all, so the stoplight loosejaw is effectively stalking them with night-vision goggles. Squid have been known to hunt with their lights off, and then strike in a blaze of brightness that stuns their prey. Some creatures squirt gobs of light, tagging their enemies. Many creatures camouflage themselves by dialing their photophores up and down like a dimmer switch to erase their silhouettes. Those are merely a few examples. Twilight zone creatures have as many different ways to wield light as there are facets on a disco ball.

*

The dive plan called for the subs to rendezvous at sixteen hundred feet—halfway to the bottom. The *Nadir* was two hundred feet below us, its lights appearing as two pale blue dots. Meeting up at close range in the mid-water is harder than it sounds, and Taylor was coaching Macdonald.

"So pop your down lights on," Taylor said. "And slow your descent rate. Remember you've got eight tons of inertia here. Eight tons of sub coming down."

"Should I flood?" Macdonald asked, sounding a bit stressed.

"Yeah, you can do."

"Five liters?"

"I don't know," Taylor said in a sly tone. "Use the force, Luke. But try not to land on them, that's never a good look."

He turned to me: "You forget sometimes that we're free-falling. And you don't realize it, but we're doing a gradual corkscrew." (It was basic hydrodynamics. The weight of the manipulator arm caused the sub to rotate around it; with no stationary objects for reference, we couldn't tell that we were turning.)

"We're at five hundred meters," Macdonald announced. The *Nadir*'s lights appeared brighter now, and level with ours.

"Perfect, just relax. Let's drive toward them."

"Control, control, *Neptune*," Macdonald called to the surface. "Depth five-zero-zero meters, holding with *Nadir*."

Taylor laughed. "Well, I hope it's *Nadir*. Could be some Russian spy submarine."

Finally the subs were hovering together, close enough to make out everyone's faces. If you ignored the hordes of creatures zipping through the lights, the scene was of two starships docking in interstellar space. Anyone nursing a phobia might have been triggered by the eerie sight of two machines suspended in oblivion, beneath megatons of salt water. But I was as happy as I can ever recall being.

"*Nadir*, *Neptune*, message from Buck," Macdonald relayed over the intercom. "You've got the lead to one thousand meters."

As the *Nadir* backed away and began to descend, a hulking shadow passed through its downward lights. Taylor and I strained to see what it was, but it had receded into the darkness. "Generally the creatures are smaller down here, but you do get some big ones," Taylor said, raising an eyebrow. "I always wonder what's out there that knows we're here and is looking at us."

We continued our free fall, life swirling around us. Two thousand feet was a catwalk of jellyfish, each one showier than the next. They sashayed by under frilly umbrellas, trailing diaphanous wisps and modeling the latest in fashionable lights. Jellies are often referred to as "simple" creatures because they lack brains, blood, bones, and a heart. Watching them, I found that description laughable. They're intricacy in motion, even though their bodies are 95 percent water—they're barely there. Despite this apparent frailty, they're the most efficient swimmers in the animal kingdom. They're also wily hunters that can eat just about anything, including fish, crustaceans, and each other. The box jellyfish has twenty-four eyes—including some with lenses, corneas, and retinas—and packs the planet's most venomous sting. Another jellyfish species is biologically immortal: it can reverse its life cycle and rebirth itself. To me, those seem like pretty advanced tricks.

Gelatinous beings have been pulsing through the ocean since the early Cambrian period, a quarter-billion years before dinosaurs walked the earth. (And these simpletons will be here, thriving, long after we're gone.) They've evolved over an unimaginably long time scale. As I could see from the sub, jellies make their own rules about body shapes. One appeared as a blazing purple ring with flowing white tentacles, the tips of which were neon violet. One was an opaque gold crown that throbbed like a heart. One was a balloon trailing multiple strings. One was a child's drawing of the sun. One was a fringed sombrero. One was a ghost.

The most exquisite jellyfish appeared as we approached the seafloor. It had a teardrop-shaped bell with luminous yellow and lilac stripes, and dozens of brilliant tentacles. Blinking its lights, it sidled right up to us. Macdonald was doing a slow 360-degree turn, sweeping the bottom with sonar to make sure we didn't smack into rocks or get jammed in a hole. "Oh, don't go through the thruster," Taylor said, swiveling in his seat to watch the creature pass. "Good, we didn't mince him."

A few seconds later the terrain came into view, hazily at first. The sub's lights reflected off a plateau of snow-white silt, but because the only wavelength down here was blue, the bottom glowed with the aquamarine hue of a Caribbean swimming pool. I'm not sure what I was expecting—maybe brown or beige ooze—but it wasn't White Sands National Park radiating a cyan halo. I was stunned into silence by its beauty.

"Remember, this bottom's going to be *super* silty," Taylor warned Macdonald. "It's better to be slightly buoyant than heavy. What's your depth now?"

"Nine-nine-seven," Macdonald said. The thrusters whined as he maneuvered the sub to touch down.

"Is that a big rock right ahead of us?" I asked.

"That's an entire rock *face*," Taylor said. "It's the base of a thousand-foot wall."

"Nine-nine-eight," Macdonald reported.

"Okay, don't drive any farther forward," Taylor advised. "Just come down, nice and gentle. We want to set it on the seabed."

"That's a thousand meters."

Now I could see the bottom clearly. We had landed on an incline, facing uphill. White limestone boulders and rock fragments poked through the white silt, which was stippled with tracks and burrows. Fifty feet in front of us, the rock face rose like a double-black-diamond ski slope. The *Nadir*'s lights were visible to my left, revealing the outlines of Stonehenge-size slabs that had tumbled from higher up the wall.

"It's a bit of a minefield," Taylor said, cautioning Macdonald to keep the sonar on so we didn't hit anything. I'd been so euphoric on the dive that I hadn't thought much about undersea rockslides or collisions or the pressure or the fact that we were more than an hour from the surface—but as I considered the heaviness of our surroundings, those thoughts were inescapable. I imagined the sphere cracking like an egg.

"Lights off so you can do your bottom report," Taylor instructed. Macdonald flicked a switch and we were plunged into blackout. A few green embers flared, courtesy of nearby fish. Taylor turned to examine some gauges. "Gases are good? Oxygen? No—" The shrill beeping of an alarm interrupted him, not jump-out-of-your-skin loud but definitely insistent. If I'd been hooked up to a heart-rate monitor, it would have registered some changes.

"Okay," Taylor said in a businesslike tone. "Look to see what's going on. What have you got there?"

"It's water in the sphere." Macdonald's voice was taut. "I'm going to turn on my headlamp so I can see what's happening."

For a minute I wondered if this was one of Taylor's simulated emergencies, but when Macdonald bent down to see if there was water by his feet, he found some.

"First things first," Taylor said. "Taste it. Is it fresh or salt?"

I held my breath.

"It's fresh."

Taylor nodded. "Have a good look around and double-check. Condensation is fresh. That big body of water that's desperate to come in and sink us is not."

"Roger," Macdonald said, sounding relieved. He muted the alarm.

"If it's fresh and you can't see any leaks, don't stress about it," Taylor said, shooting a sideways glance to make sure I wasn't freaking out. "Is everybody happy? Lovely. Put the driving lights on and let's go see what we can find."

*

We flew over to the *Nadir*, dropping in beside them. They were practicing with their manipulator arm, taking core samples of the sediment. Both subs had their lights on, illuminating the seafloor and making it possible to view animals in Technicolor—even the red shrimp and jellyfish that, without our LED assist, would never be seen in their resplendent scarlet and crimson. In the deep, red simply doesn't exist—except it does, because there are lots of red animals. But without the red wavelength to show their coloration, they're as black as the night waters around them. It's a masterful disguise. Some deep-sea fish species have taken this disappearing act to its limit. They're *ultra*black, their skin packed with a pigment that absorbs 99.5 percent of any light they encounter. I saw one skinny fish with a forked tail that was likely this type of stealth assassin, because even in the sub's brightest beam it appeared as a solid silhouette.

The *Nadir*'s sampling stirred up clouds of sediment, so we decided to leave. As we lifted off, we got a better look at the wall's colossal dimensions. "You suddenly realize how insignificant we are," Taylor said, grinning. "We're looking at a pinprick of what's down here. Not even that! And I can guarantee that we're the first people to have ever seen it."

We veered to port, back into the twinkling blizzard, heading out over rock-strewn dunes of white silt. A jellyfish shot by with impressive velocity, opening and closing like a mouth. Something that resembled a butterfly darted past a beanpole-thin fish that was hanging vertically. A siphonophore advertised the letter *J* in marquee lights. "Skim along close to the seabed," Taylor told Macdonald. "There are some nice steep gulleys we can get into."

"Is that a squid down there?" Macdonald asked.

We banked forward. It *was* a squid, but unlike any I had ever seen. From afar, it looked like a black kite fluttering across the bottom. It was maybe two feet long and moving slowly, waving elegant diamond-shaped fins, with its arms and tentacles gathered in a bouquet that rested on its mantle.

"Oh, he's very cool," Taylor said, urging Macdonald to follow it. "Keep coming down. Use your side thrusters."

We dipped into a trough and approached the creature from behind. Upon closer inspection it was a rich velvety red, and didn't seem bothered by us. (Later, after watching my video of the encounter, the Smithsonian National Museum of Natural History's cephalopod curator, Mike Vecchione, would identify it as *Pholidoteuthis adami*—also known as a scaled squid, because its mantle is covered with cartilaginous "dermal cushions" that resemble fish scales.) I was so focused on the squid that I almost missed a big jellyfish with a shocking pink center and an explosion of tentacles like a nimbus of white hair.

"This is insane," I said. "You couldn't make this up."

Taylor nodded. "It's so much fun that it's easy to forget we're in a hostile environment."

By now we'd cleared the rock field and found ourselves gliding over huge white dunes, as stark as a bleached-out Sahara. "Let's cruise along this ridge," Taylor said. "Then we can drop down on the other side."

"Turn to starboard?" Macdonald asked.

"Yep," Taylor said, and then tapped me on the knee. "Do you want to have a drive?"

A beat passed as I processed the question and determined that he was serious. He was asking if I wanted to fly the three of us around at thirty-three hundred feet.

"Sure!" I was not sure. I didn't think I'd feel great about getting into a sub that had me as its pilot. In earlier years I'd failed miserably at learning to fly airplanes and had managed to crash a boat. But I knew I'd never be able to live with myself if I said no.

Taylor looked delighted. "You'll need to do a special funky takeover," he said, squeezing to the side as Macdonald and I contorted ourselves to swap seats. The pilot's perch was slightly elevated, offering an unobstructed 340-degree view. I slid in behind the control panel, a screen swimming with numbers and graphics, and gripped the joystick with a clammy hand.

"This is it," Taylor announced. "This is your moment. You're full captain."

I hoped that wasn't actually true; that he had a second set of controls within quick reach. Already the sub was sort of tilted on its side, and we were headed straight at a steeply peaked dune. Nervously, I nudged the joystick forward. There was a loud whining noise.

"Just relax," Taylor said. "Don't twist it, push it gently ahead. You see the slope here? Try to follow that up."

"Up? Like this?" I torqued the throttle.

"Yes—that's enough. You don't have to have the death grip. Now just gradually climb up that hill."

The sub was heading forward as intended, albeit somewhat drunkenly. After a few minutes of this I was able to smooth out my line and fly along the spine of a ridge, down its flank, and back up again. Relaxation was not in the cards, but I was getting a feel for the sub as it moved in three dimensions through a landscape as fantastical as Atlantis and as elemental as the earth itself. For so long I'd dreamed of knowing the deep as a real place—a location rather than an idea— and now I was bopping around it in a little yellow submarine, driving through it as surely as I would drive to the grocery store. For once, my imagination couldn't exceed my reality.

"You're doing great," Taylor said, breaking into my utopia. "I think you're going backwards though."

*

Who knew how long we'd been in the twilight zone? It could have been ten hours or ten minutes. Time had slipped its leash. Specifically, our usual conception of time—measured with clocks and calendars on a scale of human life spans—was gone. Down here, you were never late and you were never early. In the deep there was only deep time, the infinitely slow ticking of the earth's geological clock. How long had this white limestone wonderland existed on the seafloor? Millions of years? Hundreds of millions? Earth's biosphere is 95 percent deep ocean, and that ocean is four billion years old. If you think too hard about numbers like that, terror can trickle in. But there's grace in bowing down to the deep's sublime dimensions: a peacefulness that comes from knowing your place in the true order of things.

To me, that humility felt like ecstasy. In the *Neptune,* it was as though I was meeting the world for the first time. Because that *is* the only way we can meet the world—on her terms. The late geomicrobiologist Katrina Edwards, who studied life within the ocean's crust, put it directly: "We're so biased to light because that's where we live. But in fact, most of the biosphere exists in the dark."

As the deep's first human witness, William Beebe had recognized its immensity—not just chronologically or physically, but spiritually. To meet the deep is to have your beliefs recalibrated, your perspective permanently tweaked. "When once it has been seen, it will remain forever the most vivid memory in life," Beebe wrote, "solely because of its cosmic chill and isolation, the eternal and absolute darkness and the indescribable beauty of its inhabitants." Every marine creature, every one of those quadrillion fish, was here to play its tiny role in Earth's epic production—and so were we.

This alliance with the rest of life should leave us giddy with gratitude, but it doesn't. Despite its brief appearance onstage, humanity has never seen itself as a bit player. We've created our *own* world, and it's a machine that doesn't give a damn about the indescribable beauty of the twilight zone. Show the machine world a magnificent wilderness brimming with glittering fish, and it responds in language like this: "As mesopelagic resources may represent a new seasonal fishery, sustainable capacity adaptations may be further complicated for vessels conducting combined fisheries with the same vessels for different fish stocks with a different managerial status." In other words, the twilight zone's spectacular aliveness makes it a target. Like every other ecosystem it's under threat, most immediately from industrial fishing.

In the annals of bad ideas, this is one for the hall of fame: trawling giant nets through water thousands of feet deep to catch thumb-size fish and wads of gelatinous tissue that nobody wants to eat, but that we could grind up to feed the farmed fish we're now being forced to breed because of our previous indiscriminate trawling with giant nets. And that's before you consider that tinkering with the balance of life in the twilight zone will reduce the ocean's ability to remove carbon from the atmosphere, and will disrupt (if not crash) marine food webs. As scientists race to understand the mesopelagic's many complex processes—all of which benefit us—they've learned that it's inextricably entwined with the rest of the deep, and ultimately the entire planet. (Without its carbon-pumping services, the earth's average temperatures would rise by an estimated six to eleven degrees Fahrenheit.) The idea of commercially fishing the twilight zone is off-the-charts stupid. That doesn't mean we won't do it.

Norway and Pakistan have already issued mesopelagic fishing licenses; other nations are investigating it, or have tried it in the past. To catch such peewee prey requires fine-meshed nets that scoop up horrendous amounts of bycatch—anything that happens to be swimming in the vicinity—and much of the twilight zone is located on the high seas, making regulation difficult. Currently the cost of fishing that deep outweighs the potential profits, but as the bigger fish are wiped out that equation will surely change. But again, we won't actually *eat* the lanternfish, jellies, and krill that we haul up. They'll be pulped into fishmeal, poultry feed, nutraceutical oils, and crop fertilizer. In one industry research paper, which I quoted above to illustrate the machine view of nature—charmlessly titled "Institutional Nuts and Bolts for a Mesopelagic Fishery in Norway"—the authors refer to the twilight zone creatures as "low-value species" that require "large-scale catch volumes" in order to be "economically viable." The machine considers them to be a "raw material."

*

We began our ascent with Macdonald back in the pilot's seat. I'd given up the joystick reluctantly: my fifteen minutes at the helm had not been enough. "You have to be a little bit mad to be a sub pilot," Taylor joked, knowing that I was hooked. "Do you want a job?"

Now, on the way up, we were leading, with the *Nadir* following a few hundred feet below. Taylor had positioned us this way on purpose: he was setting up a dramatic finale to our dive.

"What's our depth?" he asked Macdonald.

"Seven-zero-zero meters."

"Okay. I think we're in the zone. Let's get all these pesky bright lights off. Throw the towel over your control panel. Everybody ready? Eyes shut, and I'll count to three. On three, open them. One . . . two . . ."

On two, he'd flashed the sub's brightest lights on and off.

"THREE!"

I opened my eyes. All around us, the twilight zone creatures flashed back. It was as though we were at the center of a meteor shower, streaks and bursts and aureoles of light bejeweling the darkness to the far edges of our vision, enfolding us in a zodiac of sparks.

"*Oh my God,*" I said, because that's all you could say.

"Look at what's out there!" Taylor marveled. "Now you see the depth of them. They're absolutely everywhere."

Macdonald shook his head, laughing. "This doesn't even seem possible. It's like we're in a planetarium."

After a moment the sparks disappeared, as if a circuit had blown. We ascended another two hundred feet. "Okay, eyes shut!" Taylor commanded. "One, two, THREE!"

The water lit up again, beaming greetings.

"Ahhhhhhhhh!" I said inarticulately.

"Crazy!" Macdonald said redundantly.

"I never get bored of this," Taylor noted.

We rose through the depths, flashing and swooning as the creatures winked and flashed back. Even with the latest low-light cameras, Taylor said, this communion was impossible to film. It was too evanescent, too delicate—too inextricable from its realm and the instant in which it happened. If you tried to capture it, you'd miss it. Only by being part of the greeting could you experience it, because it was more about energy than form. We weren't on the outside looking in, or the inside looking out. We were on the inside, looking deeper within.

*

Around four hundred feet there was a reverse dusk, a subtle upshift to indigo. Again, I didn't see it happen, but as the deep released us and we crossed the threshold to sunlight, I felt it in my solar plexus. To my surprise, the emotion that hit me was grief. It started as a twinge, a minor note of longing, and I squashed it down because it felt absurd to be thrilled and heartbroken at the same time. But the feeling kept rising in my throat. I knew that I might never experience the deep again, and the thought of this being *it*—my one encounter, now over—was achingly sad, like falling in love with someone and then never seeing them again. I didn't want to get out of the submersible.

As we approached the surface the hypnotic blue reestablished itself, increasing in intensity until it was almost unbearable. We slipped through a meniscus of bubbles and then we were out, rocking in the chop. The sky was also blue, but it looked wan by comparison. I shielded my eyes against the sun's glare.

Macdonald announced our arrival. "Surface, *Neptune*, tanks are blown, breakers are off. Over."

Frey responded: "Copy, *Neptune*. Welcome back. Clear for swimmer."

Back on the ship there were hugs and high fives. I thanked everyone profusely and kept quiet about my feelings of bereavement. We'd returned in the midafternoon, and for the rest of the day my mind played its new screen saver: a vision of the seafloor, thirty-three hundred feet down, bathed in aqueous turquoise light. In the evening, after dinner, I joined Taylor and the pilots in the control van—a metal box like the shipping container where I'd logged data on the *Atlantis*—for a round of gin and tonics and a meeting. Taylor was sitting on a desk, the others lounging in chairs and leaning against the walls, still high from their dives. When it came time to critique Macdonald's piloting performance, Taylor said simply: "I think it went well." He turned to Macdonald. "Do you have anything to add?"

Macdonald smiled. "Yeah—that was fucking sick. It was *amazing*."

He was a thousand percent right, and his words stayed with me. After I left the ship and traveled home, the first thing I did was call Lahey. In a rush of detail I replayed the dive, everything that had happened and everything we'd seen and how it felt and how it affected me. I told him about the grief because I knew he would understand.

"It's a valid feeling," Lahey agreed, when I let him get a word in. "You're connecting to the ocean in a way that you never have before. You're inside the life force of the planet." I wasn't the only person to have reacted emotionally, he added. The whole point of a manned sub was to create a visceral connection; that was Lahey's—and OceanX's—central thesis. "We're sensory beings. And when you see the deep with your own eyes, it changes the way you feel about it."

In my case, it wasn't so much a change as a confirmation of my own thesis: the ocean simmers with magic, and the deeper you go, the more magical it becomes. And now, of course, I wanted to go deeper. Lahey told me that the *Limiting Factor* and the *Pressure Drop* were in Barcelona for maintenance, but would soon be leaving on the Ring of Fire expedition. McCallum had recently emailed the January 2020 to July 2020 schedule, noting, "We have a LOT of dive sites on this program." As I scanned the itinerary, an idea began to form. The next call I made was to Vescovo.

He was at home in Dallas, still digesting the Five Deeps but also itching to get back out to sea. There was much of Hades left to explore. On this next expedition, he told me, he intended to take more passengers, especially scientists, along with him in the sub. "I've done the hero dives," he said. "Now it's about sharing it."

I saw the opening and leaped into it. "I don't know if this is possible," I began nervously. "But is there any chance you'd consider letting me dive with you?"

"Possible?" Vescovo replied, and then paused. I prepared myself to hear a polite no.

Instead, I heard this: "Yes. I do think it might be possible."

CHAPTER 9

Selling the Abyss

There is still time to choose.
—SYLVIA EARLE

RABAUL, PAPUA NEW GUINEA

Who owns the abyss? It's a crazy question, like asking who owns the sky. And yet the abyssal zone is prime real estate. Half of the earth's surface lies beneath waters ten thousand to twenty thousand feet deep—and in a mercenary world, assets like that aren't usually left unclaimed. So who *does* own the abyss? The good news is that you are one of the owners. The bad news is that your interest is controlled by a group you've never heard of in Kingston, Jamaica, and they're handing the abyss over to other groups that can't wait to start strip mining it. If this sounds outrageous, wait until you hear the whole story. It gets crazier.

It started innocently enough, back in 1873. That's when the *Challenger* expedition dredged up the strange black nuggets that became known as manganese nodules. At the time, the nodules were scientific riddles: Were they volcanic in origin? Had they, perhaps, fallen from space like meteorites? Why did so many of them contain sharks' teeth? What were they *doing* down there?

Nearly a century passed before researchers glimpsed the nodules in situ, photographed on the bottom. Finally we'd seen the face of the abyss—and it had acne. There were nodules upon nodules stippling the sediment across the abyssal plains. They were as small as marbles and as big as softballs, and they flourished in the depths around fifteen thousand feet, especially in the Pacific. American geologist Bruce Heezen, a pioneer of seafloor mapping, declared the nodules "the largest mineral deposit on the planet." Along with manganese, they contained nickel, copper, cobalt, and traces of other metals. It was inevitable that someone would do the math on what trillions of these lumps might be worth.

And someone did. In 1965, another American geologist, John Mero, crunched the numbers in his book *The Mineral Resources of the Sea*. Reading it, you get the sense that in Mero's perfect world the entire ocean would be an open-pit mine. He lamented that the deep had been "little exploited relative to its potential," despite its clear purpose as "a boundless, inexhaustible storehouse of the material stuff of civilization."

He wasn't just talking about nodules. Mero wanted to suck the "mineral matter" out of salt water and use deep-sea ooze as a substitute for cement, because it was "available on a royalty-free basis." Mining the seabed would be harder, though definitely not out of the question: "With nuclear explosives . . . it may be possible to crush competent mineral-containing rock within the ocean floor." Noting that sea squirts have minerals in their mucus, he enthused about "breeding superconcentrators" and using their tissues as "a possible ore." (If Mero were an alien, you could imagine him scanning your body and telling other aliens that your teeth and bones, melted down, would make excellent construction materials.)

For a slim volume on a dry topic, his book received big attention. Industrialists gasped at Mero's claims: according to his calculations, the deep ocean contained enough metals to power "a world population of 20 billion people" for thousands of years. There was gold in the seawater and phosphorus in the rocks and tellurium in the mud, but the nodules were the obvious place to start. Mero envisioned "crawler-type bathyscaphes" pulling giant "scraper units" across the seafloor and depositing nodules into "large tonnage submarines with manned spherical control chambers." Or something like that.

What followed was a commercial race into the deep, with companies like British Petroleum, Royal Dutch Shell, Mitsubishi, and U.S. Steel jockeying to mine the abyss. None of them had ever attempted to work fifteen thousand feet underwater, so the first order of business was to invent the machines that could do it.

In 1974, the eccentric billionaire Howard Hughes appeared to take the lead when his spectacular new mining ship, the 618-foot *Hughes Glomar Explorer,* sailed to a site off Hawaii that was known to be flush with nodules. The ship had a twenty-story derrick and an interior moon pool through which a nodule-plucking apparatus could be lowered. Deep-sea mining was *on.* The *Explorer* was proof of concept—or so it seemed. In fact, Hughes's operation was an elaborate CIA ruse. His ship was seeking another type of metal: it was stationed above a sunken, nuclear-armed Soviet submarine. The Russians had lost the sub in 1968, and the Americans found it on the seafloor and hoped to raise it surreptitiously. The only thing being mined was Cold War intelligence.

Even after the press blew the *Explorer*'s cover, the allure of the nodules grew. "It is a treasure hunter's dream, a prize so magnificent as to enflame even the richest of emperors," *The Washington Post* gushed in 1981. "Envision a vast plain, hundreds of miles broad, covered with some of the scarcest and most sought-after minerals in the world."

The assumption was that the prize was there for the taking, but the question of rights was a sticky one. Who owned the abyss? Countries had fought about this; it had never truly been settled. Alarmed by the rush for nodules, the United Nations declared the deep's mineral wealth to be the "common heritage of mankind"—not a line item on some corporation's balance sheet—and set out to pass an international law to back that up.

This effort would result in the UN Convention on the Law of the Sea, a hefty treaty specifying who could do what, and where, in the waters beyond national jurisdiction. That vast expanse of deep seabed, encompassing more than a hundred million square miles—an astounding 54 percent of the world's ocean—was tagged as "the

Area." Any nation that ratified this treaty could be allocated chunks of the Area to mine. Any nation that did not would be restricted to mining in its own waters. (The United States signed the treaty but was one of the few nations that did not ratify it.)

What a job to be sheriff of the Area, overseeing the abyss on behalf of "mankind." The Law of the Sea treaty established a new organization to fill that role: the International Seabed Authority (ISA). It would be an autonomous entity with serious power, unbeholden to the UN. Like all bureaucracies, the ISA would have committees and councils and commissions, and it would spout jargon and overuse acronyms to the point where nobody but insiders knew what it was talking about. Legally, the ISA was mandated to protect the abyss from environmental harm and to divvy it up for seabed mining, as if those two things weren't in polar opposition.

None of this would be finalized until 1994, however, and in the meantime, corporations were still vying for a piece of the deep. Mero had observed that some nodules were more valuable than others: their metal concentrations and the densities of their fields varied widely across locations. Some places were packed with nodules; other places had only a scattering of nodules or no nodules at all. The richest area was thought to be the Pacific's Clarion-Clipperton Fracture Zone (CCZ), a 1.7-million-square-mile swath of abyssal plains, hills, and seamounts that stretches from Hawaii to Mexico. The CCZ was estimated to hold twenty-one billion tons of Grade-A nodules, a jackpot of metals that dwarfed terrestrial deposits.

It made economic sense to team up, so in the 1970s and early '80s—with the Law of the Sea treaty not yet in force—the corporate players formed consortiums to investigate how to mine the CCZ. With a singularity of purpose and a flair for unoriginality, they named themselves Ocean Management Incorporated, Ocean Minerals Company, Ocean Mining Incorporated, and Ocean Mining Associates. Ocean Management Incorporated invented a Zamboni-like nodule-collecting machine that plowed across the seabed. Ocean Mining Associates figured out how to shoot nodules up a pipe to the surface using compressed air. Ocean Minerals Company chartered the *Glo-*

mar Explorer, the fake mining ship, only to realize that it couldn't actually retrieve nodules.

While all this was going on, a marine scientist named Sylvia Earle arrived on the oceanic scene. Earle was petite and unstoppable, with credentials from Duke and Harvard, and a notable fearlessness: "I want to go to the deepest part of the ocean. I mean, who doesn't?" In 1970, she'd led an all-female group of aquanauts who lived in an undersea habitat for two weeks, captivating the public as thoroughly as William Beebe and Otis Barton had on their *Bathysphere* dives. Like Beebe, Earle was strikingly articulate—a natural voice for the ocean. Her passion for the abyss, and the nodule miners' need to study it so they could best exploit it, ran on improbably parallel tracks. "They were investing in exploration," she recounted to me recently, "and I kind of folded into that era of trying to understand what was down there."

To the companies that wanted to mine them, the nodules were simply inert rocks. But Earle recognized that the odd black orbs were nothing like volcanic basalt or hunks of coal: they *grow* on the seabed. The nodule-forming process is enigmatic, and even now, scientists don't fully understand it. Under certain conditions at abyssal depths, minerals from the seawater accrete in atomic layers around a nucleus—any hard scrap on the seafloor will do. Microbes are likely involved in the production, but nobody exactly knows how. Animals and other organisms live on, under, and even inside the nodules, which are habitats and microbiomes in the manner of corals on a reef.

Mero believed that the nodules sprang up with alacrity, and that deep-sea miners would "be faced with the very interesting situation of working deposits that grow faster than they can be mined." In fact, the opposite is true: nodules are born from one of the slowest reactions in nature. In a million years they might grow by a few millimeters, as an intricate ecosystem evolves around them.

None of the companies stopped to wonder if it was a good idea to remove a living layer of the seafloor that had been there, untouched, since the Mesozoic era. Mero's book didn't contain a single sentence about environmental concerns. Back in those DDT-spraying days, keeping ecosystems functioning wasn't high on the list of corporate priorities, but public perception had to be considered. It would be terrible PR to destroy some important habitat or wipe out some charismatic animal; luckily, the abyssal plains appeared to be empty. "They were using pretty crude methods for sampling and analyzing the nature of life," Earle recalled. "The whole business of microbial diversity, the ecosystem—it just hadn't clicked in. They were only thinking about things they could see. The report I remember reading back in that time said there wasn't much down there. Just a few sea cucumbers and worms, and who cares about that?"

*

These days, Earle is known as the most influential marine scientist of our age, an ocean defender so fierce she's been described as "Joan of Arc with a submarine." The abyss is known to be humming with life even a mile beneath the seabed. And people do care about it, even if they can't see it. Environmental awareness has progressed to the point where companies have to care, too, if they want to stay in business. But in the last forty years one thing has not changed at all: people who care deeply about money are still after the nodules.

The initial enthusiasm about deep-sea mining faded in the eighties when the metal markets collapsed. No one needed any more manganese; you couldn't give away a ton of cobalt. But now, the technologies we rely on—computers, smartphones, rechargeable batteries, solar panels—all require heaps of metals, including the rare earth elements with spooky names like yttrium and dysprosium. (They're not rare at all, just costly to mine. The deep contains high concentrations of them.) We need more of everything, as we always do, and with global warming tightening its grip, battery metals will be in increasingly high demand. Deep-sea mining is back on the map for a perfect storm of reasons—and this time it's likely to happen.

But not without a blistering fight. There's no stopping a country from mining the seabed on its own continental shelf—and several nations, including Norway and Japan, intend to do that—but the

ISA's abyssal mining plans are another matter. If the Area belongs to humanity at large, then it stands to reason that people should have a say in whether it gets carved up and sold for metal, and whether a small clutch of bureaucrats should be in charge of the fate of half the earth.

Since the International Seabed Authority opened its doors in 1994, its 168 members (now numbering 168 delegates from 167 nations and the European Union) have met annually at its headquarters in Jamaica to prepare for the biggest resource haul the world has ever known—or as one marine scientist put it, "the greatest assault on deep-sea ecosystems ever inflicted by humans."

The starchy formality, the impenetrable UN lingo, the thick pavement of official documents, the closed-door meetings: it's not easy to follow what's going on at the ISA. But in short, unless you're a mining contractor, the news is dispiriting. Despite its legal obligation to "protect and preserve the marine environment," the ISA's leadership lists heavily to the side of the mining industry. (Much of the decision-making power resides in its Legal and Technical Commission, an opaque group of forty-one appointees that has been conspicuously pro-extraction.) The ISA's secretary-general, British lawyer Michael Lodge, has dismissed scientists' and environmental groups' concerns as "absolutism and dogmatism, bordering on fanaticism," and described deep-sea mining as "an essential component of a vision for a sustainable world."

To date, the ISA has granted thirty-one mining exploration contracts covering about six hundred thousand square miles, a seabed footprint the size of Alaska. Nineteen of those contracts are for nodule mining, but with an expansiveness that would have delighted Mero, the other twelve contracts would allow miners to investigate scalping the tops off seamounts and grinding up hydrothermal vents.

Apparently it's not hard to get an exploration contract, because so far the ISA has approved every application. Any member state that pays a $500,000 fee and follows procedure can soon have exclusive rights to its own patch of seabed. There's no stinting on size: an average CCZ contract area spans about thirty thousand square miles. Some nations already hold multiple contracts (China has five) and there's nothing to prevent any ISA member state from obtaining more, usually by sponsoring contracts on behalf of private-sector mining companies. (In that case, the mining company puts up the capital, runs the show, and would pocket most of the profits—theoretically, billions. The sponsoring nation receives a small royalty, and could be liable for damages if anything goes wrong.)

Ultimately the ISA stands to benefit from every contract it grants, with royalties rolling in from each mining operation. Some of that cash will be distributed to developing nations, but a portion flows to the ISA itself—a jarring conflict of interest for a group that also serves as the industry's regulator. Absurdly, there are even plans for the ISA to develop its own nodule mining concession called "the Enterprise."

At the moment, the ISA and its delegates are still negotiating the "exploitation" rules, known as the "Mining Code," that must be in place before commercial mining can begin in the Area. But the exploration phase has been under way for years, with mining contractors surveying their sites, testing their equipment, and preparing their mandatory environmental baseline studies (what's down there; who lives there) and impact assessments (how badly will the region be damaged; how many of the deep's inhabitants will be snuffed).

Under any circumstances this is hard information to pin down: the impact of an industry that's never existed before, on thousands of square miles of seafloor that nobody's ever seen before, that's home to God knows how many mobile, sessile, burrowing, super-tiny, and microscopic creatures. Even a dedicated scientific institution would need years to produce a comprehensive report. But the mining contractors are self-reporting to the ISA—with varying degrees of compliance and data quality—so it will be difficult to check their homework. One part of their impact, at least, is easy to predict. "I don't think there's any reasonable expectation that any life that exists in the [direct mining] area will survive," MIT oceanographer Tom Peacock declared in a recent talk. "That will be lost."

*

Deep-sea mining first appeared on my radar in 2015, when I was on a ship in Papua New Guinea. In a sort of double coincidence, the world's first commercial deep-sea mine was slated to open in that nation's waters—and I happened to be traveling with Sylvia Earle. She was hosting a conference for a hundred ocean scientists, artists, and conservationists that I'd been invited to attend. The gathering was called Mission Blue II, and it was held on the *National Geographic Orion*, a sporty 340-foot cruise ship. The TED organization was running the trip, so in the evenings we assembled on deck to listen to TED-style twenty-minute talks. During the days, as we sailed through a region known as the Coral Triangle—a crown jewel of marine biodiversity—we stopped to dive on reefs that were like artworks of the divine.

Earle wanted to draw attention to the area because it was threatened by everything from pollution to global warming to fishermen whose tools included cyanide and dynamite. That list of threats would now include deep-sea mining. When we boarded the *Orion* in Papua New Guinea's port of Rabaul, we were only thirty miles away from a hydrothermal vent field known as Solwara 1—the first of many local seabed sites that a company called Nautilus Minerals hoped to obliterate.

Rabaul is perched on the northeastern tip of New Britain, a skinny crescent of an island that faces the Pacific's Bismarck Sea. This is not a great place to put a town. New Britain looks like the rim of a submerged crater, because it is: Rabaul is inside the caldera of a pyroclastic volcano. More than once during explosive eruptions, its buildings have collapsed under ash fall. The Bismarck Sea is part of a volatile subduction zone that includes the thirty-thousand-foot-deep New Britain Trench; its seabed is pocked with submarine volcanoes and hydrothermal vents.

I learned more about this a few days into the voyage, when deep-sea biologist Cindy Van Dover presented her TED talk. Along with her position as the director of Duke University's Marine Laboratory, Van Dover was the first—and to date, only—female *Alvin* pilot, and she flew the sub on trailblazing dives to black smokers. She'd stared legendary vents in the eye: Godzilla in Canada, Rose Garden in the Galápagos, Broken Spur on the Mid-Atlantic Ridge. "When you're down there," she'd reflected, "you really see how this environment could have been the cradle of life on earth."

Van Dover was a soft-spoken woman in her early sixties. Few people know hydrothermal vents as intimately as she does, so it was logical that her talk focused on them. But there was a twist: for the past decade she had been working as an advisor to Nautilus Minerals.

It was a common, if soul-numbing, arrangement. Mining companies need scientists to inform their projects; scientists need funding to do field research (especially expensive deep-sea research). By signing on with Nautilus, Van Dover got an all-access pass to Papua New Guinea's vent fields; a chance to conduct baseline studies and determine guidelines that could potentially reduce the industry's damage. "I don't think it's sensible or right to not try to contribute scientific knowledge that might inform policy," she explained in an interview. "We can't just stick our heads in the sand and complain when it goes wrong." Van Dover was familiar with these waters, and so as we floated in the Bismarck Sea, she described the seafloor below us—and what was about to happen to it.

Like nodules, sulfide vents are a rich source of metals, including copper, gold, silver, and zinc. Unlike nodules, these vents don't occupy much horizontal space: The deep's known active vent fields, clustered together, could fit into Iowa City. (Solwara 1 is the size of ten football fields and contains about forty thousand vents.) But a vent field's vertical structure means that its mineral deposits extend far beneath the seabed. As the hydrothermal fluid pumps, its metal sulfides form chimneys and meld with the underlying rock, so extraction of the ore is a violent affair. "It's open-cut mining, basically," Van Dover said, pulling up a slide that Nautilus had created to depict its undersea work site.

It was a cross-section illustration like something you'd find in a child's storybook, with a toy ship rendered in happy primary colors, sitting peaceably on the surface. Below it, a pipe dropped like a drinking straw to the seafloor, which was gilded in gold (in case anybody forgot what this was about). Cute ant-like machines plowed a tidy trail along the bottom, surrounded by little castles of rock. The scene looked downright bucolic.

The reality would be different. Once the mine was operational—in 2018, according to Nautilus—it would be a maelstrom of pulverized rock and sediment, a riot of noise, vibration, and light. A furious, thousand-foot storm cloud would engulf the site, assailing every creature around. Van Dover showed us a photo of the cute machines being built: they were two-hundred-and-fifty-ton beasts designed to tear apart everything in their path, grind it into a slurry, and blast it five thousand feet up the pipe. This slurry would be a mix of animal and mineral, with fish, clams, worms, crustaceans—the entire vent community—rocketing to the surface along with the ore. (And since each vent is home to its own crowd of strangers, many species would be gone before we'd ever met them.) On the ship, the metals would be sieved out. Then the remains of everything else—that is to say, most of it—would be pumped back down to the seafloor and spewed out in a continuous plume that would drift through the water for miles before settling into smothering heaps on the bottom.

In this manner, for twenty-four hours a day, 365 days a year, over the course of three to five years, Nautilus would extract two million tons of ore, leaving behind a crater up to seven hundred feet deep where the vent field used to be. That was the idea, anyway. No one had ever tried this before, so the whole thing was a blind experiment.

You wouldn't have guessed that from the next image Van Dover showed, a map of the Bismarck Sea. It was covered with flags that marked future seabed mine sites: Nautilus had obtained multiple exploration licenses from the Papua New Guinea government. The company also planned to expand into the waters of Tonga, Fiji, the Solomon Islands, and other South Pacific nations. "You know, as scientists we don't want to advocate for or against mining," Van Dover said, sounding a bit hesitant. "But if it is going to take place, then we have to figure out how can they do it best."

I looked around to see if everybody else was buying this. Nautilus's scheme sounded insane to me. The only thing worse was the idea of helping them execute it. It's true that science demands a certain neutrality, and there's something to be said for minimizing harm, but I couldn't fathom the emotional whiplash of discovering the wonders of chemosynthesis, seeing unimagined animals feasting on six-hundred-degree vent fluid, being humbled and astonished by the revelation that life had far more dimensions than we'd known—and then watching a mining company annihilate it.

When the talk was over, I headed straight for the bar. Earle was right behind me. She was wearing a flowing blue and green jacket and chic black pants, but her preferred garment even at age eighty was a wet suit. I asked her what she thought about the mining issue and she rolled her eyes. Her organization, Mission Blue, was working to create a global network of marine protected areas, with the goal of placing 30 percent of the ocean off-limits to commercial exploitation by 2030. (Currently, only about 8 percent is protected.) A new deep-sea mining industry was the last thing she needed.

"Why would we do this when we don't have to?" she said, shaking her head. Earle's voice is magnetic, deep and husky and resonant, and she speaks slowly, measuring every word. "It's not just going to affect the seafloor, it's going to affect the water column. The chemistry and the biology of the ocean are truly on the line with this massive new way of stirring things up." The bartender handed her a glass of wine, and she set it down and turned to me, her eyes conveying that she knew how I felt. "I mean, *come on*—these are the very systems that keep us alive. The deep sea has kept us relatively safe. We should leave it the hell alone."

*

In 2019, Nautilus Minerals went bankrupt. It had burned through nearly $700 million, its stock was worth a penny, and it hadn't mined a single vent. The company needed such a gargantuan ship for its humongous mining machines that it couldn't afford to pay for the vessel's construction. Also, the Papua New Guinea people had organized against the project. "We will not accept this seabed mine," a group

called the Alliance of Solwara Warriors proclaimed. "We put Nautilus and its investors on notice that we will resist this venture and
you will bear the costs!" The group noted that Nautilus had assured
coastal villagers that its mine would have "no impact" on marine life,
while disclosing to its shareholders that "the actual impact of any . . .
mining operations on the environment has yet to be determined."
"What a bullshit!!" one woman wrote in an online forum, expressing
the local sentiment. "Why don't these destroyers experiment in their
own backyard?"

Independent science reviews of Nautilus's plans were equally
scathing. Reviewers wrote that Solwara 1 "would result in severe,
prolonged, and perhaps region-wide impacts to a globally rare and
poorly understood biological community," and cause harm "propor-
tionately far greater" than the "impacts of terrestrial mines on global
forest habitat." Ten million tons of contaminated effluent would be
disgorged into the deep annually; the machines' din would resound
through the Bismarck Sea for hundreds of miles, driving whales, dolphins, and other marine creatures away from their feeding and mat-
ing grounds. The mine's sediment plumes would choke filter-feeding
animals, interfere with bioluminescence, and expose the marine food
chain to toxic particles. The list of hazards went on for pages.

The Papua New Guinea government had borrowed money to
buy a 15 percent stake in Solwara 1 and it lost $120 million, which
would've gone a long way in a place where a large percentage of the
population has no electricity. To this day, Nautilus's mining machines
are rusting near a harbor in the country's capital city, Port Moresby.
They're large enough to be seen on Google Earth.

*

This debacle should have been a red alert for the International Seabed
Authority, which was busy doling out its own contracts, and whose
leader, Michael Lodge, had lauded Solwara 1 as "an exciting opportunity." It should have been a stake through Nautilus's heart—and it
was, but only on paper. The enterprise was awash in subsidiaries, and

some of them survived, including two CCZ nodule-mining ventures sponsored by ISA member states Tonga and Nauru. Meanwhile, a crew of former Nautilus executives and investors—who'd cashed out before it all came crashing down—had launched DeepGreen Metals, a company that quickly became the industry's most aggressive player.

To nobody's surprise, Nautilus's Tonga and Nauru subsidiaries were soon among DeepGreen's holdings. DeepGreen also partnered with a third ISA member state, the Republic of Kiribati, which sponsored yet another nodule contract—meaning that a privately owned mining company had now gained exclusive rights to eighty-seven thousand square miles of seabed, in nodule-rich areas of the CCZ reserved for developing nations. Clearly the ISA was fine with this: Michael Lodge appeared in DeepGreen's promotional video and photographs wearing a hard hat emblazoned with the company's logo.

Even now, with deep-sea mining poised to change the abyss in perpetuity (at least on human time scales), the public knows little about it—and when they do learn about it, they don't tend to like it very much. "This is a terrible, terrible, terrible idea and I don't know how it is being allowed to go forward," read one typical comment when *The New York Times* ran a story on the subject. "LEAVE THE OCEANS ALONE," "This will not end well," and "So gross" were a few others. An industry this immense, this unprecedented, this terrifyingly risky and heartbreakingly sad would be hard to sell. It needed some spin. And DeepGreen's CEO, Gerard Barron, was just the man to do it.

Barron, a voluble Australian in his mid-fifties, had an unusual talent for a mining executive: he was a marketing expert. Before he reportedly invested $226,000 in Nautilus and walked away with $31 million, he'd founded an online advertising agency. Barron set about rebranding his deep-sea mining company; he even rebranded himself. A 2013 photo shows him as a close-shaven, clean-cut suit. A few years later, as the face of DeepGreen, he was as shaggy and bewhiskered as an Aerosmith guitarist after a rough night, dressed in a tight T-shirt, leather jacket, and chunky man-bracelets.

Now manganese nodules were "polymetallic" nodules, and they

came with a slogan. "It's an electric vehicle battery in a rock," Barron would say during interviews, pulling a nodule out of his pocket with a magician's flourish. On a *60 Minutes* segment about deep-sea mining, Barron mused, "I don't call myself a miner." Rather, DeepGreen would be "harvesting" metals for the "green transition." "We want this to be cool," Barron told reporters.

The story got bigger: DeepGreen's pursuit of nodules was about nothing less than saving the world. And Barron was everywhere, telling it at length. It was a blur of land-based mining destroying rain forests Africa child labor toxic tailings fossil fuel human rights violations versus renewable energy closed-loop recycling no-waste sustainable abundant nodules that "literally lie on the ocean floor like golf balls on a driving range." "I'm doing this for the planet and the planet's children," Barron said.

Also, for an 8.1 percent stake in a company that would soon be listed on the NASDAQ. In 2021, DeepGreen went public in a SPAC merger, changing its name to the Metals Company (ticker symbol: TMC). It was valued at $2.9 billion, which was impressive given that it had no revenue, and its only assets were seabed rights that supposedly belong to the rest of us.

The TMC stock debuted at $11.05 a share, spiked to $15.39, and fell to $3.48 within a month. (It has since dropped below a dollar.) Accounting problems emerged; lawsuits followed. Shareholders joined a class-action suit, alleging the Metals Company had made "materially false and misleading statements"; the Metals Company sued private equity investors over a missing $200 million. It was a Nautilus-style mess, complete with its own version of the Solwara Warriors: "Deep sea mining is not needed, not wanted, not consented!" warned the Pacific Blue Line, a coalition of environmental activists from island nations. "To TMC, we say, you will face fierce opposition from our growing public mobilization."

At the same time, the International Union for Conservation of Nature—a powerful UN-affiliated network—called for a moratorium on deep-sea mining until its impacts were understood, a halt to further mining contracts, and a reform of the ISA. A petition signed by hundreds of marine scientists seconded this, citing the "risk of large-scale and permanent loss of biodiversity, ecosystems, and ecosystem functions." Google, Samsung, BMW, Volvo, and other major companies pledged not to use deep-sea metals in their supply chains. (And it's debatable whether any businesses will be clamoring for nodule supplies in the future: lithium-ion batteries are fast being replaced by new battery chemistries that don't require cobalt or nickel.)

With its cash flow pinched and resistance building, the Metals Company needed to start mining soon, or likely miss its chance. It needed the ISA to wrap up the Mining Code, pronto, and hand over the exploitation contracts. But the negotiations had been under way for years with no consensus—and the process was complicated by the fact that it's impossible to protect the deep from harmful effects, as required by the Law of the Sea, while opening it up to industrial-scale mining.

And how can anyone guarantee it won't be harmed when we don't know what's down there to begin with, or how it all works? The CCZ's leading expert, University of Hawaii oceanographer and professor emeritus Craig Smith, estimates that scientists have sampled less than 0.1 percent of it. Along with the seabed, the waters above will also be affected. After the nodules and the sediments they're embedded in and all nearby creatures come hurtling up to the ship, the nodules will be separated out, and then (as with vent mining) the remains will be fire-hosed back into the deep: the Metals Company plans to release this material at the bottom of the twilight zone. "It is frightfully clear that the impact of this drifting plume on open-water ecosystems will be severe, varied, and global in scale," deep-sea researchers Steven Haddock and Anela Choy wrote in an editorial.

The sane thing to do, the smart thing, the ethical thing, was to slow down, not speed up. As Earle argued, "Just *stop*. Just *wait*. Just give us a chance to think about the alternatives. It's not just rocks and water out there. It's alive—*all* of it. And we can't put it back once it's gone."

*

The Metals Company couldn't afford to wait, and it had a partner willing to make sure it didn't have to: the Pacific Island nation of Nauru. In a brassy move, Nauru invoked an obscure Law of the Sea trigger clause that allows a member state to give the ISA a deadline: two years on the clock to finalize the Mining Code. At the end of that time— July 9, 2023, in this case—if the rules aren't in place, exploitation contracts can be issued anyway. It was an ultimatum that would affect the future of the global ocean, coming from a country the size of the San Francisco Airport, on behalf of a commercial mining company.

Letting Nauru drive deep-sea mining policy is about as wise as letting Hunter S. Thompson drive you to Las Vegas. This eight-square-mile island, population eleven thousand, is the poster child for bad environmental decisions. It has been so ruthlessly mined for phosphate that it's basically a husk. Eighty percent of the island is simply *gone*, its guts ripped out, only the gnawed bones left of what was once a densely forested, tropical atoll. Author Jack Hitt, who wrote about his visit to Nauru in *The New York Times Magazine*, called it "one of the scariest things I have ever seen."

Obviously tourism isn't an option. There's no locally grown food because there's hardly any topsoil. Or birds. Or native wildlife. Without tree cover, the bald limestone becomes infernally hot; the rising heat shunts the clouds away, causing perpetual drought. Drinking water has to be imported.

Some of this isn't Nauru's fault—colonialism played its rapacious role—but for more than a half century, since gaining its independence in 1968, the country has continued to mine itself into oblivion. For a minute, this may have seemed like a good idea: in the 1970s and '80s Nauruans were among the world's wealthiest people, second only per capita to Saudi Arabia. But Nauru swiftly blew its $1.2 billion phosphate royalties trust through a mix of naiveté, graft, and ill-advised investments, like backing a London theater production, *Leonardo the Musical: A Portrait of Love*, in which Leonardo da Vinci gets the Mona Lisa pregnant. (The play closed after five weeks.) Australian economist Helen Hughes, who studied Nauru, noted that it had a "long history of being taken to the cleaners by crooks."

By the nineties, Nauru had decided to become a crook itself, as the world's most prolific money-launderer. Seventy billion dollars in Russian money alone sluiced through Nauruan shell banks in 1998. The United States accused Nauru of extending "an open invitation to financial crime" and slapped it with the heaviest sanctions available, forcing it to reform. Nauru also sold passports—some to members of Al Qaeda—and considered letting its phone lines be used for 1-900 sex calls, before taking money from Australia to host a brutal refugee detention center on its blasted-out phosphate fields. (Amid protests from human rights groups, and some refugees begging to be sent back to the war-torn countries they'd fled, Nauru closed its borders to journalists.) And now, as a sponsoring state for the Metals Company, it was turning to deep-sea mining.

"Nauru has proudly taken a leading role in developing the international legal framework governing seafloor nodules in the international seabed area (the Area)," the country announced in a press release. It added: "Nauru is comfortable with being a leader on these issues."

*

What could possibly go wrong? What happens when Nauru's two-year deadline arrives and it's open season on the Clarion-Clipperton Zone and the Metal Company's nodule-mining machines get out there and start vacuuming up the seafloor and spraying the mid-waters with a haze of sediment and nodule debris? It's obvious what they have to gain from this (dollar-sign-face emoji), but what do we stand to lose? Since commercial deep-sea mining will have unknown and enduring effects, that question will remain long after it has begun.

Nature doesn't do things fast in the abyss. It's our largest, most stable ecosystem, and it swaddles the earth like a blanket. Deep-sea ooze is fine and sticky; the CCZ's bottom waters are pristinely clear. Sediment plumes kicked up by the mining machines will stay in suspension longer than they would in other parts of the ocean. "We'll create a bit of a storm down there as we collect [the nodules]," Barron has said. "A little bit of dust. We don't think that's going to be a big issue."

Perhaps it won't be a big issue—to the Metals Company. Three miles down in the CCZ, the natural sedimentation rate is extremely low, less than a centimeter every thousand years. None of the exquisitely adapted animals—sponges, anemones, corals, crinoids, brittle stars, crustaceans—that reach upward from the nodules to feed on the vestiges of marine snow will survive in turbid conditions. Fifty percent of the CCZ's creatures are there *because* of the nodules: they coexist symbiotically. They may be attached to the nodules' hard surface, or reliant on the nodules for food or shelter—but in all cases, it's not like they can swim away. When mining begins, their fate is to be choked, buried, crushed, or shredded. And like the nodules they live on, they're not coming back soon—if ever.

To hear Barron and his team tell it, the CCZ is as bleak as it comes already, a loser of a place that nobody will miss—a "monotonous" "lifeless desert," "the most common area on the planet," "a very challenging place for organisms to live," where "the abundance of life is up to fifteen hundred times less than in the vibrant ecosystems on land." In a Metals Company report thick with authoritative-looking statistics and infographics, I came across the following paragraph:

> Despite life likely originating in the ocean and oceans accounting for 70% of our planet's surface, most life on our planet lives on land. Only 3% of biomass resides in the oceans while 97% resides on land. While oceans are vast, most of their area lacks vegetation and, as a result, the opportunities for life to evolve are limited.

It's hard to know where to start with this, but probably by pointing out that ecological value is not a factor of weight. If it were all about biomass, then a hundred cows would be far more precious than a hundred hummingbirds or a million lanternfish or zillions of phytoplankton. The terrestrial biosphere is heavy because trees compete for sunlight: they have an evolutionary reason to grow as tall as they can.

Deep-sea animals—and, of course, microbes—have no such need: in the deep, it often pays to be gossamer-thin or small. (You don't have to be big to have a profound impact in the scheme of life—ask a virus.) What's more significant is biodiversity—the kaleidoscope of species, the vast genetic reservoir within an ecosystem. Unlike biomass, biodiversity is irreplaceable. And the more scientists explore the CCZ, the more apparent it has become that this "lifeless desert" is one of the most spectacularly biodiverse places in the deep.

Ironically, we may not have learned this if it weren't for the ISA. Mining contractors are required to fund research in their contract areas, and every science cruise has returned with astonishing finds. An estimated 90 percent of the animals collected have been new species: corals like pearl necklaces and anemones like dripping flowers and pearl-white brittle stars and acid-yellow sea cucumbers with huge swishy tails; urchins that gallop across the bottom and glass sponges that house so many residents they're like apartment buildings designed by Dr. Seuss. The CCZ is a wonderland of xenophyophores—single-celled protozoans that build elaborate shells of various shapes from bits they forage on the seafloor. That would seem ambitious enough for a cell, but xenophyophores think big: their shells can be the size of grapefruits.

The CCZ's landscape presents its own mysteries, more chapters in the Deep Sea Book of Weirdness and Surprises. One group of researchers found a fossil bed glittering with metal-encrusted whale skulls; another found thousands of divots in the sediment that were likely made by beaked whales diving to epic depths to hunt. Rather than an ecological wasteland, the CCZ is a cabinet of curiosities; rather than homogenous plains, it's studded with seamounts, ridges, and hills—and even on the flats, it's a mosaic of unique fauna. "In terrestrial environments it's really remarkable if you find an animal, especially a larger animal, that's new to science," the University of Hawaii's Craig Smith told me. "In the CCZ in particular, every box core we take, every sediment sample we take—we have *hundreds* of species that haven't been described."

And that's before you consider the galactic circus of microorganisms. The abyssal sediments (and the nodules themselves) are teeming with extraordinary life. Recently, researchers analyzed genetic sequences from four hundred and eighteen deep-sea sediment samples and identified nearly one hundred thousand DNA variants, *60 percent* of which belonged to seafloor-dwelling critters that didn't fit into any known taxonomic group. These weren't merely new species—they represented entirely new branches on the tree of life. "We're talking about small animals less than a millimeter in size," the study's coauthor, marine scientist Andrew Gooday, explained, "and probably a lot of protozoans, a lot of single-celled organisms."

Considering that deep-sea ooze covers half the earth, it's boggling to think of the scope of its inhabitants. Whoever they are, they're doing some heavy lifting: cycling nutrients, sequestering carbon, keeping the earth's geochemistry in balance, and even more remarkably, serving as life's database—an almost infinite archive of genomic innovation reaching deep into the past. (In 2020, Japanese scientists cored into abyssal sediments and found hundred-million-year-old microbes alive and well, in a state of hibernation.)

"Just as scientists have discovered through ever more powerful telescopes that stars number in the billions," the evolutionary biologist Mitchell Sogin has noted, "we are learning through DNA technologies that the number of marine organisms invisible to the eye exceeds all expectations and their diversity is much greater than we could have imagined."

It's dizzying to think of the discoveries that could be made in this realm, the genetic resources that could help us solve our most intractable problems. Deep-sea microbes and other organisms (sponges in particular) have already yielded potent antiviral, anticancer, antimalarial, antifungal, and antibacterial drugs—and the research is just getting started.

Razing the seafloor will disrupt life within the sediments—and that's a euphemism. In a 1989 experiment, researchers dragged a plow back and forth across a nodule field in an attempt to approximate mining's effects. When scientists revisited the field in 2015, they found something akin to a ghost town. Microbial activity was reduced four-fold, while microbial cell numbers were diminished by nearly 30 percent. The twenty-six-year-old plow tracks looked like they could've been made yesterday. Scientists don't know when—or if—the site will fully recover.

By wrecking the sediments we also lose the ability to study the ocean's paleoclimatology, the layers of fossilized microorganisms, laid down over eons, that reveal temperatures, currents, wind patterns, and chemistry across deep time. Every geological event has left its signature. "The sediments are a sort of epic poem of the earth," Rachel Carson reflected in *The Sea Around Us*. "When we are wise enough, perhaps we can read in them all of past history. For all is written here."

When we're unwise enough, we trash the book before we've even cracked the cover. If the deep's ancient nodule fields are strip-mined, the damage will be unquantifiable and irreversible. The Metals Company's 2021 Annual Report spells it out in the "risk factors" section, where it's legally bound to disclose the actual facts: "Given the significant volume of deep water and the difficulty of sampling and retrieving biological specimens, a complete biological inventory might never be established. Accordingly, impacts on CCZ biodiversity may never be completely or definitively known."

*

In an attempt to offset deep-sea mining's ecological mayhem, the ISA has set aside large swaths of seafloor that it calls "Areas of Particular Environmental Interest (APEI)." These "protected" zones form a choppy perimeter around the CCZ, and are meant to be refuges for some abyssal creatures, and, one assumes, serve as examples of what the region was like before we got there. The tragedy is that the entire CCZ is an Area of Particular Environmental Interest.

Lodge often asserts that the deep's most ecologically sensitive areas will be avoided, but it's hard to take that claim seriously. In 2017, scientists reeled in disbelief when the ISA granted Poland an exploration contract for hydrothermal vents along the Mid-Atlantic Ridge—in an area that includes Lost City.

Seventeen years after Deb Kelley and her colleagues on the *Atlantis* discovered that singular vent field, and first laid eyes on Lost City's transcendent white spires, and one year after UNESCO proposed it as a "World Heritage Site of Outstanding Universal Value in the High Seas," and praised it as "a sample of the truly iconic treasures our deep oceans harbor"—and even as the top microbiologists study its otherworldly life-forms and chemistry and make declarations like "This is an example of a type of ecosystem that could be active on Enceladus or Europa right this second" and "Lost City is one of those places that's got the origin of life written all over it"—the ISA sold it out.

In 2020, Alan Jamieson and fellow hadal biologist Thom Linley launched *The Deep Sea Podcast (A Punk Take on a Science Podcast About Everything Deep Sea),* and in an episode about mining, Jamieson interviewed Michael Lodge. "So how does that work, then?" he asked Lodge about the ISA's decision to hand over Lost City to Polish mining interests. "If it qualifies as a World Heritage Site, what would we have to bring [for you] to not grant the [mining] license? Because that's raising the bar pretty high."

Lodge made a noise in his throat that may have been some kind of laughter. It was definitely a sound that meant: *I'm tired of talking about this.* "First of all—what is the Lost City?" he lectured Jamieson, a professor of deep-sea science at a major university. "The Lost City is one of *many* hydrothermal vent sites around the world." He allowed that "yes, the Lost City is a site that is kind of charismatic, one could say," and then implied that time had run out for it. "They've been studying the Lost City for years and years," Lodge said in a patronizing tone. "Scientists have made their whole careers studying hydrothermal vents. Which is *great,* it's *fine.* And in the process they become very attached to them." He batted away the significance of the World Heritage Site designation: "UNESCO doesn't actually have any jurisdiction in the deep sea."

It was ridiculous to worry about Lost City being harmed, he

added, when all Poland was going to do—for the time being—was "study the mineralogy." "So you know, this is good for science!" Lodge barked. "This is not supposing in *any* way that *anyone* is *ever* going to mine the Lost City!" "First of all, it's absolutely unlikely that the Lost City itself has any mineral resources. So why would anybody mine it?"

Afterward, Jamieson and Linley discussed the interview. "We're absolutely exhausted," Linley told listeners. "Exhausted," Jamieson agreed. "I must admit, Thom, it hasn't been that pleasant an experience." The two spent a few minutes correcting Lodge's misstatements. Lost City, Jamieson clarified, is *not* like any other hydrothermal vent field: "It is of paramount scientific interest because it's thought that it may be one of the only analogs that represents the conditions of the primeval earth. So there you go."

*

"Disgraceful" was Earle's one-word verdict, when I brought up Lost City's potential future as a mine site. I'd called her to talk because my reporting on this subject had made me heartsick, and as the years had passed I could see the industry marching closer to reality, in defiance of many in the scientific community, public opinion, common sense, and even the rules laid down by the Law of the Sea—and that seemed to represent something bigger and darker about where we were headed, into a future where nothing in the natural world, not even in the farthest reaches of the deep ocean, would be left alone if it could be converted to cash.

In the media, at conferences, and even within the ISA itself, the arguments about deep-sea mining were growing louder and more heated. "We need to convince our cousins of the Pacific to stop this craziness," French Polynesia's minister for marine resources told the press, after his country voted for a ban. Costa Rica, Chile, Spain, France, Germany, New Zealand, Ecuador, Panama, Fiji, Palau, Samoa, Vanuatu, and the Federated States of Micronesia have argued that Nauru's two-year deadline should be disregarded, and declared their support for a moratorium or "precautionary pause."

The New York Times and the *Los Angeles Times* ran investigative features that alleged corruption and mismanagement at the ISA, quoted whistleblowers who'd worked there, and accused its leadership of cozying up to the Metals Company. "Interviews and hundreds of pages of emails, letters, and other internal documents show that the firm's executives received key information from the Seabed Authority . . . giving a major edge to their mining ambitions," *New York Times* reporter Eric Lipton wrote. The stories were damning, and the ISA hired a law firm to rebut them.

Soon, we will reach an inflection point. The Metals Company and others may get their exploitation contracts from the ISA, but what they will not have is the social license to start deep-sea mining: society's agreement that this is necessary now; that we understand the risks and benefits in full; that they're the ones who should do it; that it's being regulated properly; that the technology being used is the best it can possibly be, to do the least harm. Right now we're nowhere near that conclusion, and until we are—if we *ever* are—deep-sea mining will be a rogue operation, the antithesis of something that's being done for the benefit of humanity. And all the greenwashing in the world won't change that fact.

"If we're serious about dealing with climate change, we're not going to deep-sea mine," Earle told me when we spoke, her voice low and emphatic. "This is the last best chance we will ever have to hold the planet steady. Our highest priority must be to safeguard whatever remains of the natural carbon-capturing systems, whether it's an old-growth forest, an intact desert, a grassland—and by far the largest, relatively undisturbed, intact part of the planet is the deep sea. And it should be hands off."

One day when we were in Papua New Guinea, standing on deck in our wet suits, I'd asked Earle if she had a favorite dive site. "Anywhere, fifty years ago," she shot back with a dry smile. More than anyone, Earle felt the pain of the ocean's decline—but somehow she had maintained her optimism, her focus on the glimmers of what could be, rather than what seemed inevitable.

Even over a phone line, I could feel that energy coming through like a laser. "We're at a point in time where we have a choice," she said. "Here's the thing: We didn't know before. Fifty years ago, there was *so* much we didn't know. We are now better prepared, armed with the superpower of knowledge. I tell kids, 'Be glad that you're a twenty-first-century human. Imagine if you didn't have the facts about what the problems are. Imagine if you didn't have the solutions. But you have both.'"

CHAPTER 10

Kamaʻehuakanaloa

(Red Child of the Deep)

The child of Haumea (earth) and Kanaloa (sea) is born. Kamaʻehu,
the red island child, rises from deep in the ocean floor.

——HĀLAU O KEKUHI

18.70° N, 155.17° W, 57 MILES
SOUTHEAST OF HILO, HAWAII
CENTRAL PACIFIC OCEAN

The water was black and the night was black, the moon off duty and the stars missing in action. Rob McCallum steered the Zodiac (also black) toward the *Pressure Drop*, floating offshore, its lights a beacon in the black waves. It wasn't the best time to be shuttling people around Hawaii in a tiny boat, but McCallum had swung into shore effortlessly, Victor Vescovo and I and a few others had jumped in with our gear, and it all felt very clandestine, speeding out to the ship under cover of darkness. (The truth was more mundane: we were boarding at night because that's when we'd arrived.) As the Big Island receded behind us, Vescovo leaned forward in anticipation of being reunited with his deep-sea machines. A year had gone by since he had agreed to take me into the abyss, and only now was that plan coming to pass. The delay had not been for lack of trying.

Originally I'd been scheduled to dive with Vescovo in June 2020 on his Ring of Fire expedition. What's more, he had offered to let me accompany him to an unexplored site in the Mariana Trench. The site didn't have an official name. It did have a depth, however: 32,300 feet. It was as though I had asked to go for a flight and ended up with a ticket to the moon. And even though when he'd invited me in January 2020 the dive was six months away, I thought about it every minute of every day and felt exhilarated about it and made lists of the gear I'd need and booked a plane ticket to Guam. But 2020 was about to impose its own agenda.

When the year began, the *Pressure Drop* and the *Limiting Factor* were in the Mediterranean Sea, where Vescovo had dived to the 16,762-foot Calypso Deep (with Prince Albert II of Monaco); located the wreck of *La Minerve*, a French military submarine lost with its fifty-two crew in 1968; and scouted brine pools on the bottom of the Red Sea with Alan Jamieson and scientists from Saudi Arabia's King Abdullah University of Science and Technology. In March, the *Pressure Drop* was beginning its long transit into the Pacific when everything tipped sideways due to a little virus called COVID-19.

Could any hysteria compare to the first spun-out weeks of a global pandemic? Cruise ships were viral nightmares; ports were wary of letting vessels dock. Flights continued, but the list of destinations that weren't accepting incomers was growing fast. Other places imposed draconian quarantines, which resulted in bleak arrangements such as people living in cardboard boxes at Tokyo's Narita Airport. To pursue a deep-sea expedition in the midst of COVID seemed like the height of futility, but Vescovo had no intention of tapping out. "We're still on," he announced in late March. "Until I know differently, that is the plan." "I will swim to Guam if I have to," I'd emailed McCallum.

I was expecting the worst, but in June I was able to fly to the island and board the ship in Guam's Apra Harbor. Before I arrived, Vescovo had made six dives to the Challenger Deep, including one on which he'd taken Don Walsh's son, Kelly Walsh, to the spot where the *Trieste* had touched down sixty years earlier. It seemed like the expedition was going well despite the world's ambient chaos, but behind the scenes the submersible was malfunctioning badly. The electrical issues that had flared up in Tonga were occurring on every dive. One junction box had burst into flames on deck. After each meltdown, the

Triton crew labored nonstop to replace the charred components, but ultimately the damage was beyond shipboard repair. The sub would have to be dry-docked and rewired, a process that could take months. When I learned that my dive was postponed, I had been in Guam for less than twenty-four hours. I turned around and flew home.

*

Now it was January 2021, and the *Limiting Factor* was ready to dive again—this time, in Hawaii. Patrick Lahey was running sea trials around Kona to test the sub, and then Vescovo planned to ascend Mauna Kea, the volcano that happens to be the tallest mountain on earth. It measures 33,500 feet from seafloor to summit, but inconveniently for climbers the bottom half is underwater. A full ascent would require diving seventeen thousand feet to Mauna Kea's base, coming back up to the ship, paddling an outrigger canoe twenty-seven miles to shore, biking thirty-seven miles up the volcano's flank, and then hiking six miles to its snowcapped peak. No one had ever done this before, because no contender had ever showed up with his own full-ocean-depth submersible.

My rescheduled dive was slotted in among those endeavors. In a twist of fate that I couldn't have predicted, Vescovo and I were going to Lōʻihi. True, in terms of sheer depth it wasn't the Mariana Trench, but this location was personal. Terry Kerby's stories about the volcano's mystique had stayed with me, and my visions of it—conjured from his words, a handful of photos, and my own imagination—were never far from the surface of my subconscious, like an eerie but irresistible dream. There was something complete about experiencing the abyss in the waters where I had learned to love the ocean in all its moods, and even after years of swimming with Hawaii's large animals and playing in its big waves and feeling like I knew the place, I really didn't, because I had not met the Hawaiian deep.

Although now I knew that showing up for a dive was no guarantee that it would happen. I was already sweating the marine forecast, which called for twenty-knot winds and a ten-foot swell. As we approached the *Pressure Drop,* whitecaps flashed in the darkness. "Not exactly tropical, is it?" McCallum said, maneuvering the Zodiac into a lee on the ship's starboard side. "It's winter," I replied, and we all knew what that meant: squalls, waves, stress. These were dicey conditions for launching the *Limiting Factor,* and we had a tight four-day window to complete three ambitious dives. Initially the plan had included only two dives—Lōʻihi and Mauna Kea—but on a test dive earlier in the day, one of the landers had failed to resurface. "Our big fear is that it's stuck in fishing debris," McCallum said, delivering the news. Unless Vescovo wanted to leave a half-million dollars on the seafloor, he would have to go down and retrieve it.

This wasn't great to hear, because I knew if anything dropped off the schedule it would be Lōʻihi. Vescovo had spent months prepping and training for Mauna Kea, and the logistics were complicated: special permits, cultural approval, escort boats, a support team, the loan of an outrigger canoe. Hawaiian marine scientist (and pro surfer) Cliff Kapono would be accompanying him. The whole thing was choreographed down to the hour and when it was done, Vescovo was flying home, and the ship was sailing to Australia for a chartered science cruise. First the pandemic, then Guam—and now I was being squeezed between a small fortune in lost gear and a volcano larger than Everest. There was nothing to do but hope that Pele was feeling benevolent.

*

With McCallum holding the Zodiac steady, we climbed a ladder and boarded the *Pressure Drop.* The ship was familiar to me from Tonga, but after months of hiatus it felt less lived-in, less like a traveling show. The pirate flag still hung on the wall of McCallum's office but without the same swagger. Jamieson's nook was empty, his lab equipment stowed. No one was pulling a giant isopod out of the freezer or blasting Korn's "Narcissistic Cannibal" at top volume or walking around with an egg that had been to the bottom of the Mariana Trench. Without all of the Five Deeps' large personalities in residence, the ship seemed lonely, the way a house does after the movers have just left.

Lahey had been on board for two weeks, but his departure coincided with my arrival. McCallum had dropped him onshore when he picked us up, and with the Zodiac's motor running there was no time to talk. "Call me after your dive!" Lahey had shouted to me in the wind. He was headed to Barcelona to deliver a six-person sub to a yacht owner who wanted to be able to dive with his entire family. But Kelvin Magee, Frank Lombardo, and others from the Triton team were here, including my twilight zone guide, Tim Macdonald. Since our dive in the Bahamas, there had been a major development in Macdonald's career: Lahey had trained him to fly the *Limiting Factor*, and Vescovo had hired him as a second pilot. "This one's more like a spaceship," Macdonald said, grinning, when I asked how it compared to a thousand-meter sub.

In the dry lab I met Tomer Ketter, the scientist in charge of sonar operations on this trip. Ketter was a bit of a ringer: an oceanographer, seafloor-mapping specialist, dive master, and skipper, who had served as a navigator in Israel's military. "I've always been a map nerd," he told me. He was studying Lōʻihi's bathymetry, his three computer screens displaying the acoustic images. Even as an abstraction rendered in acid-bright colors, the volcano was a monumental presence. Lōʻihi tops thirteen thousand feet in height and sprawls for twenty miles across the bottom; it's bigger than Italy's Mount Etna, the tallest active volcano in Europe. "I want to do our own sweep on it with the sonar," Ketter said, frowning. He pointed to a screen: "The waypoint they gave me for the summit is wrong." I leaned over his shoulder to take a closer look. If the coordinates were off, we could easily end up in the wrong place on our dive—and there were definitely some places that we needed to avoid.

Among Hawaii's deep sites, Lōʻihi had been an obvious choice. But it was harder to determine *where* to dive on the volcano, a decision that Vescovo had delegated to me. At the summit we could descend into Pele's Pit, the deep collapse crater with its basalt spires, venting chimneys, and ghost sharks; or we could drop onto the south rift, a steep, fissured freeway of lava; or we could investigate the Shinkai Ridge, a colossal slab of rock that sheared off the volcano's east flank and toppled to the seafloor at least six thousand years ago. And then there was Lōʻihi's enigmatic base, seventeen thousand feet down.

When I'd asked Kerby for advice, he had warned against going into Pele's Pit. "Fascinating place, but it's unstable. We would hear landslides on the communications hydrophones. Rocks are still coming down in there." He also did not recommend the south rift zone: "That is *really* rugged terrain. It's all box canyons and boulders the size of buses. The sub might need a paint job afterward." Although his *Pisces* subs didn't descend below two thousand meters, Kerby had once dived to Lōʻihi's southeastern base in Russia's six-thousand-meter *Mir* submersible, and he was more encouraging about that area. "We found a hydrothermal vent field with little chimneys, and these incredible pillow lava formations," he recalled. "Yeah, you could see some really dynamic stuff."

I wanted to go deep, so the base was ideal. But it, too, occupied a huge area—and the dive plan called for specifics. Vescovo had requested the "exact coordinates" for where the submersible and each of the three landers should be dropped, "the path you want us to take sideways, up, or down," and a "vector" (whatever that was). In our three hours of bottom time, the sub could travel about two miles horizontally. The trick was to identify the most exciting stretch. A good selection might mean crossing paths with a deep-sea shark, or meandering through a sculpture garden of lava, or finding hydrothermal vents. A bad selection might result in a lander falling into a crevasse, or the sub getting bashed up, or scenery that consisted of nothing but rubble.

There's no guidebook to the abyss, no Tripadvisor reviews for Lōʻihi's deepest reaches. "It's frustrating to me that our one big Hawaiian submarine volcano, we know so little about," University of Hawaii marine scientist Ken Rubin told me—and Rubin is one of the world's leading experts on submarine volcanoes. He has studied scores of them in the deep's farthest-flung places, and is extremely enthusiastic about them (especially the ones in Tonga), and can describe them in encyclopedic detail, yet right in his own backyard Lōʻihi had stubbornly resisted his scrutiny. After its attention-getting eruption in 1996, Rubin had attempted to install a cabled observatory at its base, only for the sediment to swallow the instruments like quicksand. On numerous occasions he had sailed to Lōʻihi with Kerby and the *Pisces*, and not once had the wind conditions been calm enough to launch the subs.

One of Rubin's colleagues, oceanographer Brian Glazer, had experienced better luck. Glazer had dived into Pele's Pit with Kerby, and investigated areas around Lōʻihi's base with the ROV Jason and another Woods Hole robot called Sentry. "Lōʻihi is one of my favorite underwater volcanoes on the planet," he raved, when we connected on a Zoom call. "There are some really cool things going on there." But then he began to describe *why* Lōʻihi was so intriguing—its distinctive biogeochemistry—and I could barely understand a word he was saying.

Glazer has his own lab at the University of Hawaii—though he looks young enough to be a student—and he began his career at NASA's Astrobiology Institute, which gives you a sense of his brain's horsepower. According to his web page, he studies "the relationship between redox disequilibria and microbes living in proximity to, or even mediating, steep redox gradients and pronounced geochemical interfaces." When I asked what that meant, he paused, unsure of where to even start. "I guess it's been a while since you've taken chemistry?"

In simple terms, Glazer studies how the ocean works, how its life and chemistry have coevolved with its geology over billions of years. Lōʻihi is compelling to him (and NASA) because its ecosystem is iron-rich—rather than sulfide-rich, like Axial and other volcanoes at mid-ocean ridges—and it hosts weird, iron-eating microbes that binge on Lōʻihi's vent fluids like it's the last days of Rome. Iron plays a key role in ocean chemistry, one that scientists don't fully understand, and this iron-spewing volcano is the perfect place to study it—not only to understand the alchemy of our own planet, but also to inform the search for extraterrestrial life. If microbes are found on Mars or

Europa, Glazer explained, they may well be similar to the ones that are thriving on Lōʻihi.

He recalled sitting in the control van with Jason's pilot as the robot explored an area of the volcano's base that was covered with bulky black mounds. No one knew what these were, so Glazer asked the pilot to poke one with Jason's manipulator arm: "And all of a sudden we were in this giant orange snowstorm." Under a thin manganese crust, the mounds were made entirely of bacteria. The swirling organisms were fluffy and flame-colored, because they were ingesting iron and excreting rust. In some places the bacteria were heaped more than seven feet high. "There were acres and acres of the stuff," Glazer enthused. "It's a massive microbial playground."

Dunes of bacteria? That did sound cool, but if they were all we saw for three hours the dive would be pretty monotonous. I asked Glazer what else was around—and if there were any deep features he was particularly curious about. He nodded, and switched his screen to a seafloor map. "See these two mounds to the southeast? These are pillow basalts that erupted in place. They come straight out of the bottom and go straight up. We never got an opportunity to visit them with Jason."

The structures were easy to make out. They appeared on the map as two prominent pinnacles; from above, they appeared to be shaped like volcanic cones. Each was about a thousand feet high. It was possible, Glazer said, that they were gigantic vents. "I'm not guaranteeing that, of course," he added. "But it could be. If you get to the top and you see shimmering water, that's the smoking gun, so to speak. And you can't miss them—they're huge. I'll give you the coordinates."

*

Finally I knew where we should go, but that didn't mean I knew how to make a dive plan. So I'd called Jamieson. He had intended to come to Hawaii, but the pandemic had marooned him in England. The two pinnacles were interesting, he agreed, examining the bathymetric map: "They do look like they've boiled to the surface. Hmmm. I'm not sure you should be putting landers on top of that."

"I don't have the slightest clue where to put the landers," I said. "Will you do it?"

Jamieson laughed. "Sure. I'll place one to the south as a start point, and then you can head north and go straight up the hill. It's nice to cruise along the flat bottom first—you can see all the animals out doing their thing. Then you'll get to the rugged stuff, and that's amazing geology. It'll be a great dive. It'll be brilliant. I'll write up a plan and send it to Victor."

I thanked him, and asked: "Anything else I should know?"

"Yeah," Jamieson said. "You'll get cold. There's only a sliver of neoprene between your feet and the titanium hull, right? I normally wear about three pairs of Antarctic socks and I've never once stopped my toes from freezing solid."

After we hung up I felt a rush of relief—but there was one thing I still needed to do. Beyond the science, beyond the technicalities of the dive, beyond how thick my socks should be, there was a more ancient kind of intelligence about Lō'ihi, and a code of respect that accompanied it. The Hawaiian volcanoes are sacred territory, and you don't just barge in without permission. I knew exactly who to ask.

Dr. Pualani Kanaka'ole Kanahele is a Hawaiian matriarch: Her ancestral lineage descends directly from Pele. Now in her eighties, she is a *kupuna* (wisdom keeper), a *kumu hula* (hula master), an author, a teacher, a pilot—among other distinctions. She is also my friend. When I explained to Pualani that I hoped to dive to Lō'ihi's base, she told me that Lō'ihi was not the volcano's name. The Hawaiians know it as Kama'ehu, for short, or by its full name, Kama'ehuakanaloa: the "Red Child of the Deep."

In 1954, it had appeared on a bathymetric survey of the Big Island's seafloor, and been dubbed Lō'ihi—"the Long One"—because of its elongated shape. Back then, the scientists who "discovered" it believed that it was merely the "parasitic" appendage of another volcano, Kilauea. They had no idea that it would one day become the next Hawaiian island. Despite this case of mistaken identity, Lōʻihi was the name that ended up on the maps. Meanwhile, in their traditional *mele* (chants), the Hawaiians had been describing Kamaʻehu for hundreds of years. Although they couldn't see it beneath the waves they knew precisely what it was up to, and that it was tinged rust-red with iron. "The red child that is born is the volcano," Pualani said. "*Puka Kamaʻehu* means to erupt. It shows itself. It comes to life."

She asked if our dive involved the federal government, and I said that it didn't—this was about pure exploration. "I don't see any problem with going down deep if you're just discovering," she said.

"So we have your blessing?" I asked.

"You've got it," she confirmed.

*

Early the next morning Vescovo and Macdonald dived to retrieve Skaff, the stranded lander. The rescue had to be made quickly, while Skaff still had the battery power to respond to the sub's sonar interrogations. Locating the lander would be like a game of Marco Polo, with Vescovo sending out pings, waiting for Skaff to ping back, and then adjusting the sub's heading until he got close enough to spot the lander's blinking strobe light.

If that sounds simple, it isn't. Finding a lost object in the deep is like rummaging through the Grand Canyon at night wearing horse blinders. The ocean is so good at hiding things that even when a target's position is *known* it can still elude searchers—who lack GPS, landmarks, light, or a vista by which to orient themselves. Only when Skaff was within sight would Vescovo be able to assess its predicament. If it was mired in silt, a tug from the sub's manipulator arm could free it. But if it was entangled or wedged under rocks or stuck in a hole, a rescue might not be feasible. Skaff had gone missing in a rocky area at a depth of only about a thousand feet, so fishing lines and tight crevices were a real possibility.

I waited out the dive on the top deck, where I could monitor the weather. The day was sunny with a few loose clouds, a rolling swell, and an undeniable wind. The sea state was fine for diving—but we were still in the lee of the Big Island. That would not be the case tomorrow at Kamaʻehu, where the launch site was thirty miles offshore, completely exposed, and subject to currents that collided around the island's south point. I leaned over the railing to watch a pod of spinner dolphins that had come by to inspect the ship. Then I heard a shout at the stern: Skaff had popped up. An hour later, the *Limiting Factor* was back.

Vescovo and Macdonald climbed out of the sub. "Where is he?" Vescovo asked, scanning the deck for Skaff and laughing when he saw it sitting in the corner, tagged with a sign that read, "I HAVE BEEN A BAD LANDER." It was true that Skaff had a reputation for trouble. It had disobeyed its acoustic orders before—most notably in the Challenger Deep, where it had once resided for two days before Lahey, piloting the sub on the world's deepest salvage operation, went down and yanked Skaff from the ooze. For all their bland boxiness, Vescovo swore the landers had distinct personalities. "Skaff is my recalcitrant child. Flere's a good kid, he's just a bit wild. Closp is . . . kind of dull."

Today it hadn't been hard to nudge Skaff to the surface. Oddly, the lander hadn't even been stuck. It was hovering above the seabed with its weights dropped, inexplicably staying put. (Without ballast it should've automatically floated to the surface.) Maybe there *was* a ghost in the machine. Or maybe Skaff liked the scenery on the bottom. Macdonald told me that he'd seen a stunning wall of black and gold corals, and a pregnant tiger shark.

As I listened to everyone talking on deck, puzzling over Skaff and discussing possible fixes, I realized that I was nervous. *Now it's getting real*, I thought. By this time tomorrow I could be three and a half miles down in the abyss, on an active volcano—if only the conditions would cooperate. I added Macdonald's shark to the dolphins I'd seen and the rainbow that had arched across the ocean at sunrise, and convinced myself that those were three good omens.

*

Turning south, the *Pressure Drop* began its thirteen-hour sail to
Kamaʻehu. Judging from the collective mood, I was the only one who
was looking forward to the voyage. McCallum had set out his tray of
seasickness meds, warning that when the ship crossed to the island's
windward side, "We're going to get pounded. The wind forecast is
twenty-five to thirty-five knots." In the dry lab, Magee, Lombardo,
and Triton technician Shane Eigler sat around the table with an air of
resignation.

Vescovo didn't appear deterred by the rough seas. As the ship
steamed on, we climbed into the *Limiting Factor* so he could give me
a safety briefing. Crouching atop the hangar, I stepped into the hatch,
set my feet on the ladder, squeezed myself through the hatch tunnel,
and inched my way down to the sphere. It was snug but not uncomfortably so, small but not impossible. It was like being in a rocket ship,
with every inch of surface housing some kind of instrumentation, and
three viewports the size of dessert plates. There was no mistaking the
fact that we were sitting inside a serious feat of engineering.

I lowered myself into the body-hugging passenger seat. "So we'll
go over the things that can go wrong in the sub," Vescovo began.
"The big three are entanglement, fire, and runaway oxygen."

"What's runaway oxygen?"

Vescovo nodded. "We'll get to that. It's what killed Apollo One."
(The tragedy he was referring to occurred in January 1967, when a
flash fire in their lunar module—oxygen igniting in a sealed capsule
lined with combustible Velcro—overwhelmed astronauts Gus Gris-
som, Ed White, and Roger Chaffee during a test launch. The men had
also been clad in nylon space suits, a grave materials error that led to
the adoption of hardier fireproof suits, like the Nomex ones that we
would be wearing.)

"Okay, entanglement," Vescovo continued. "There's really nothing you can do about that." Fortunately, the sub was designed for
Houdini-like escape. If any of its protruding parts got snagged—its
thrusters or batteries or manipulator arm—he could burn through
bolts to eject them, thus freeing (hopefully) the vehicle: "Your job is
not to panic."

That I could promise. There would be no panicking in the abyss. Not because I'm immune to fear—far from it—but because there was nothing in my life that I had ever wanted to do more than this dive, and if, against all odds, it was also the last thing I did, that was an outcome I was willing to accept. And I knew Vescovo felt the same way. "I don't discuss it a lot, but if there's a catastrophic failure with the submarine and I turn into plasma in less than a second, I'm okay with that," he'd said in Tonga. "I've been okay with that for decades in mountain climbing. If the worst happens, you know what? I'm doing what I want to do and there's no better way to go than that." Talk like that can sound cavalier, but only if it comes from the reckless. Vescovo wasn't a risk junkie—he was a risk analyst. I trusted him, the crew, the sub, every part of this endeavor. For all its extremes, the *Limiting Factor* was hyperengineered for safety.

"All right," Vescovo said. "Smoke or fire." He handed me a canister topped with a regulator: if we needed auxiliary air, we would use this scuba kit while assembling a more durable breathing apparatus that was stashed behind our seats. "In the direst emergency—if there's a fire inside the sphere—I'll spray a halon mixture to put it out," he explained. "Then I'll shut off the oxygen to starve any fire. And we have a burn bag for anything lithium. There should be nothing flammable in here. Okay?"

"Got it," I said, and we moved on. "Now, runaway oxygen. Which is one of the scariest." He pointed to the oxygen cylinders racked like artillery above our heads. "These bottles are under extremely high pressure, and they're in a very reliable NASA-tested system," he prefaced. "But it's possible, however unlikely, that something could start leaking, and no matter what I do I can't stop the leak."

"So we get flooded with oxygen?"

Vescovo nodded. "Which is fine. It's medical grade. But if the oxygen reaches a certain level, it can actually combust—and this is all electronics." He waved a hand at the instrument panels in front of us, behind us, lining the sphere. "If I can't stop the leak, I'll then shut down everything. Our phones, *everything* goes off. We won't be able to talk to the surface anymore. We'll be coming up dark. And the oxygen levels will keep rising. We'll be okay—we'll be breathing pure oxygen." He chuckled darkly. "The problem is that now we're kind of sitting in a bomb."

I digested that, and then pushed it out of my mind. The chance of it happening was next to zero, but he was right: it was scary.

"Otherwise during the mission, your job is simply to observe as much as you can," Vescovo added in a brighter tone. "The key is to make sure I don't hit anything at high speed. And that's basically it. Any questions?"

*

We rounded the island's south point shortly after midnight, and as predicted the ocean turned belligerent. In my bunk, I felt the ship surging and thought, *There's no way*. But we plowed on to the launch site, and around four o'clock I heard the landers being craned over the side. The dive was on.

By sunrise I was out on deck in my long underwear and my flame-proof suit, and the launch preparations were under way, although nobody looked too happy about it. Lombardo stood at the railing with his arms crossed, stoic beneath his whiskers. Eigler, who always had a smile on his face, wasn't smiling. Magee appeared to be grinding something to a pulp with his back molars, his mouth set in a grim line. The three men were watching the waves barreling over the horizon and the ship's flags snapping in the wind.

"We're trying to tame the ocean," Eigler told me, grinning slightly.

"What we have right now is a confused sea," Magee said, shaking his head. "As soon as I pick up that sub and a big set comes in and makes the ship roll, it's gonna smash right into the side of that A-frame." He shrugged tersely. "But if Victor tells me to put it in the water, I'll put it in the water."

I felt frazzled and had to keep moving, so I rotated into the dry lab. Ketter and Vescovo were poring over seafloor maps, considering alternate dive sites—but everything out here was exposed to the same wind. Macdonald and McCallum stood by, assessing the conditions and checking the forecast on a laptop. "We've got a north swell and a south swell and trade winds coming from the east," Macdonald said, summing it up.

"It's not so much about the height of the swell," McCallum noted. "It's about the frequency of the waves." (I'd already checked it: ten feet at nine seconds. Hammering.)

There was a call to be made and I knew it wasn't mine, and I didn't want to lurk behind Vescovo with my eagerness to get into the sub radiating like a microwave oven while he was making the decision. I left the dry lab and went down to my cabin to pace around. If we didn't dive, it was unclear when I'd get another chance to go this deep—or if I ever would. Vescovo was actively trying to sell the sub, and that could happen at any time. COVID was still raging. The schedule, everything, was up in the air.

I gave it a half hour, and then came back up. Vescovo was standing outside McCallum's office and I could tell from his face that we weren't going anywhere. "We've scrubbed it," he said, putting a hand on my shoulder. "I'm sorry, Susan." He and McCallum were trying to find another slot for me—maybe in the Sirena Deep—but of course that meant more months of uncertainty and waiting, and another trip to Guam. "I'll be there," I told Vescovo, and assured him that I was fine. And I was—kind of. My adrenals were blown and my soul felt like roadkill, but intellectually I knew this was the way it had to be. This was the smart call.

The two landers that were sent down had been summoned from the bottom, and once they were back on board we would depart for Mauna Kea, which was sixty miles north and thus more sheltered. On my way to remove my many layers of inflammable clothing, I stopped by the mission deck to thank everyone, and noticed that the wind was blowing even harder, the waves feathering with spray. Lombardo was outside the hangar smoking, and he looked at me sympathetically: "The ocean is a rough mistress sometimes."

I needed to be alone but I didn't want to sulk in my cabin, so I changed and went up to the Sky Bar. It was empty. I parked myself in a captain's chair, put my feet up on the railing, and stared numbly at the endless Pacific. The ship rose and fell with a metronomic rhythm, rocking in swells that had run three thousand miles from Alaska. Today, the ocean had a duller face. A silvery sun poked weakly through a low ceiling of bruised clouds. Someone had wound a lei of ti leaves around the railing, the Hawaiian navigator's protective talisman.

An hour went by, maybe two. Whatever. I heard footsteps on the stairs and turned to see Vescovo. He walked over and sat down in the chair next to me. "Closp is back, but Flere didn't come up," he told me. Flere had been our target lander—it was stationed at the dive's deepest point, south of the pinnacles. "Oh wow," I said, as the implications of that dawned on me.

"Yeah," Vescovo replied. "You and I are diving tomorrow."

*

"All checked and cleared, we have a green board. Life support engaged."

Magee's voice crackled through the radio. "Thrust to starboard, Victor."

"Thrust to starboard. Roger."

A tenser Magee: "ALL STOP on the thrusting."

The sub yo-yoed in the waves, which were knocking it back toward the ship. Vescovo was focused on his instruments; I was focused on not vomiting. Macdonald and another swimmer—a massively athletic six-foot-seven Aussie named Swain Murray—were struggling to unhook the lines. These were the same conditions that had scrapped our dive yesterday, but here we were. Thanks to a wayward lander, our next stop would be seventeen thousand feet down.

"We have been here before," McCallum had told the group last night, in a pre-dive pep talk. "We have been here for Tonga. We have been here for the *Titanic*. This is like Challenger on a big day.

We're not asking anybody to do anything heroic. It's all about being methodical in our approach."

Outside my viewport, something started flapping. "What's that?" I asked Vescovo.

"Oh, it's the depth sensor," he replied and then radioed: "I've got an RBR sensor that's detached next to the manipulator. Can the swimmer come and secure it?" Macdonald soon arrived in goggles and wrestled with the cable. "That's really dangerous, what he's doing," Vescovo noted. By now, Swain was out: he had injured his back, though we didn't know that at the time. On the surface, the sub was an eleven-ton bucking bronco. I understood why Lahey had felt the need to install seat belts.

McCallum radioed from the Zodiac. "Okay, LF is clear. You're free to pump."

"Roger that. Cable is secured. I'm pumping in," Vescovo responded.

Magee acknowledged the sub's release: "Roger that, Victor. Have a good one."

"Thank you, gentlemen," Vescovo signed off, and flicked a few switches. That was it for the radio: now we would communicate by acoustic modem, checking in with the control room every fifteen minutes. With a loud whooshing noise, the sub's ballast tanks filled with seawater. Pulled down by the weight, we sank beneath the waves. "And we're off," Vescovo announced. "Look at that beautiful blue. Now watch how quickly it gets dark."

*

Dropping at a meter per second, we plummeted through the twilight zone. Creatures whizzed past my viewport, bounced by like tumbleweed. "Squiggles, motes, flares," I scribbled in my notebook, the closest I could come to identifying them. We would be free-falling for the next two and a half hours, the pressure slowing our descent rate as we neared the bottom.

Unbuckling my seat belt, I settled myself. Sparks flashed in the

mid-water. Vescovo was logging data from the sub, looking up peri-
odically to check his instruments, reaching occasionally to tap a touch
screen. "Don't mind the cracking noise," he advised me. "That's just
the fiberglass under pressure." I nodded, but I hadn't noticed it: I'd
been thinking about how stellar everything was. It would take a much
louder noise to jolt me from my bliss.

"SURFACE, LF, DEPTH 222-NINER, HEADING 180, LIFE
SUPPORT GOOD," Vescovo relayed from the midnight zone,
shouting because the connection was feeble. "HELLLOOOO? SUR-
FACE, HOW DO YOU READ?" He was hunched over his control
panel, pinging the lander for a response. "C'mon, Flere—where are
you, buddy?" Flere was mute, making no commitments.

Outside the sub I heard a high-pitched moaning sound: "Are those
whales?"

"Those are the thrusters."

I checked the depth meter and proclaimed: "We are officially in
the abyss."

"We are!" Vescovo replied. "Three thousand meters. Two tons of
pressure per square inch."

I pressed my face against the viewport, and felt a subtle chill emanating through the sphere. The abyss looked back at me: it was silent,
peaceful. Vescovo had described the feeling of going deep as "almost
an embrace," and now I understood what he meant. There was a pulse
outside the sub, with real substance and weight, and the deeper we
plunged the more pronounced it became—the long, slow heartbeat
of something very big and very serene. I adored the chiaroscuro
quality of it, the way anything might materialize out of the velvety
darkness—and whatever it was, you wouldn't be expecting it.

"Thirty-five hundred and fifty meters," Vescovo said. "Two-
thirds of the way."

We fell past an S-shaped siphonophore that was glimmering like
Cassiopeia. "At Challenger Deep, a lander caught one that was in a
double-helix shape," Vescovo recounted. "Which was wild." Blowing people's minds is a siphonophore specialty: on a research cruise

off Western Australia, scientists encountered one that had arranged its chain of cloned cells into a glowing 150-foot spiral that could have easily passed as a portal to another dimension.

A squeegee sound erupted from the modem. "SAY AGAIN?" Vescovo replied. "SAY AGAIN LOUDER. SPEAK UP, BOYS. ROGER. DEPTH 3,882, HEADING 105, LIFE SUPPORT GOOD." He turned to me: "I can hear them, but it sounds like *mahhhwahhhhh*. It all mooshes together."

A sizable fish flew past, glinting indigo and gold. It was followed by a jelly that looked like a donut in a fright wig. Vescovo, bent over a notebook jotting down numbers, suddenly snapped upright and touched his headset: "Flere! He finally spoke! He's twelve hundred meters away. Our odds of finding him just went way up."

Two hours had passed like ten minutes; in the sub, time was as slippery as quicksilver. At four thousand meters I played a recording of the Hawaiian chant to Kama'ehu, which was beautiful and fierce, with shouting and drumming that seemed to drill into a primordial part of the brain.

"Wow," Vescovo said.

"Yeah, it's intense. But then, so are these volcanoes."

We continued our descent. The sphere had become frigid, the condensation from our breath beaded on the walls. I popped a piece of candy into my mouth: Macdonald had recommended a steady flow of sugar to kindle body heat. I had decided to eat nothing but chocolate on the dive, in homage to Don Walsh and his Hershey-bar lunch on the *Trieste*.

"Three hundred meters to go," Vescovo declared. "Now I'll start prepping to make us neutrally buoyant so we can navigate on the bottom. I think we need to drop about ten weights." I heard a *chu-chunk* noise as some steel bars released from the belly of the sub, slowing our descent noticeably. "Thrusters are armed. VBTs are both armed. Okay, now the altimeter's reading stable," Vescovo said, running through his checklist. "Let's put on the sonar."

Outside my viewport there was a faint ebbing of darkness, our beams reflecting off terrain that wasn't completely black. "There's the seafloor," Vescovo pointed out. "You can see it now—four meters away."

I leaned forward and watched the reflection sharpen into focus: a rough, sandy plain that was pale gold, dotted with white anemones and obsidian rocks, dashed with splashes of orange: unearthly tertiary colors that were, simultaneously, the dreamy hues of the earth. In our lights, the water glowed a crystalline jade green. The sub touched down with a soft hop and a puff of sediment. We had landed on the flats south of the pinnacles.

"Welcome to the bottom," Vescovo announced. "We're at 5,017 meters."

"*Amazing,*" I whispered. This was the abyss. My God, I loved it. Everything shimmered with a languid beauty, an uncanny gentleness, an amniotic calm. Yet at the same time, there was an enveloping aura of gravitas. The abyss was profound, in the literal sense: depth itself had deepened its enchantment. I felt as though I were melting into it, surrendering—as though I had made it home, and now I could finally rest. "When, at any time in our earthly life, we come to a moment or place of tremendous interest, it often happens that we realize the full significance only after it is all over," William Beebe had observed on a deep dive. But he was quick to add: "In the present instance the opposite was true." I took a moment to bask in the reality of where I was, to let it sink into my cells, so I would always be able to summon the memory and its emotional heft.

Vescovo cut into my reverie. "Okay, this is where the work begins." He gestured to the sonar, which was sweeping like a windshield wiper across a black screen. Hard objects, like rocks or a lander, showed up in red, with yellow highlights indicating major features. Three hundred feet ahead of us, there was an explosion of red and yellow that filled the sonar screen.

"What the hell is *that*?" Vescovo said.

"Maybe it's a UFO."

"I would love that. Then this would all be worthwhile."

"Actually I think that's lava. Like, a lot of lava."

"Well, we're gonna find out. We're heading right for it."

We set off, gliding across the seafloor. It was crisscrossed with tracks and confettied with burrows; labyrinths shuffled into the sediment. A bigheaded fish rushed by, chased by a white spider-like creature. "Do you think we're going in the right direction?" I asked.

Vescovo sighed. "I don't know. I'm kind of guessing right now. Once I start closing the range, then I'll know. Range doesn't lie." He listened for a ping from the lander. "Oh, there's Flere! *Shoot.* He's farther away."

I looked down. There were animals everywhere, darting, wafting, pulsing. Holothurians grazing on the sediment like tiny translucent-purple cows. Shrimps doing wheelies. Jellies blinking a welcome. Patches of neon-orange microbial mat webbing the rocks— Kamaʻehu's signature.

"Let's see what a heading of 180 does," Vescovo muttered to himself. "Yay, I'm closing! Not by a lot, but . . . let's go to 165."

"Hey, there's something really bizarre beneath us," I pointed out. "It's fluttering! And it's *big.*"

Vescovo ignored me. "Argghh, this makes no sense. No matter what I do we get farther away." He grimaced in frustration, and then perked up: "Okay, four hundred and seventy-five meters! *Closing.* Now we're in the ballpark. We're fifteen minutes away from Flere."

"Great!" I hoped we could deal with the lander swiftly and move on. "Fifteen minutes—that's not long."

"I mean, maybe."

Now the seafloor was more ragged, with knife-edged shards of basalt poking through rumpled hillocks of sediment. "This is what Challenger Deep looks like," Vescovo said. "It undulates. The major thing is, let me know if we're going to hit anything."

"There's some dark stuff ahead," I said, straining to see its contours. The light dimmed abruptly, turning shadowy.

"*Whoa!*" Vescovo exclaimed, indicating a yellow burst on the sonar. "Look at this bad boy! I've never seen anything like that— ever. What the *hell* is it?"

"Okay, there's a big field of black up here," I narrated. "There are ridges . . ."

"Those are *boulders*," Vescovo corrected me, squinting through his viewport. "Wow. Flere got himself into a rough neighborhood."

"Something bioluminescent just flashed," I said. "It was incredibly bright."

"Is it blinking?"

I stared through the viewport. "Yes—it's a strobe! Ten degrees to the left."

Flere loomed out of the darkness, a white box on a battlefield of black lava. Around it, jagged black rocks jutted like daggers. The lander was tilted diagonally, as though it had tried to rise but couldn't get quite enough momentum for its buoyancy to kick in. "It looks like he dropped his weight," Vescovo noted. "He's floating. I'm going to do a pirouette around him. *Oh, damn it!* I used too much thrust. I stirred up the silt."

The water outside our viewports was suddenly a fiery, billowing orange. "That's not silt," I said. "That's bacteria." Clots, strings, and hunks of it—living embers whirling in vortices around the sub. The effect was spectacular, like surfing through the clouds of Jupiter. But in the tumult we got turned around, and lost sight of Flere.

"*Shit*," Vescovo said.

"Wait—he's dead ahead," I said, pointing. "We're practically on top of him."

"Get the manipulator arm off its pedestal," Vescovo directed me. "Can you unhook—"

With a thunk, the sub rocked to the side. I couldn't see anything: my viewport was shrouded in campfire orange.

Vescovo looked surprised. "Oh! There he goes! He's headed up."

"What?"

"I rammed him," he said, laughing. "Just between you and me." He checked the lander's range: "Yep, he's up. Okay! Let's go do some exploring."

*

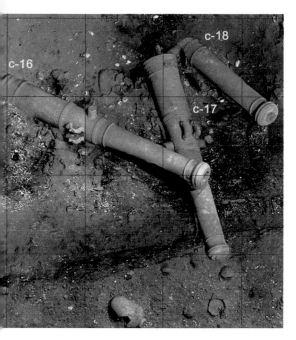

The *San José*'s cannons, with handles cast in the shape of dolphins (*left*); ceramic jars and Chinese porcelain on the seafloor (*center, left*); a box of enema syringes (*bottom, left*). Roger Dooley and Garry Kozak (*below*).

The *San José* in her glory: an illustration commissioned by Dooley, based on extensive historical research (*above*).

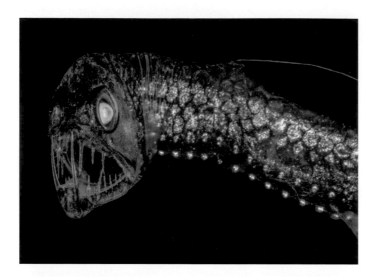

Toothy, glittery predators: a Sloane's viperfish (*top*), elongated bristlemouth (*center*), and luminous lanternfish (*bottom*). They may have fearsome teeth and scary faces, but many twilight zone fish are surprisingly small.

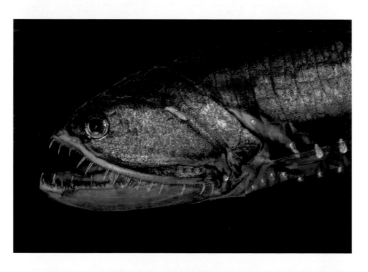

A cosmopolis of creatures: a glass squid (*top*), a tiny, exquisite jelly (*center*), and a ctenophore, or comb jelly (*bottom*). In the deep's mid-waters, bioluminescence is a central fact of life and the key to survival.

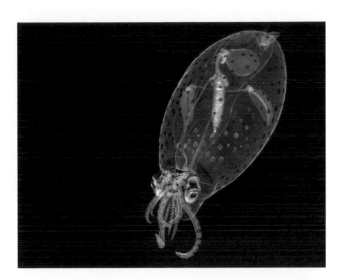

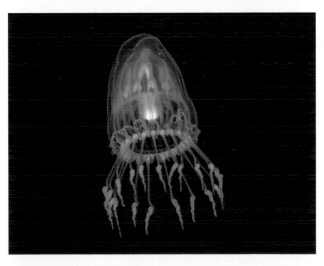

The blue that "admitted no thought of other colors": OceanX's Triton 3300/3 sub, the *Nadir* (*above* and *below, right*) in the Bahamas.

Pilot Buck Taylor, a veteran of four thousand submersible dives (*above*).

The barreleye's tubular eyes are sealed inside its transparent head (*above*).

A garden of glass sponges and bamboo corals (*above*). The long-nosed chimaera, a primitive, cartilaginous fish that is sometimes referred to as a ghost shark (*left*).

Extraordinary and subtle life: a deep-sea anemone on a manganese nodule field (*above*). A particularly vibrant holothurian (*below*, *left*).

Anemones, sponges, and other creatures taking advantage of a rock to reach upward for particles of food (*above*).

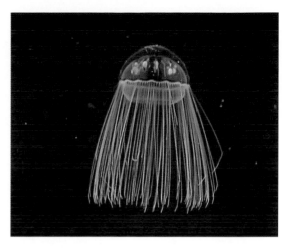

Her Deepness: the voice of the ocean, marine scientist Sylvia Earle (*left*); a jellyfish in the twilight zone (*above*).

A new species from a new order: an anemone-like *Relicanthus* with eight-foot-long tentacles perches on a nodule in the Clarion-Clipperton Fracture Zone (*above*); a holothurian grazing on the sediment (*left*).

Vescovo and Casey in the *Limiting Factor,* diving into the abyss, January 31, 2021 (*left*); the bathymetry of Kama'ehua (*below*); pillow lava in Pele's house (*below, left*); a jelly blinking a greeting (*bottom, right*).

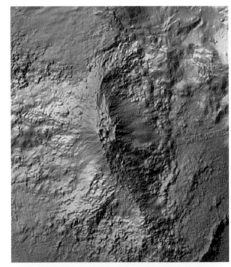

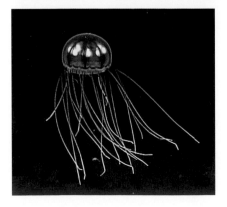

Now we had entered Pele's house. There was a sense of trespassing, of tiptoeing into a giantess's lair while she was sleeping. "This is *all* pillow lava," I said, gaping at the twisted and tortured formations that surrounded us: pillars and knolls of black basalt rose from the seafloor like an eerie maze. It was as though the lava had been squeezed from a giant toothpaste tube, under the glow of a giant lava lamp. It wound and looped and S-turned like the intestines of a giant beast. It dripped like the wax of giant candles. This was surely one of the pinnacles.

"It's going up pretty steeply," I advised. "I think there's a wall ahead . . . Oh *no*! Pull up! Pull up!" Too late—we smacked into the wall, pitching forward, and in that instant I was acutely aware that we had the weight of five hundred atmospheres on our heads.

"Sorry," Vescovo said. "That does tend to unnerve passengers."

At the top of the rise, the terrain leveled out momentarily. Some of the lava was cracked and swollen, as if sunburned. In other places, it looked like it had been torn apart in a fit of anger and vandalized with orange paint. Heaps of pillow lava were dusted with ochre sediment, and brought to mind piles of ash-covered bodies at Pompeii—a sea of heads, outstretched arms, legs tangled with other legs, backs curled into the fetal position. We flew slowly across the plateau, tracing the pinnacle's flank. A ghostly creature with a filamentous body hung motionless, regarding us as we passed. In our titanium orb we felt the metallic chill, breathed the metallic air.

William Beebe wrote that he had perceived a "compounding of sensations" in the deep, likening his dives to a headlong tumble through the looking glass that grows increasingly fantastical by the hour. "More and more complete severance with the upper world follows, and a plunging into new strangenesses, unpredictable sights continually opening up, until our vocabularies are pauperized, and our minds drugged." Whatever this drug was, I craved it—I had *always* craved it. I was immersed in awe, and experiencing awe is like mainlining the truth. Above the surface, life can feel like it's spinning apart, fracturing into pixels, dispersing like dust by the day. But here was solidity and eternity, and a reality bigger than anything we could imagine.

"People have this intrinsic fear of the abyss," Vescovo mused, glued to his viewport. "But they've got it all wrong. It's not some deep, dark, automatic death sentence."

"It's magnificent," I whispered, spellbound, as a ctenophore twinkled by. You could live for a thousand years and never see this, and never know what you were missing. You might never know that this elemental wildness is what keeps your spirit's pilot light aflame. We always want more, but descending into the depths is a process of subtraction. Subtract air, light, weather, horizon. Subtract ego. Subtract human illusions of supremacy and control. When those things are gone, others can be added: real humility, new visions of beauty, altered perceptions, expressions of life that include unfamiliar schemes. (Why have one heart when you can have three? Who says that blood can't be blue?) In the abyss, you don't glimpse the mystery—you *enter* it, and your consciousness is the only fixed point. Subtract time and you're left with presence. In the deep, you lose your bearings and you find yourself.

The Deep Future

The sea is my sacred place . . . Even now in the winter of my life
I am still going out there, as sacred has no expiration date.
—CAPTAIN DON WALSH,
U.S. NAVY DEEP SUBMERGENCE PILOT NO. 1

THE GLASSHOUSE
MANHATTAN, NEW YORK
APRIL 23, 2022

The Madagascar hissing cockroaches were roasted extra-crispy, glistening on wooden skewers. They were bigger and shinier than the Costa Rican cave cockroaches and the Argentinian wood cockroaches, with longer antennae and a formidable set of horns. If you were hoping for a black-tie occasion to sample a variety of roaches, this was the only game in town: the 118th Explorers Club Annual Dinner. Archaeologists, anthropologists, geologists, field biologists, polar explorers, marine scientists, endurance athletes, space travelers, wildlife photographers, ocean conservationists, extreme climbers, single-handed sailors, and assorted peers: this was their party, a global gathering of adventurers.

For the last two years, the annual dinner had been canceled due to COVID, so the night was both a reunion and a celebration, the rooms overflowing with the club's members and guests. The dress code was evening wear gone wild. There were women in tiaras, leis, and beaded tribal headbands; men in mountaineering garb, astronaut suits, tiger-claw necklaces, and Himalayan robes. Along with the cockroaches, people were eating mealworms and Asian carp, and washing them down with fine Cabernets and single-malt scotch. The downtown Manhattan skyline glinted through walls of windows.

I stood at the "appetizers" table with Patrick Lahey and Alan Jamieson, who were wearing tuxedos. "CAUTION! THESE ITEMS ARE CONSUMED AT YOUR PERSONAL DISCRETION AND RISK," a sign warned. I had just eaten iguana on a cracker and I regretted it; a fishy, metallic tang lingered in my throat. "That's probably the mercury," Jamieson noted. Large insects—even those fried in tempura—were not on my menu, but he and Lahey seemed to be considering it. Peer pressure was a factor here: If you weren't willing to ingest a bug or two, then you weren't much of an explorer, were you?

"I don't know . . . ," Lahey said, shaking his head.

"Why did they import these cockroaches?" I asked. "There are plenty of them right here in New York."

While we were debating this, Jamieson downed a Madagascar hissing cockroach the size of a Lego brick, chewed it furiously, retched for a second, and then wiped his mouth with a napkin.

"What's it taste like?" Lahey asked.

Jamieson winced. "Like a fucking cockroach."

Perhaps recalling the many times he'd seen Jamieson tweezing apart amphipods in the *Pressure Drop*'s wet lab, Lahey pointed out: "Well, you of all people know what these things eat."

I turned to see Kelvin Magee and Tim Macdonald, also clad in tuxedos. Magee had flown in from Barcelona, Macdonald from Perth. Behind them, Victor Vescovo was talking to a group of well-wishers. Tonight he would be accepting the Explorers Club Medal, the group's highest honor, bestowed for "extraordinary contributions" in the field of exploration. Previous recipients included fellow deep-sea alumni Sylvia Earle, James Cameron, and of course, Don Walsh.

Typically Walsh himself would be here, but he was laid up with a broken hip after slipping during an appearance at SeaWorld. "The irony is not lost," he told me wryly, when I called him. "I went to San Diego for a four-day business trip and ended up being incarcerated in the hospital for a month. I'm not used to being cooped up, staring at the walls." Walsh reminded me that he had turned ninety during the pandemic: "At my age, I don't have time for this. I've got so much I want to do." Now that he was on the mend, Walsh's forthcoming plans included trips to China, Spain, and the Fort Lauderdale International Boat Show, not to mention a cruise around Cape Horn.

I knew that Walsh had wanted to see Vescovo receive his medal, and that he viewed the Five Deeps and its aftermath as a turning point in our relationship with the deep. From their first conversation, Walsh had been steering Vescovo away from the "hero dives" and toward a more enduring contribution. "I told him bluntly," Walsh recalled, " 'Look, so you got to X thousand meters and you found a plastic bag on the seafloor—that's news at eleven, Victor. Thirty years from now everyone will have forgotten it, because that's the way the media is, they just devour that stuff. But if you mobilize all the science you're doing—and the amount of seafloor that you've mapped is phenomenal—*that's* your legacy. You'll go down in history as one of the great expeditions, like *Challenger* and some of the others.' And he took that path. I'm just delighted that he's come into our community. He's fully into it now."

If anything, that was an understatement. Since the Five Deeps, Vescovo had racked up so many firsts that even he couldn't keep track of them. He had now dived into seventeen hadal trenches. He had been to the Challenger Deep fifteen times, spending a total of thirty-six hours there, surveying every corner of it, taking scientists and even paying clients down with him. (The cost of bragging rights to the world's deepest spot was $750,000—money that went into keeping the ship operations and science program going.) "It's an old saw in the media that more people have been to the moon than to the bottom of the Mariana Trench," Vescovo had pointed out. "Well, not anymore."

The numbers told the story: 1.2 million square miles of deep seafloor mapped at high resolution. Hundreds of lander deployments. Hundreds of thousands of biological samples. Scores of new species.

A steady river of scientific publications from Jamieson, Heather Stewart, and other marine researchers who'd received data from the *Pressure Drop,* and its sub, sonar, and landers. The world's two deepest shipwrecks (to date) found, at 21,180 feet and 22,621 feet, respectively.

None of this was breaking news, as it would've been if these achievements had occurred in space. ("As NASA and SpaceX Prepare to Fly Another Crew to the Space Station, Engineers Are Fixing a Leaky Toilet on the Spacecraft," *The Washington Post* had trumpeted in a recent headline.) The deep was still largely out of sight and out of mind—but with each awesome discovery more people's curiosity was piqued, and fewer perceived it in that old, tired way, as a hellscape populated by monsters. Or a barren wasteland that has nothing to do with us. With each dive, science paper, eyewitness account, video— each revelation—more people understood that the deep's mysterious obscurity isn't something to fear: it's the essence of life itself. They recognized its sublime beauty, and the knowledge and wisdom it had to offer. All of the scientists and submersible creators and explorers I'd met were like publicists for the deep, sharing their own fascination as though sprinkling stardust.

Earlier in the evening, when my Uber driver had pulled up to the Glasshouse, the dinner's venue, and seen women entering the building in gowns blinking with LED lights and men milling on the sidewalk in lederhosen and sarongs, he'd asked me what the event was that drew such a colorful crowd. When I told him—the Explorers Club—he wanted to know: "What do you explore?"

"The ocean."

The driver nodded his approval. "Isn't it amazing that we know more about space than what's at the bottom of the ocean?" he said, and then paused. "What *is* at the bottom of the ocean, anyway?"

*

The answer to that question is endless, because we're always adding to our master list of what's down there. But as I had learned, there was no better time to be interested in the deep than right now. Technology had cracked the vault open. Woods Hole Oceanographic Institution was building a new generation of intelligent autonomous robots like the stealthy, red-lit Mesobot, designed to spy on the twilight zone—scanning DNA fragments in the water, following and filming animals without scaring them off—and Nereid Under-Ice, a hybrid remote-operated vehicle capable of traversing twenty-five miles beneath ice cover; and Orpheus and Eurydice, a pair of free-swimming robots that will venture to the bottom of hadal trenches.

Now there are tiny swarming robots, and soft-bodied robots shaped like snailfish, and humanoid mer-bots with stereoscopic eyes and touch-sensitive hands. In the near future, fleets of drones will monitor the deep cooperatively, sharing information, responding to conditions, and transmitting data to the surface—realizing the vision that John Delaney put forth when he created the Regional Cabled Array. The need for us to have eyes and ears and cameras and smart instruments and DNA analyzers and sensors and robots in the deep ocean—so that we can protect it and understand it and survive in a time of harrowing climate change—will only grow. Ignoring the aquatic realm that dominates this planet is no longer an option.

As technology enables us to see through the veil of the abyss, it will become easier to find lost things (including, most likely, the elusive wreckage of flight MH370). In March 2022, robots filmed Ernest Shackleton's ship, the *Endurance,* ten thousand feet down in Antarctica, eerily well preserved and sprouting a vibrant garden of sea anemones. And of course, robots had documented the resting place of the Spanish galleon *San José.* Seven years after Roger Dooley found it, the ship was still on the seafloor, but a new Colombian president, Gustavo Petro, seemed determined to change that, publicly vowing to "return that galleon to the city of Cartagena in the form of a museum." Dooley had met with Petro; the two men connected over their mutual desire to do justice to the *San José.* If all went as planned, the project would proceed using an innovative economic model that would enable Colombia to pay for the ship's excavation, conservation, and display. "Now I can sleep," Dooley told me, sounding cautiously optimistic. "I spent four years almost without sleep."

In this robotic age, manned deep-sea exploration is also on the rise (thanks to private funding), no longer sidelined as too expensive, too risky, and too time-consuming. OceanX and its storytelling expertise, Triton and its covetable subs, Vescovo and his ambitious expeditions: they had rekindled the public imagination. "Imagine your wife is about to give birth to your child," Lahey had challenged an interviewer, making the distinction clear. "And you could be in the room, present, or you could hire a videographer to record it so you could watch it on TV later. Which do you think would have a more lasting impact?"

"There were a lot of people who said manned subs were a thing of the past," Terry Kerby agreed. "I think the science community has learned that's not true. There's nothing that can ever replace getting the human brain on the bottom." Plus, he added, much deep-sea terrain is too complex for tethered ROVs, which simply aren't nimble enough. Narrow or twisty canyons, rock walls with vertical overhangs, rugged volcanic areas, anywhere near discarded fishing nets or lines—only a pilot in situ can navigate such a three-dimensional obstacle course.

But alas, we are still short on research submersibles. The *Alvin* would soon return to sea after an eighteen-month overhaul that extended its range to sixty-five hundred meters. The venerable sub has taken scientists into the deep for nearly sixty years, and it has never been a more capable machine. So that's progress. But at the same time, to the dismay of countless researchers, the University of Hawaii shuttered the *Pisces* program. As always, it came down to money. The *Pisces*' mother ship, the *Ka'imikai-o-Kanaloa*, had needed a three-million-dollar investment to remain seaworthy, and the university opted to scrap it instead.

Understandably, nobody took the decision harder than Kerby. "Those subs could serve science and exploration for decades more," he told me, dejected. "Now they're rotting away in a warehouse. I haven't seen them since I was laid off." On the day the *Ka'imikai-o-Kanaloa* left Oahu for the final time, towed by tugboat to a shipbreaker in Mexico, Kerby had climbed to the top of the Koko Head crater and watched the vessel disappear over the horizon. Not long after that, as though the Pacific Ocean were amplifying his feelings, a ferocious windstorm tore the roof off the Hawaii Undersea Research Lab's now-empty hangar.

*

Amid this flurry of gains and losses, it was easy to forget that as recently as 2018, no full-ocean-depth passenger sub had existed. Now, in 2022, there were two diving repeatedly. In November 2020, China's new three-man submersible, the *Fendouzhe*, joined the *Limiting Factor* in the eleven-thousand-meter club, making its debut in the Challenger Deep. The two subs could not have looked more dissimilar. If the *Limiting Factor* was a Lamborghini, the *Fendouzhe* was a Winnebago: it weighed thirty-five tons and traveled with a crew of eighty technicians. In one photo, three Chinese aquanauts were shown eating a hot meal in the pressure hull, which looked big enough to pass for a dorm room. "Dear friends, the bottom of the sea is incredible," the crew broadcasted through their acoustic modem.

The truth of that statement is increasingly evident, with the deep demonstrating its flare for constant astonishment. In 2021, scientists on a research cruise in Antarctica's Weddell Sea flew a towed camera above the seafloor, and by pure chance they found a nesting colony of *sixty million* icefish—sleek, sizable predators with eyes like head-lights, flat triangular heads, translucent blue skin, and white hearts that pump gin-clear antifreeze blood. The nests were bowl-shaped and perfectly uniform, and each contained its architect: a single male icefish, guarding the eggs. The colony spanned a hundred square miles. Meanwhile near the North Pole, deep-sea researchers happened upon sprawling gardens of giant sponges—and myriad other fauna—thriving atop extinct submarine volcanoes. This supposedly defunct region was bursting with life. The sponges were feeding on the fossilized remains of animals that had lived and perished eons ago, when the sites had contained active hydrothermal vents. Neither of these ecosystems resembled anything seen before.

That same year, taxonomists introduced new deep-sea species with names that could've been coined by William Beebe, including the Jurassic pig-nose brittle star, the balloon backpack isopod, the hidden Horniman mysid shrimp, and the Yokozuna slickhead. They came on the heels of 2020's E.T. sponge, feisty Elvis worm, and the most press-ready deep-sea creature of all: the amphipod known as *Eurythenes plasticus*.

Plasticus was discovered in the Mariana Trench by Alan Jamieson and one of his doctoral students, Johanna Weston; its name reflected the fact that every specimen had plastic microfibers embedded in its guts. There was no baseline, no plastic-free version of this species. It had become one with its contaminant, and it stood as stark proof that we had managed to saturate the ocean with plastic down to its greatest depths and its tiniest creatures. Which is tragic enough, but that isn't the extent of it.

Along with microplastics and synthetic fibers, in recent years scientists have found that the hadal trenches are thick with every toxin we've ever unleashed: persistent organic pollutants like PCBs (industrial poisons), PBDEs (flame retardants), DDT (pesticide), phthalates (plasticizing chemicals), along with lead, mercury, pharmaceutical waste, and radioactive carbon from nuclear bombs. From the ocean's surface to its deepest sediments, all the way up the marine food chain, we have left our mark.

This burden isn't one that marine life bears lightly. When it doesn't outright kill them, it affects their immune systems, reproductive success, ability to digest food—their longevity and chances of survival. "The more you think about it, the more depressing it is," Jamieson said. When I asked if he thought there were more hybrid-plastic creatures like *plasticus* in the deep, he replied: "All of them."

The microscopic fug of this pollution is its most insidious form, but on dives in the *Limiting Factor*, Vescovo, Lahey, Jamieson, and Macdonald—like Terry Kerby and Buck Taylor and every other submersible pilot—had been floored by the visible signs of human degradation. The Philippine Trench had been the worst: it contained so much plastic debris that the sub's sonar was rendered virtually useless, because every target turned out to be some piece of long-lived junk. "I saw a teddy bear," Vescovo recalled, when I asked him about it. "My favorite was the plastic bag marked 'eco-friendly' that floated by the sub at ten thousand meters," Jamieson told me. "It was insane. It was just horrid."

That defilement was like a shadow falling across every magnificent sight that Jamieson had witnessed on what was now approaching twenty dives: the meadow of yellow crinoids ("Normally with crinoids, you see one or two. There were *thousands* of them") and the pristine manganese nodule field ("We're not telling anyone where it is") in which every nodule was crowned with its own exquisite animal ("If you look at the anemones, there are no two the same, and every species of sponge is different than the last") and the indigo-colored snailfish that showed up in the Japan Trench. ("It's *blue*. It's completely blue. It's like someone's gotten a snailfish and sprayed it blue. And it has an elderly feel to it, like an old basset hound.")

Along with the amazing visuals, Jamieson's sub dives and lander deployments had yielded amazing insights and facts. As the videos were analyzed and the DNA samples sequenced and the new species described—and as time allowed him and other scientists to put the terabytes of data into perspective—more jigsaw pieces would be slotted into the ever-evolving puzzle of the deep.

∗

After we were summoned to our tables by a man wailing on bagpipes, and after the standard three-course rubber-chicken dinner was served and mostly ignored, the awards began, and because there were three years' worth of pent-up honors to distribute, the whole thing passed in a lively parade of Mars Rover designers and cheetah rescuers and cave divers and space artists and lion cinematographers, and with so many honorees there was no time for acceptance speeches. Vescovo was photographed onstage with his medal; the Five Deeps team received a Citation of Merit, and then—bam!—it was over.

I was sitting with McCallum and Macdonald, and since I'd last seen them in Hawaii, they had dived together to the Challenger Deep on Macdonald's first hadal piloting mission. "So how was it?" I asked.

"Epic," Macdonald offered.

McCallum grinned. "We went deeper than Victor, but don't tell him that."

"*No* we didn't," Macdonald said.

"Nah, right," McCallum agreed, laughing. "We didn't."

"We had vegemite sandwiches and beer on the way up," Macdonald added.

"Victoria Bitter," McCallum specified. "The deepest-diving beer."

Macdonald sat back and took a swig of his martini. A lapel pin glinted on his tuxedo, and I knew what it was because I had one, too. Don Walsh had given us our own gold dolphins, mine for my dive to Kamaʻehuakanaloa—now the volcano's official name, to the joy of Hawaiians—and Macdonald's, more legitimately, for being a *Limiting Factor* pilot. Not long after I'd returned from the Big Island, a bulky envelope had arrived at my house containing the pin and a letter.

"Now that you are a deep diver, I wanted you to have this bauble," Walsh had written. "It's the official USN uniform badge that designates those officers and enlisted men who have qualified as pilots and crewmembers in manned submersibles. I consider you to be fully qualified to possess and use the badge." A separate page gave instructions on how to wear it properly: "The pin is worn on the left breast of the uniform just below the ribbons for medals and campaigns."

*

Our plates were cleared, and the live auction began. There was a $30,000 glacier trek in Greenland and a $20,000 cruise up the Nile and a $25,000 glamping week in Bhutan—and stiff competition for all of them. As I watched a woman lunge from her chair to bid, I was reminded that the other ingredient in adventure and exploration (besides the adventurers and explorers) is the patrons (who were often adventurers and explorers themselves). In his quest to dive in all the deepest spots, Vescovo had become a major sponsor of deep-sea technology. Without him: no sub. No vegemite sandwiches in the Mariana Trench and no indigo snailfishes and no 1.2 million square miles of desperately needed seafloor maps, and no one to report that the Philippine Trench was a dump. No all of this.

The U.S. government's funding for deep-sea science is a fraction of what it should be. Its funding for deep-sea exploration is basically nonexistent: if not for privately funded efforts, it would barely occur at all. After their stunning discovery of Lost City, and hoping to test their theory that similar sites were out there, Deb Kelley told me, "Jeff Karson and I tried multiple times to get an expedition funded to find another one." You'd think this would be a slam dunk: their proposal to the National Science Foundation included annotated maps and was bluntly titled "In Search of New Lost Cities." But when they couldn't guarantee their hunt would be successful—because no one had ever *been* to these places—they were turned down, and told that eventually someone would stumble across a second Lost City. "We were dumbfounded," Kelley said.

Since the U.S. Navy currently has zero vehicles that can operate at full ocean depth, it seems fortunate that an intrepid explorer in Texas had commissioned one. Vescovo was open about the fact that this endeavor had tested the limits (admittedly, high) of his net worth, and that he now needed to sell. "Financially, I knew that I couldn't sustain it indefinitely," he'd explained. "It's unbelievably expensive to own and operate a deep-ocean research vessel and submersible. Four years at full burn was a very heavy commitment—which I met. But I'm not a billionaire. I can't do what a Ray Dalio can: fund this thing for years and not even notice it. I notice it."

It was hard to fathom why the Navy hadn't leaped at the chance to buy the *Limiting Factor*. ("In the military world, Victor's system is dead cheap," Don Walsh told me. "They spend fifty million dollars before their first coffee break.") A sub that can prowl anywhere in the deep and be launched from a small, unobtrusive ship would seem like a handy item for a country to have. But instead, a private buyer stepped in.

As of November 2022, the new owner of Vescovo's sub, ship, sonar, and landers would be a software entrepreneur and video game developer named Gabe Newell, who began his career at Microsoft in the early days of Windows 1.0. Newell loved boats and submersibles and technology and, quite obviously, the ocean. The *Limiting Factor* would be his third sub; the *Pressure Drop* would be his fourth ship. To put these vehicles to good use, Newell had formed a marine research group called Inkfish, with a division dedicated to the deep. Its purpose? To explore and study the unknown abyss. Its chief scientist? Alan Jamieson. Its lead geologist? Heather Stewart. Its submersible team leader? Tim Macdonald. Stuart Buckle would remain as captain, along with his handpicked crew. The band would play on.

"This is fantastic," I said to Lahey when I heard about Inkfish, and he agreed: "It's the best of all possible outcomes. Gabe will continue the legacy of exploration that Victor began, but expand on it and focus more on the science aspect. And it's even more exciting to know that Jamieson is going to be part of that scenario."

Then, in December 2022, Lahey would announce that Ray Dalio and James Cameron had taken equity stakes in Triton. The two would be joint owners with Lahey, infusing the company with resources and enabling Lahey's team to build even more futuristic subs—next-generation aquatic machines that would fit right into, well, a James Cameron movie.

*

When the *Limiting Factor*'s sale was made public, I'd ask Vescovo: "Do you think you'll ever get another sub?"

"Oh, absolutely."

"Full ocean depth?"

"Yeah. You can't go back. There are seven or eight trenches I still haven't dived, and you have to imagine that they're definitely calling me."

Clearly, there was no lack of options for Vescovo's next chapter. And clearly, Gabe Newell's Inkfish group was a dream home for the *Limiting Factor*. Yet even so, for Vescovo the handoff would be bittersweet. "When you work with a machine and you get to know it, it becomes like a life-form," he would reflect on the day the sale went through. "With every little sound the *LF* made, I could predict her behavior. And I knew one thing: she would always get me home. There were tense times, and there were times like 'Okay, I have a situation here'—the fire in Tonga, or the cables around the *Titanic*— but I knew she was a great craft. And she never let me down. She *never* let me down. When things really mattered. She pissed me off now and then like anyone, but . . ." He would leave the sentence unfinished.

*

That night at the Explorers Club, after-dinner dancing and cocktails were held in a room that showcased a Blue Origin crew capsule—the six-person module that perches on the tip of Jeff Bezos's New Shepard rocket. I saw Vescovo standing by the capsule and walked over to congratulate him. "Thanks," he said, and then whispered: "Can you keep a secret?" I nodded, and he pointed to the spacecraft: "I'm going up in it on May 20. I'll be the first person to go to the bottom of the ocean, the top of the Himalayas, and into space."

Before I could respond—and what could you really say except, "Cool!"—Vescovo was commandeered by a group of young men wearing Jacques Cousteau–red toques with their tuxedos. I turned to Lahey, standing nearby. We sipped our wine and stared at the space capsule. "If someone said you can have a trip to space or a trip to the bottom of the ocean, I'd pick the ocean every time," Lahey noted.

"For sure," I said.

"I know they talk about all the billions of planets and galaxies and there's probably dozens of places like Earth," he continued. "But guess what? As far as I know, this is the only one. And we should take better care of it. That's why it's so important to get people into the water— you send someone down in a sub and they come back changed."

I nodded. "Why would anyone buy an airplane when they could have a submersible?"

"Well, of course, you think like I do."

It was true: Lahey and I did think alike. So did the scientists and explorers I'd gone to sea with, and the trailblazing aquanauts I admired. There is an aquatic tribe on this earth, people drawn to the mysteries beneath the waves, drawn to water itself, happier at sea than on land—"If it floats, I want to be on it," Walsh had declared—and if you're reading this book, it's an excellent bet that you're one of them.

In which case, good news: the future is aquatic. If you want to immerse yourself in the deep, you'll have opportunities like never before. "Things are happening at almost a dizzying rate," Lahey said. "I mean, look at anything—whether it's a battery, a light, a sonar system—*everything* is moving just as quickly as the technology in your cell phone. And we're always pushing the envelope. We're always thinking about what else we could do."

I toasted that, and Lahey, and Triton—and Vescovo, leaning on the bar beside us, engaged in another conversation.

"Hey, one of these days we'll get to Europa's ocean," I teased Lahey. "And we'll need a sub."

Vescovo wheeled around, suddenly interested. "Europa? I would totally do that! I would *totally* go to Europa."

Acknowledgments

Nobody does anything by herself in the deep ocean, a statement that's also true in book publishing. I owe thanks to many, but none more than Don Walsh, Patrick Lahey, and Victor Vescovo. Don welcomed me into his home, shared his time and incredible wealth of experience, connections, stories, and wisdom. I cherish the conversations we've had and look forward to those yet to come. (And hope that some will take place over a "jar of Malbec," Don's signature drink.)

When I first met the team from Triton Submarines, one of the sub technicians told me, "We all bleed Triton blue." It didn't take long to understand why: the inspired leadership of Patrick Lahey. His vision, integrity, and commitment to the ocean were evident from our first phone call. My gratitude to Patrick is profound—equaled only by my admiration of who he is and what he has achieved. If I could have one wish for ocean awareness, it would be that everyone gets the chance to experience the deep's wonders in one of Triton's revolutionary submersibles.

Don and Patrick introduced me to Victor Vescovo, a man who over the course of four years changed the face of deep-sea exploration. To be present during his history-making run of dives was about as lucky as a writer could get, but Victor gave me far more than that. His generosity and graciousness are epic, his spirit of adventure unrivaled. The opportunity to dive with him was a gift beyond measure, and one that I am honored to share in these pages. Mahalo nui loa, Victor. A tsunami of thanks doesn't begin to cover it.

Victor's ship, the *Pressure Drop*, was a hub of great people who were integral to this book. For three years, starting within minutes of when I boarded in Tonga, I asked Alan Jamieson nonstop questions about deep-sea science, and he answered them all with his trademark mix of brilliance and humor. When he moved from Newcastle to Perth—where he became a professor at the University of Western Australia, and the founding director of its Minderoo-UWA Deep-Sea Research Center—he took hours from his packed schedule to speak with me on Zoom. I'm immensely grateful to Alan. There is no better guide to the deep.

Thanks to Rob McCallum for his steady, friendly support, and for sharing his knowledge about ocean exploration, shipboard psychology, dive planning, risk management, and life in general. Rob makes running a global expedition look easy, though it's the opposite of that. Additional thanks to Karen Horlick, Kelvin Murray, and the team at EYOS Expeditions.

What a bolt of great fortune it was to sail with Joe MacInnis. Thank you, Joe, for your wisdom, warmth, and encouragement, for your sage counsel and inspiring example, for sharing your words, photos, and company, and most of all, for being a kindred aquatic spirit.

Heartfelt gratitude to Kelvin Magee, Frank Lombardo, Shane Eigler, Steve Chappell, Hector Salvador, Ticer Pfeifer, Jarl Stromer, and everyone at Triton. John Ramsay and Tom Blades: thank you for your amazing creations. Special thanks to Tim Macdonald, swimmer-engineer-pilot-friend: it was a joy to spend time with you in the underworld.

I'm indebted to Captain Stuart Buckle, Captain Alan Dankool, P. H. Nargeolet, Heather Stewart, Cassie Bongiovanni, Jonathan Struwe, Tomer Ketter, Swain Murray, Enrique Alvarez, Henschel Orsino, Kath D'arville, Tiziana Lahey, Rachel Jamieson, and my fantastic cabin-mate, the Five Deeps' official artist, Alexandra Gould. (To see Alex's artwork from the expedition: https://www.behance.net/alexandragould.) Many thanks to everyone on the crew: Charlie Ferguson, Fraser Retson, Peter Coope, Erlend Currie, Andrew Welsh, Noli Garcia, Leo Sinoro, Melvin Lucido, Marcos Benavides, Manfred Umfahrer, Peter Barlow, Scott Cherry, Brendhan Thompson, Narciso Sagitarios, Panfilo Lanchinebre, Rolando Belmonte, Randolph Quiton, Jessie Eusebio, Roger Divinagracia, Ali Benarabi, Veselin Botev, Whern Carvajal, Daryl Majarocon, Kyle McDowell, Sunny Regolis, and John Wallace—all of whom made my reporting at sea a pleasure.

I send my deep appreciation to Sylvia Earle. Her devotion to the wild marine world is legendary, and we are all its beneficiaries. I wish I could think of an adequate way to thank Sylvia for the generosity she has shown me—and I hope she recognizes her influence suffused throughout this book. Actually, there is one sure way we can all thank Sylvia: by reveling in the ocean's beauty, and doing everything we can to protect it.

As I write this, commercial deep-sea mining appears to be imminent. It will be up to an outraged public to stop it: I urge readers to follow the issue and make their voices heard. The Deep Sea Conservation Coalition is an excellent source for news and information about how to take action. Thank you to everyone who is working to raise awareness about this devastating threat.

No single volume could do justice to Terry Kerby's deep-sea stories, but I'm grateful for the chance to try. To see Terry light up as he talks about his *Pisces* dives is to glimpse the thrill and power of the abyss, the spell it casts and the gifts it bestows. His experiences illuminated the deep for me. Additional thanks to my friends Paul Atkins and Grace Atkins for introducing me to Terry and HURL.

I owe huge thanks to Deborah Kelley, whose warmth and intelligence sparked conversations that continue to this day. Sailing with her was a highlight of my reporting, and a master class in marine geology. Thanks to Mike Vardaro, Katie Gonzalez, Julie Nelson, Janel Hershey, Mitch Elend, Trina Litchendorf, Orest Kawka, James Tilley, Eric Olson, Rachel Scott, Eve Hudson, and everyone on the *Atlantis*. Thanks also to the Woods Hole/Jason team: Scott McCue, Webb Pinner, Drew Bewley, Chris Judge, Chris Lathan, Korey Verhein, Summer Farrell, James Convery, Matthew Heinz, and Nile Akel Kevis-Stirling.
    Deb introduced me to John Delaney, a true deep-sea visionary. I'm grateful to John for his time and encouragement, and for everything he has done, and continues to do, in science and education, on behalf of a greater understanding of the ocean.

What an extraordinary experience it was to dive with OceanX—and what an extraordinary group of people. Deep thanks to Ray Dalio, Mark Dalio, and James Cameron for creating such a great organization, and for sharing it. Thanks also to Mark Kirby, the *Alucia*'s crew, submersible pilots Dave Pollock, Lee Frey, Alan Scott, Toby Mitchell, Colin Wollerman, Chris May, Matt Awty, Alex Gottschall—and especially, Buck Taylor.

Roger Dooley and Garry Kozak opened up the world of deep-sea search and archaeology to me. I thank them for that, and thank Roger for entrusting me with the details of his successful hunt for the galleon *San José*. I look forward to the day when I can visit the *San José* museum in Cartagena, and see the result of his decades of impassioned work.

Many deep-sea scientists and experts welcomed my inquiries and helped me. This book relied on them, and I greatly appreciate their contributions: Thom

Linley, Johanna Weston, Kim Picard, Samantha Joye, Andy Bowen, Ken Rubin, Brian Glazer, Craig Smith, Jeff Drazen, Patricia Fryer, Andrew Gooday, Duncan Currie, Bruce Robison, Steve Haddock, Antje Boetius, Matthias Haeckel, Lisa Levin, Mike Vecchione, Nerida Wilson, Doug McCauley, Glenn Moore, Sally Ley, Will Kohnen, Andy Sherrell, and James Delgado.

For her meticulous fact-checking, research, and insights, I thank Naomi Barr. Philip Keifer and Martha Corcoran also contributed research to *The Underworld* with their characteristic expertise. Any errors are mine.

It's been a pleasure to write for *Outside* magazine over the years. Before I chronicled my time with the Five Deeps expedition in these pages, I wrote about it in a 2019 *Outside* feature shepherded by editors Chris Keyes, Mary Turner, and Michael Roberts. I am grateful to them for the opportunity.

Doubleday is my publishing home, and I couldn't be more fortunate. My editor, Bill Thomas, is the partner and guide that every author hopes for. His intelligence, patience, and skill are precious to me (as is his willingness to do things like answer a panicked email, sent at three a.m. from Guam). Thank you, Bill, for traveling with me on this journey. You are simply the best.

Many thanks to the exceptional team at Doubleday: Khari Dawkins, Nora Reichard, John Fontana, Maria Carella, Daniel Novack, Michael Goldsmith, Milena Brown, Anne Jaconette, Vimi Santokhi, Andy Hughes, and Kathy Hourigan; and to Kristin Cochrane and Amy Black at Doubleday Canada.

As always, I'm grateful for the invaluable advice, support, and friendship of my agent, Eric Simonoff—and to Criss Moon, Elizabeth Wachtel, and everyone at WME.

Thank you to my friends: Sharon Ludtke, Martha Beck, Andy Astrachan, Maria Moyer, Brooke Wall, Kelly Meyer, Cristina Carlino, Sara Corbett, Laird Hamilton, Gabby Reece, Donna Shearer, Elizabeth Lindsey, Pualani Kanaka'ole Kanahele, George Crowley, Jeff Shapiro, Hinako Shapiro, Terry McDonell, Stacey Hadash, Jennifer Buffett, Peter Buffett, Amber Rubarth, Hila Katz, Bob Dandrew, and my Munay, Finca Mia, and Hatunsonqo tribes (you know who you are). Thanks to Angela Casey, and to Shele Letwin, Sharlane Letwin, Lorna Walkling, Bob Casey, and Mike Casey.

A deep ocean of thanks to Deborah Caulfield Rybak and Michael Rybak—ohana forever—to whom this book owes much, as do I. Love and gratitude that words can never express go to my family: Rennio, Leo, and Mia Maifredi.

Notes

PROLOGUE

1   "My dear child": Homer, *The Odyssey,* trans. Stephen Mitchell (New York: Atria Books, 2013), 141.

2   "The passion caused": Edmund Burke, *A Philosophical Enquiry into the Sublime and Beautiful* (Oxford, UK: Oxford University Press, 2015), 47.

2   "a sort of delightful horror": Ibid., 60.

5   "Who has known the ocean?": Rachel Carson, "Undersea," in *Lost Woods: The Discovered Writing of Rachel Carson,* ed. Linda Lear (Boston: Beacon Press, 1998), 4.

5   Yet the deep ocean: Roberto Danovaro, Cinzia Corinaldesi, Antonio Dell'Anno, and Paul V. R. Snelgrove, "The Deep-Sea Under Global Change," *Current Biology* 27, no. 11 (June 2017): R461–R465.

5   The Pacific alone: The Pacific's immensity is hard to visualize, especially when you consider that it's also the deepest ocean basin, with a maximum depth of 10,935 meters, or 35,876 feet—nearly seven miles at its nadir. My back-of-the-envelope calculation of how many continents would fit into the Pacific is based on its area of 64 million square miles; a measurement for the earth's continental landmass of 57 million square miles; and South America's footprint of 7 million square miles. (Obviously, to be accurate, cubic miles would be preferable as a basis for measurement.) In any case, the Pacific is *big.*

7   I pored over any deep-sea news: Ean Higgins, "Search for MH370 Unveils a Lost World Deep Beneath the Ocean," *The Australian,* July 21, 2017.

7   more like estimates than facts: Larry Mayer, Martin Jakobsson, Graham Allen, Boris Dorschel, Robin Falconer, Vicki Ferrini, Geoffroy Lamarche, Helen Snaith, and Pauline Weatherall, "The Nippon Foundation–GEBCO Seabed 2030 Project: The Quest to See the World's Oceans Completely Mapped by 2030," *Geosciences* 8, no. 2 (2018): 1–18, https://www.mdpi.com/2076-3263/8/2/63.

8   crisply detailed three-dimensional maps of the deep: To see a graphic representation of the newly revealed Indian Ocean seafloor detail: "The Data Behind the Search for MH370," https://geoscience-au.maps.arcgis.com/apps/Cascade/index.html?appid=038a72439bfa4d28b3dde81cc6ff3214 (accessed January 4, 2023).

8   the *West Ridge*: Ross Anderson, *Maritime Archaeological Analysis of Two Historic Shipwrecks Located During the MH370 Aircraft Search*, Report No. 322, Department of Maritime Archaeology, Western Australian Museum (April 2018), https://museum.wa.gov.au/maritime-archaeology-db/sites/default/files/no-322-mh370-shipwreck-analysis.pdf.

8   A hundred million years ago: K. Picard, B. Brooke, and M. F. Coffin, "Geological Insights from Malaysia Airlines Flight MH370 Search," eos.org, https://eos.org/science-updates/geological-insights-from-malaysia-airlines-flight-mh370-search (accessed January 4, 2023).

8   a huge mass of igneous rock: Kim Picard, Walter Smith, Maggie Tran, Justy Siwabessy, and Paul Kennedy, "Increased-Resolution Bathymetry in the Southeast Indian Ocean," hydro-international.com, https://www.hydro-international.com/content/article/increased-resolution-bathymetry-in-the-southeast-indian-ocean (accessed January 4, 2023).

9   In one adjacent area: Kim Picard, Brendan P. Brooke, Peter T. Harris, Paulus J. W. Siwabessy, Millard F. Coffin, Maggie Tran, Michele Spinoccia, Jonathan Weales, Miles Macmillan-Lawler, and Jonah Sullivan, "Malaysia Airlines Flight MH370 Search Data Reveal Geomorphology and Seafloor Processes in the Remote Southeast Indian Ocean," *Marine Geology* 395 (January 2018): 301–19, https://doi.org/10.1016/j.margeo.2017.10.014.

10   If all 3.6 nonillion marine microbes: These estimates come from the International Census of Marine Microbes (ICoMM), a project that was part of the International Census of Marine Life, a global initiative to assess the diversity and abundance of marine life, conducted from 2000 to 2010, http://www.coml.org/international-census-marine-microbes-icomm/ (accessed January 4, 2023). Also: https://www.calacademy.org/explore-science/microbe-census.

10   new diagnostic tests: Elise Hugus, "Finding Answers in the Ocean," whoi.edu. https://www.whoi.edu/oceanus/feature/finding-answers-in-the-ocean/ (accessed January 4, 2023).

10   "To sense this world of water": Carson, "Undersea," 4.

10   the foundation of the planet: A. R. Thurber, A. K. Sweetman, B. E. Narayanaswamy, D. O. B. Jones, J. Ingels, and R. L. Hansman, "Ecosystem Function and Services Provided by the Deep Sea," *Biogeosciences* 11, no. 14 (July 2014): 3941–63.

11   "If the stars should appear": Ralph Waldo Emerson, "Nature," in *Nature and Selected Essays*, ed. Larzer Ziff (New York: Penguin Books, 2003), 37.

CHAPTER 1: MAGNUS'S MONSTERS

13   "Indeed, I should also add": Olaus Magnus, *A Description of the Northern Peoples*, vol. 3, trans. Peter Fisher and Humphrey Higgens, ed. Peter Foote (London: The Hakluyt Society, 1998), 1089.

13   a hub of feisty Norse pagans: Joshua Mark, "Temple at Uppsala," worldhistory.org, https://www.worldhistory.org/Temple_at_Uppsala/ (accessed January 5, 2023).

14   with unprecedented accuracy: Leena Miekkavaara, "Unknown Europe: The

Mapping of the Northern Countries by Olaus Magnus in 1539," *Belgeo Revue Belge de Géographie* 3, no. 4 (December 2008): 307–24.

15   Mariners vanished into its maw: *Folktales of Norway,* ed. Reidar Christiansen (Chicago: The University of Chicago Press, 1964), 30–32.

15   from the Greek word *kētŏs,* or sea monster: John K. Papadopoulous and Deborah Ruscillo, "A Ketos in Early Athens: An Archaeology of Whales and Sea Monsters in the Greek World," *American Journal of Archaeology* 106, no. 2 (April 2002): 187–227.

15   Throughout the Middle Ages: Charles Singer, *A Short History of Biology: A General Introduction to the Study of Living Things* (Oxford, UK: The Clarendon Press, 1931), 26. Singer notes: "Aristotle appears to have been the first to illustrate a biological treatise. In his works such diagrams are often referred to. Unfortunately the figures have long since disappeared, but his descriptions are not infrequently of a character that enables us to reconstruct them."

16   "a conglomeration of monsters": Magnus, *Description of the Northern Peoples,* 1081.

16   It was the Age of Discovery: Lindsay J. Starkey, "Why Sea Monsters Surround the Northern Lands: Olaus Magnus's Conception of Water," *Preternature* 6, no. 1 (2017): 31–62.

16   "sundry monstrous appearances": Friar Jordanus, *Mirabilia Descripta: The Wonders of the East,* trans. Colonel Henry Yule (London: The Hakluyt Society, 1863), 1–3.

16   "murder on the seabed": Walter Isaacson, *Leonardo* (New York: Simon & Schuster, 2017), Chapter 20, iBooks.

17   The other one: In Munich, the original print of the *Carta Marina* is in the Bayerische Staatsbibliothek's map collection.

18   I'd seen reproductions of it: For reasons unknown, most contemporary reproductions of the *Carta Marina* are colorized, and as a result all the fine detail is lost. In its original black-and-white the map is stunning; the colorized versions are awful.

19   "It has a ghastly head": Magnus, 1096.

19   "like sea-robbers or ill-disposed visitors": Ibid.

19   "a serpent of gigantic bulk": Ibid., 1128.

19   "The whole head of this creature": Ibid., 1087.

20   "Paulus Orosius declares": Ibid., 1089.

20   "a sea-monster with a huge horn": Ibid., 1097.

20   Magnus cited one onlooker's impression: Ibid., 1095–96.

21   "Such great wonders have their position": Starkey, 37–38.

21   "Even the deadliest": Edward O. Wilson, *Biophilia: The Human Bond with Other Species* (Cambridge, MA, and London: Harvard University Press, 1984), 84.

22   "large melon-thistle": Philip Miller, *The Gardeners Dictionary Containing the Methods of Cultivating and Improving All Sorts of Trees, Plants, and Flowers, for the Kitchen, Fruit, and Pleasure Gardens, As Also Those Which Are Used in Medicine* (London: John and James Rivington, 1754), n. 238.

22   Over the years, taxonomy: Marta Paterlini, "There Shall Be Order: The Leg-

acy of Linnaeus in the Age of Molecular Biology," *Embo Reports* 8, no. 9 (September 2007): 814–16.

23   "In the deeps of the sea": Magnus, 1089.

24   "It is no trifling evil": Charles Darwin, *The Voyage of the Beagle* (New York: P. F. Collier & Son), 528.

24   "Through want of instruments": Helen Rozwadowski, *Fathoming the Ocean: The Discovery and Exploration of the Deep Sea* (Cambridge, MA, and London: The Belknap Press of Harvard University Press, 2005), 5.

25   "endless novelties of extraordinary interest": Charles Wyville Thomson, *The Depths of the Sea* (London: Macmillan and Company, 1873), 49.

25   "The wonders of the ocean floor": H. Noel Humphreys, *Ocean Gardens: The History of the Marine Aquarium* (London: Sampson, Low, Son, and Co., 1856), 9.

25   "Yet they are not flowers": Ibid., 13–14.

26   "exhibited in deadly conflict with human divers": Ibid., 111.

26   British mariners observed one with a giraffe's neck: A. C. Oudemans, *The Great Sea-Serpent: An Historical and Critical Treatise* (London: Luzac & Co., 1892), 271, 347.

26   "I saw no less than three sea-serpents": J. Cobbin letter to *Annals and Magazine of Natural History*, dated January 22, 1872.

27   "There is no a priori reason": Thomas Henry Huxley, letter to *The Times* (London), January 1893, quoted in The Huxley File, http://aleph0.clarku.edu/huxley/, http://aleph0.clarku.edu/huxley/UnColl/LonTimes/SeaSerp.html (accessed January 6, 2023).

27   "I can no longer doubt": Louis Agassiz, 1849 lecture, quoted in Sherrie Lynne Lyons, *Species, Serpents, Spirits, and Skulls* (Albany, NY: State University of New York Press, 2009), 17.

27   "The sea saurians of the secondary periods": Richard Owen, letter to *The Times* (London), November 11, 1848, quoted in Oudemans, 284.

27   "the living, and perhaps sole": Richard Owen, *Memoir on the Pearly Nautilus* (London: Richard Taylor, 1832), 2.

28   "Everything is corroded by the brine": Plato, *The Collected Dialogues of Plato Including the Letters*, ed. Edith Hamilton and Huntington Cairns, trans. Lane Cooper et al. (Princeton, NJ: Princeton University Press, 1961), 91.

29   "Britons, whether scientific or unscientific": Edward Forbes, *The Natural History of the European Seas*, ed. Robert Godwin-Austen (London: John Van Voorst, 1859), 3.

29   Its scoop was too small: Thomas R. Anderson and Tony Rice, "Deserts on the Sea Floor: Edward Forbes and His Azoic Hypothesis for a Lifeless Deep Ocean," *Endeavour* 30, no. 4 (November 2006): 131–37.

30   "its inhabitants become more and more modified": Forbes, 26.

30   at least nineteen species: Thomson, *The Depths of the Sea*, 270.

30   "teeming with animal life": James Clark Ross, *A Voyage of Discovery and Research in the Southern and Antarctic Regions During the Years 1839–1843* (London: John Murray, 1847), 202.

31   "This sounding far exceeds": G. C. Wallich, *The North-Atlantic Sea-Bed: A Diary of the Voyage on Board H.M.S. Bulldog, in 1860* (London: John Van Voorst, 1862), 68.

31   a mythical undersea kingdom: Ibid., 63–67.

31   They named it *Bathybius*: Philip F. Rehbock, "Huxley, Haeckel, and the Oceanographers: The Case of *Bathybius haeckelii*," *Isis* 66, no. 4 (December 1975): 504–33.

32   a specimen so bizarre: Thomson, *The Depths of the Sea*, 423–24. Thomson's impression of the glass sponge was tempered by skepticism: "Anything very strange coming from Japan is to be regarded with some distrust. The Japanese are wonderfully ingenious, and one favorite aim of their misdirected industry is the fabrication of impossible monsters by the curious combination of the parts of different animals."

32   materials engineers study glass sponges: Matheus C. Fernandes, Joanna Aizenberg, James C. Weaver, and Katia Bertoldi, "Mechanically Robust Lattices Inspired by Deep-Sea Glass Sponges," *Nature Materials* 20 (September 2020): 237–41.

32   Their fractal-like skeletons: Vikram C. Sundar, Andrew D. Yablon, John L. Grazul, Micha Ilan, and Joanna Aizenberg, "Fibre-Optical Features of a Glass Sponge," *Nature* 424 (August 2003): 899–900.

33   "Certainly it was always the first": Thomson, *The Depths of the Sea*, 59–60.

33   The weather calmed: Ibid., 279.

34   "I had to summon up some resolution": Ibid., 155–57.

34   "In some places nearly everything": Ibid., 98.

34   "We concluded": Ibid., 93.

34   He spent the rest of his career: A. L. Rice, "G. C. Wallich M.D.—Megalomaniac or Mis-used Oceanographic Genius?," *Journal of the Society for the Bibliography of Natural History* 7, no. 4 (February 2011): 423–50.

36   which the sailors referred to as "insects": *At Sea with the Scientifics: The Challenger Letters of Joseph Matkin*, ed. Philip F. Rehbock (Honolulu: University of Hawaii Press, 1992), 29.

36   In a letter to his family: Ibid., 39.

36   "ghastly objects": Lord George Campbell, *Log-Letters from the Challenger* (London: Macmillan and Co., 1877), 5.

38   "There was light enough": Ibid., 89.

38   The sailors called it: Rehbock, ed., 81.

39   "This is the greatest depression": John James Wild, *At Anchor: A Narrative of Experiences Afloat and Ashore During the Voyage of H.M.S. Challenger from 1872 to 1876* (London: Marcus Ward and Co., 1878), 160.

39   "So, hey for the *Challenger*": Campbell, 420.

39   "It is possible even for a naturalist": H. N. Moseley, *Notes by a Naturalist on the Challenger: An Account of Various Observations* (London: Macmillan and Co., 1879), 578.

39   "He bothered me much": Ibid., 593.

40   "center of creation": Forbes, 8.

41   "the most magnificent contribution": "Work of the Challenger Expedition— III. Geologically Viewed," *Science* 6, no. 132 (August 1885): 138.

41   "seems to have needed a live harpooner": "Work of the Challenger Expedition—II. From a Zoological Standpoint," *Science* 6, no. 128 (July 1885): 54.

41   Thomson quoted in Doug Macdougall, *Endless Novelties of Extraordinary Interest: The Voyage of H.M.S. Challenger and the Birth of Modern Oceanography* (New Haven, CT, and London: Yale University Press, 2019), 25.

CHAPTER 2: AQUANAUTS

42   "These descents of mine beneath the sea": William Beebe, *Half Mile Down* (New York: Harcourt, Brace and Company, 1934), 148.

47   wasn't pursued with any rigor: For an overview of the development of the U.S. Navy's research sub *Alvin*, at the Woods Hole Oceanographic Institution, see National Research Council (US) Ocean Studies Board, *50 Years of Ocean Discovery* (Washington, DC: National Academies Press, 2000), https://www .ncbi.nlm.nih.gov/books/NBK208815/ (accessed January 7, 2023).

49   He liked to end his journal: Carol Grant Gould, *The Remarkable Life of William Beebe: Explorer and Naturalist* (Washington, DC: Shearwater Books, 2004), 65.

49   He was a prolific writer: Ibid., 33. William Beebe was born on July 29, 1877. His first article was published in the January 1895 edition of *Harper's Young People* magazine.

50   he charmed some financiers: Ibid., 241.

50   the use of a private island: Ibid., 272.

50   he and a friend, President Theodore Roosevelt: Beebe, 90.

50   On Manhattan's Upper East Side: Brad Matsen, *Descent: The Heroic Discovery of the Abyss* (New York: Pantheon Books, 2005), 10–12.

54   He guessed, correctly: Beebe, 108.

54   "Only dead men": Ibid., 100.

54   "It was of an indefinable translucent blue": Ibid., 109.

54   Beebe's inner alarm bells: Ibid., 112.

54   "The window to a wholly new world": Ibid., 65.

55   "I sat crouched": Ibid., 134.

55   "I watched one gorgeous light": Ibid., 169.

55   "Twenty feet is the least possible estimate": Ibid., 219.

56   "it will live throughout": Ibid., 212.

56   Beebe appointed these newcomers: Beebe's awe at what he was seeing is evident in his lengthy and vivid descriptions of deep-sea fauna, but he also expresses frustration at the difficulty of conveying his transcendent experiences with mere words: "This entire volume would not contain the detailed recital of even a fraction of all the impressive sights and forms I saw, and nothing at these depths can be spoken of without superlatives." Ibid., 206.

56   Twice during unmanned test dives: Ibid., 189.

56   the Chicago World's Fair: This event was also known as the "Century of Progress" International Exposition. It was held in Chicago from May 1933 to October 1934, with forty million people attending.

57   "This book ought to create": Florence Finch Kelly, "Exploring the Depths of the Ocean," *New York Times*, December 9, 1934.

60   the earth's longest-lived vertebrate: Julius Nielsen, Rasmus B. Hedeholm, Jan Heinemeier, Peter G. Bushnell, Jørgen S. Christiansen, Jesper Olsen et al.,

"Eye Lens Radiocarbon Reveals Centuries of Longevity in the Greenland Shark (Somniosus microcephalus" *Science* 353, no. 6300 (August 2016): 702–4.

61   During one notorious tantrum: R. S. Dietz and M. J. Sheehy, "Transpacific Detection of Myojin Volcanic Explosions by Underwater Sound," *Geological Society of America Bulletin* 65, no. 10 (October 1954): 941–56.

62   Rocks the size of bungalows, buildings, and city blocks: J. G. Moore, D. A. Clague, R. T. Holcomb, P. W. Lipman, W. R. Normark, and M. E. Torresan, "Prodigious Submarine Landslides on the Hawaiian Ridge," *Journal of Geophysical Research* 94, no. B12 (December 1989): 17465–84.

62   Some of the slides would have caused mega-tsunamis: Peter W. Lipman, William R. Normark, James G. Moore, John B. Wilson, and Christina E. Gutmacher, "The Giant Submarine Alika Debris Slide, Mauna Loa, Hawaii," *Journal of Geophysical Research* 93, no. B5 (May 1988): 4279–99. See also Robert Irion, "The Case for Monstrous Hawaiian Waves," *Science*, December 9, 2003, science.org. https://www.science.org/content/article/case-monstrous-hawaiian-waves (accessed February 11, 2023).

63   No one has died: There is an asterisk to this assertion. On July 1, 2019, a Russian nuclear-powered military submarine called the *Losharik* suffered an undersea fire and the loss of fourteen crew. In general, I am not including military submarines in my account of manned deep-sea exploration, for the simple reason that they don't go very deep. But while the *Losharik* resembled a small military submarine on the outside, under its steel exterior it was a different beast. Its interior consisted of seven titanium pressure hulls strung together like a pearl necklace, giving it the ability to dive to twenty thousand feet—more than ten times deeper than America's manned military submarines are thought to operate. When the *Losharik* was not making deep dives, it traveled by latching onto the underside of a larger submarine. Unsurprisingly, the Russian government hasn't divulged much about the *Losharik*, but Western intelligence agencies believe that its purpose was, at least in part, to survey (and potentially sabotage) deep-sea telecommunication cables. For more on the incident, see James Glantz and Thomas Nilsen, "A Deep-Diving Sub. A Deadly Fire. And Russia's Secret Undersea Agenda," *New York Times*, April 20, 2020.

64   The sub's captain: Stephen Chen, "Underwater Tornadoes Found Near China's Nuclear Submarine Base by Paracels That Could Sink U-Boats in Treacherous Abyss," *South China Morning Post*, December 10, 2015.

65   By the time Chapman and Mallinson: The story of the *Pisces III* rescue is told in harrowing detail by pilot Roger Chapman in his book, *No Time on Our Side* (New York: W. W. Norton & Company, 1975).

66   ten tons (or more) of ballast weights: The *Trieste*'s ballast system consisted of hoppers of steel pellets that were fitted within the float; the pellets could be released through an opening in the float's underside. This ballast was held in place by an electromagnet: to drop pellets, the pilot would flip a switch and cut the magnet's power. The amount of ballast weight the *Trieste* carried depended on the depth of a dive. For the Challenger Deep, the bathyscaphe was laden with seventeen tons of steel pellets.

67   "Could we sink and disappear": Jacques Piccard and Robert S. Dietz, *Seven*

Miles Down: The Story of the Bathyscaph Trieste (New York: G. P. Putnam's Sons, 1961), 171.

67   "Was it sheer madness": Ibid., 162.

67   As one submersible expert put it: James Hamilton-Paterson, *Three Miles Down: A Hunt for Sunken Treasure* (London: Jonathan Cape, 1998), 167.

68   "We were venturing beyond": Piccard and Dietz, 170.

68   "a waste of snuff-colored ooze": Ibid., 172,

69   The scientists in the *Pisces:* Hubert Staudigel, Stanley R. Hart, Adele Pile, Bradley E. Bailey, Edward T. Baker, Sandra Brooke et al., "Vailuluʻu Seamount, Samoa: Life and Death on an Active Submarine Volcano," *Proceedings of the National Academy of Sciences* 103, no. 17 (April 2006): 6448–53.

70   the aftermath of forty-two atomic tests: Gary M. McMurtry, Randi C. Schneider, Patrick L. Colin, Robert W. Buddemeier, and Thomas H. Suchanek, "Redistribution of Fallout Radionuclides in Enewetak Atoll Lagoon Sediments by Callianassid Bioturbation," *Nature* 313 (February 1985): 674–77.

71   Their plan was to slip into Pearl Harbor: James Delgado, Terry Kerby, Hans K. Van Tilburg, Steven Price, Ole Varmer, Maximilian D. Cremer, and Russell Matthews, *The Lost Submarines of Pearl Harbor: The Rediscovery and Archaeology of Japan's Top-Secret Midget Submarines of World War II* (College Station, TX: Texas A&M University Press, 2016).

71   Nine of the ten pilots died: The lone surviving Japanese midget submarine pilot wrote an account of his experiences: Ensign Kazuo Sakamaki, *I Attacked Pearl Harbor*, trans. Toru Matsumoto (Honolulu: Rollston Press, 2017).

72   Globally, millions of tons: Jacek Beldowski, Matthias Brenner, and Kari K. Lehtonen, "Contaminated by War: A Brief History of Sea-Dumping of Munitions," *Marine Environmental Research* 162 (December 2020): 105189.

72   there were chemical weapons down there: For more information about the weapons that linger on Oahu's seafloor, see the Hawaii Undersea Military Munitions Assessment Project (HUMMA) website: http://www.hummaproject.com/.

72   No one knew if the bombs: David M. Bearden, *U.S. Disposal of Chemical Weapons in the Ocean: Background and Issues for Congress* (Washington, DC: Library of Congress Congressional Research Service, January 2007).

73   "The purpose of today's dive": "Navy's Bathyscaph Dives 7 Miles in Pacific Trench," *New York Times*, January 23, 1960.

CHAPTER 3: POSEIDON'S LAIR

76   "There are environmental worlds": Loren Eiseley, *The Star Thrower* (San Diego, CA, and New York: Harcourt Brace & Company, 1978), 39.

76   the world's most advanced deep-sea observatory: The main information hub for the Regional Cabled Array is https://interactiveoceans.washington.edu/about/regional-cabled-array/.

77   It's like a puzzle: Jeffrey A. Karson, Deborah S. Kelley, Daniel J. Fornari, Michael R. Perfit, and Timothy M. Shank, *Discovering the Deep: A Photographic Atlas of the Seafloor and Ocean Crust* (Cambridge, UK: Cambridge University Press, 2015), 36–37.

78   If an earthquake exceeds magnitude 8.5: Andreas M. Schafer and Friede-
mann Wenzel, "Global Megathrust Earthquake Hazard—Maximum Magnitude Assessment Using Multi-Variate Machine Learning," *Frontiers in
Earth Science* 7 (June 2019), https://www.frontiersin.org/articles/10.3389
/feart.2019.00136/full.

79   What they didn't account for: Bruce C. Heezen and Charles D. Hollister, *The
Face of the Deep* (New York: Oxford University Press, 1971), 552–57.

79   the longest-lived parts of the seafloor: Paul Voosen, "Mediterranean Sea May
Harbor Piece of Oldest Ocean Crust," *Science*, August 15, 2016, science.org.
https://www.science.org/content/article/mediterranean-sea-may-harbor
-piece-oldest-ocean-crust (accessed January 31, 2023).

79   How could *continents* move: Hezeen and Hollister, 540–41.

79   At its spreading center, the Axial Volcano: Robin George Andrews, "A Deep-
Sea Magma Monster Gets a Body Scan," *New York Times*, December 3, 2019.

80   The Juan de Fuca's real hazard: Robert S. Yeats, "Cascadia Subduction Zone,"
chap. 4 in *Living with Earthquakes in the Pacific Northwest* (Corvallis, OR:
Oregon State University Press, n.d.), https://open.oregonstate.education
/earthquakes/chapter/cascadia-subduction-zone/.

80   bulging upward by a few millimeters: Roy D. Hyndman and Garry C. Rogers,
"Great Earthquakes on Canada's West Coast: A Review," *Canadian Journal of
Earth Sciences* 47, no. 5 (June 2010): 801–20.

80   the Cascadia barely whispered: Yeats, chap. 4 in *Living with Earthquakes in the
Pacific Northwest.*

80   Eighteen of them were caused: Chris Goldfinger, C. Hans Nelson, and Joel E.
Johnson, "Holocene Earthquake Records from the Cascadia Subduction Zone
and Northern San Andreas Fault Based on Precise Dating of Offshore Turbi-
dites," *Annual Review of Earth and Planetary Sciences* 31 (May 2003): 555–77.

81   Japan, the world's most tsunami-aware: Brian F. Atwater, Musumi-Rokkaku
Satoko, Satake Kenji, Tsuji Yoshinobu, Ueda Kazue, and David K. Yamagu-
chi, *The Orphan Tsunami of 1700: Japanese Clues to a Parent Earthquake in North
America* (Seattle, WA: University of Washington Press, 2015), 54.

81   geologists finally correlated: Kenji Satake, Kelin Wang, and Brian D. Atwater, "Fault Slip and Seismic Moment of the 1700 Cascadia Earthquake Inferred
from Japanese Tsunami Descriptions," *Journal of Geophysical Research* 108,
no. B11 (2003): 7.1–7.16.

81   The survivors chronicled the disaster: Ruth S. Ludwin, Robert Dennis, Deborah Carber, Alan D. McMillan, Robert Losey, John Clague, Chris Jonientz-
Trisler, Janine Bowechop, Jacilee Wray, and Karen James, "Dating the 1700
Cascadia Earthquake: Great Coastal Earthquakes in Native Stories," *Seismo-
logical Research Letters* 76, no. 2 (March–April 2005): 140–47.

81   the energy of twenty-three thousand atomic bombs: Associated Press,
"Numbers That Tell the Story of 2004 Tsunami Disaster," apnews.com,
https://apnews.com/article/4bf54ae8134a47718e8314e883b8074c (accessed
January 10, 2023).

84   It's an interactive marine laboratory: Karson et al., *Discovering the Deep*, 29–32.

84   Slope Base is striking: For an overview of the Slope Base site, and a graphic
representation of the seafloor instruments deployed there, see: https://

interactiveoceans.washington.edu/research-sites/oregon-slope-base/ (accessed January 10, 2023).

85   a shockingly ugly fish: Weirdfish was identified as *Genioliparis ferox*, a rare species of snailfish.

90   we arrived at Axial Volcano: Karson et al., *Discovering the Deep*, 177–86.

91   another astonishing type of vent called a snowblower: Julie L. Meyer, Nancy H. Akerman, Giora Proskurowski, and Julie A. Huber, "Microbial Characterization of Post-Eruption 'Snowblower' Vents at Axial Seamount, Juan de Fuca Ridge," *Frontiers in Microbiology* 4, no. 153 (June 2013), https://www.frontiersin.org/articles/10.3389/fmicb.2013.00153/full.

91   Known as the deep biosphere: Deep Carbon Observatory, *A Decade of Discovery* (Washington, DC: n.p., 2019), 40–43, https://fsmap-images.s3.amazonaws.com/dco-pr/A+Decade+of+Discovery-DCO+Decadal+Report.pdf.

91   Unless the ocean: Giorgia Guglielmi, "This Is What It Would Take to Kill All Life on Earth," *Science*, July 14, 2017, science.org/content/article/what-it-would-take-kill-all-life-earth (accessed February 15, 2023).

91   Microbes that can reside: Anais Cario, Gina C. Oliver, and Karyn L. Rogers, "Exploring the Deep Marine Biosphere: Challenges, Innovations, and Opportunities," *Frontiers in Earth Science* 7 (September 2019), https://www.frontiersin.org/articles/10.3389/feart.2019.00225/full.

91   How life emerged: Kevin Peter Hand, *Alien Oceans: The Search for Life in the Depths of Space* (Princeton, NJ: Princeton University Press, 2020), 139–50.

92   a zoo of strange animals: Evan Lubofsky, "The Discovery of Hydrothermal Vents," *Oceanus*, June 11, 2018, whoi.edu, https://www.whoi.edu/oceanus/feature/the-discovery-of-hydrothermal-vents/ (accessed January 10, 2023).

92   Black smokers were discovered: Woods Hole Oceanographic Institution, "1979—The Smoking Gun," whoi.edu, https://www.whoi.edu/feature/history-hydrothermal-vents/discovery/1979-2.html (accessed January 10, 2023).

93   "They seemed connected to Hell itself": Robert D. Ballard, *The Eternal Darkness: A Personal History of Deep-Sea Exploration* (Princeton, NJ: Princeton University Press, 2000), 187–202.

95   it has gulped down: Woods Hole Oceanographic Institution, "Ocean Warming," whoi.edu, https://www.whoi.edu/know-your-ocean/ocean-topic/climate-weather/ocean-warming/ (accessed February 5, 2023).

95   The state of ecological balance: Andrew K. Sweetman, Andrew R. Thurber, Craig R. Smith, Lisa A. Levin, Camilo Mora, Chih-Lin Wei, Andrew J. Gooday et al., "Major Impacts of Climate Change on Deep-Sea Benthic Ecosystems," *Elementa* 5, no. 4 (February 2017), https://online.ucpress.edu/elementa/article/doi/10.1525/elementa.203/112418/Major-impacts-of-climate-change-on-deep-sea.

96   NASA had recently come on board: For more detail about the INVADER project, see https://invader-mission.org/.

98   the fracturing had uplifted: Deborah S. Kelley, "From the Mantle to Microbes: The Lost City Hydrothermal Vent Field," *Oceanography* 18, no. 3 (September 2005): 32–45.

100 They named the magnificent structure: Deborah S. Kelley, Jeffrey A. Kar-

son, Gretchen L. Früh-Green, Dana R. Yoerger, Timothy O. Shank, David A. Butterfield, John M. Hayes et al., "A Serpentinite-Hosted Ecosystem: The Lost City Hydrothermal Field," *Science* 307, no. 5714 (March 2005):1428–34.

100   filmy layers of archaea: Susan Q. Lang and William J. Brazelton, "Habitability of the Marine Serpentinite Subsurface: A Case Study of the Lost City Hydrothermal Field," *Philosophical Transactions of the Royal Society A* 378, no. 2165 (February 2020), https://royalsocietypublishing.org/doi/10.1098/rsta.2018.0429.

101   central to the emergence: Carl Zimmer, "Under the Sea, a Missing Link in the Evolution of Complex Cells," *New York Times*, May 6, 2015.

101   our most distant relative may be a Lokiarchaea: Anja Spang, Jimmy H. Saw, Steffen L. Jørgensen, Katarzyna Zaremba-Niedzwiedzka, Joran Marijn et al., "Complex Archaea That Bridge the Gap Between Prokaryotes and Eukaryotes," *Nature* 521 (May 2015), 173–79.

102   Lost City's sturdy vents: Kristin A. Ludwig, Deborah S. Kelley, David A. Butterfield, Bruce K. Nelson, and Gretchen Früh-Green, "Formation and Evolution of Carbonate Chimneys at the Lost City Hydrothermal Field," *Geochimica et Cosmochimica Acta* 70, no. 14 (July 2006): 3625–45.

102   This type of vent system: Giora Proskurowski, Marvin D. Lilley, Jeffery S. Seewald, Gretchen L. Früh-Green, Eric J. Olson, John E. Lupton, Sean P. Sylva, and Deborah S. Kelley, "Abiogenic Hydrocarbon Production at Lost City Hydrothermal Field," *Science* 316, no. 5863 (February 2008): 604–7.

102   UNESCO designated Lost City: David Freestone, Dan Laffoley, Fanny Douvere, and Tim Badman, "World Heritage in the High Seas: An Idea Whose Time Has Come," *World Heritage Reports* 44 (August 2016): 32–33, https://unesdoc.unesco.org/ark:/48223/pf0000245467.

104   Ever since tubeworms: Tjorven Hinzke, Manuel Kleiner, Corinna Breusing, Horst Felbeck, Robert Hasler, Stefan M. Sievert, Rabea Schlüter et al., "Host-Microbe Interactions in the Chemosynthetic *Riftia pachyptila* Symbiosis," *American Society for Microbiology* 10, no. 6 (December 2019), https://journals.asm.org/doi/10.1128/mBio.02243-19.

CHAPTER 4: WHAT HAPPENS IN HADES . . .

106   "I could tell myself": Jacques Piccard and Robert S. Dietz, *Seven Miles Down: The Story of the Bathyscaph Trieste* (New York: G. P. Putnam's Sons, 1961), 160.

109   "I'd never get into that thing": Don Walsh, "Going the Last Seven Miles: The Bathyscaph *Trieste* Story." Personal essay.

109   others might have been chosen: Walsh's command of the *Trieste* struck me as fated when I heard the story of how it unfolded. After Jacques Piccard claimed his seat for the Challenger Deep dive, and it became clear that only one U.S. Navy representative would be going to the bottom of the Mariana Trench, Walsh had tried to give up his own seat to his mentor, the *Trieste* program's chief scientist, Andy Rechnitzer—only to be countermanded by Admiral Arleigh Burke, the U.S. chief of naval operations. "Walsh and Piccard will make the dive" was the edict from Burke. "There was no going back to it," Walsh recalled. "Andy was a PhD from Scripps, naval reserve officer, com-

mander. He was teaching me about the world of oceanography. He deserved [to make the dive], he really did. He handled the situation with great charm and dignity, but it had to have bothered him throughout his whole life."

112   only four robots: Alan Jamieson, *The Hadal Zone* (Cambridge, UK: Cambridge University Press, 2015), 65–67.

112   Kaiko, a forty-million-dollar Japanese robot: William Broad, "Japan Plans to Conquer the Sea's Depths," *New York Times*, October 18, 1994.

112   In 2014, it imploded: Daniel Cressey, "Ocean-Diving Robot Will Not Be Replaced," *Nature*, December 10, 2015.

113   it had plainly suffered: Cameron donated his sub to Woods Hole Oceanographic Institution, but its post-dive fate was sealed when a truck that was transporting it caught fire: Madeleine List, "WHOI Sues Over Deep-Sea Submarine Fire," *Cape Cod Times*, January 15, 2018.

117   There are approximately twenty-seven: Jamieson, *The Hadal Zone*, 25. The number of hadal trenches, trench faults, and troughs is an evolving one, as high-resolution mapping reveals more detail about seafloor features. A hadal trench is defined as a single elongated area deeper than six thousand meters generally (approximately twenty thousand feet), formed by tectonic subduction or faulting. Hadal troughs are depressions that are not located at converging plate boundaries, but occur as deeper basins within an abyssal plain. In his 2015 book, Jamieson enumerates twenty-seven hadal trenches (twenty-three of which are located in the Pacific), six hadal trench faults, and thirteen hadal troughs.

118   At the Tonga Trench's north end: Diana Lutz, "Release of Water Shakes Pacific Plate at Depth," *The Source*, Washington University in St. Louis, January 11, 2017, https://source.wustl.edu/2017/01/release-water-shakes-pacific-plate-depth/ (accessed January 11, 2023).

118   It's a buffet of geological havoc: Michael Bevis, F. W. Taylor, B. E. Schutz, Jacques Recy, B. L. Isacks, Saimone Helu, Rajendra Singh et al., "Geodetic Observations of Very Rapid Convergence and Back-Arc Extension at the Tonga Arc," *Nature* 374 (March 1995): 249–51.

118   a majority of the world's deepest quakes: Simon Richards, Robert Holm, and Grace Barber, "When Slabs Collide: A Tectonic Assessment of Deep Earthquakes in the Tonga-Vanuatu Region," *Geological Society of America* 39, no. 8 (August 2011): 787–90.

118   all three earthquakes shook simultaneously: Thorne Lay, Charles J. Ammon, Hiroo Kanamori, Luis Rivera, Keith D. Koper, and Alexander R. Hutko, "The 2009 Samoa-Tonga Great Earthquake Triggered Doublet," *Nature* 466 (August 2010): 964–68.

118   two-hundred-and-ninety-foot tsunami: "Wave Created by Tonga Volcano Eruption Reached 90 Meters," *Eco*, Autumn 2022, 12.

118   this same volcano would be historical in its fury: Alexandra Witze, "Why the Tongan Eruption Will Go Down in the History of Volcanology," *Nature* 602 (February 2022): 376–78.

118   another Tongan island called Lateiki: Robin George Andrews, "This Volcano Destroyed an Island, Then Created a New One," *New York Times*, November 14, 2019.

125   to measure Hades: Heather A. Stewart and Alan J. Jamieson, "The Five Deeps:

The Location and Depth of the Deepest Place in Each of the World's Oceans," *Earth-Science Reviews* 197 (October 2019), https://doi.org/10.1016/j.earscirev .2019.102896.

125   a definitive book on the hadal zone: Jamieson, *The Hadal Zone.*

126   When he published a paper: A. J. Jamieson, D. M. Bailey, H.-J. Wagner, P. M. Bagley, and I. G. Priede, "Behavioral Responses to Structures on the Seafloor by the Deep-Sea Fish *Coryphaenoides armatus:* Implications for the Use of Baited Landers," *Deep Sea Research Part 1: Oceanographic Research Papers* 53 (April 2006): 1157–66.

127   It was the first confirmed record: A. J. Jamieson, T. Fujii, M. Solan, A. K. Matsumoto, P. M. Bagley, and I. G. Priede, "Liparid and Macourid Fishes of the Hadal Zone: In Situ Observations of Activity and Feeding Behaviour," *Proceedings of the Royal Society B* 276 (December 2008): 1037–45.

127   Who could find the deepest: Thomas D. Linley, Mackenzie E. Gerringer, Paul H. Yancey, Jeffrey C. Drazen, Chloe L. Weinstock, and Alan J. Jamieson, "Fishes of the Hadal Zone Including New Species, *in Situ* Observations, and Depth Records of Liparidae," *Deep Sea Research Part 1: Oceanographic Research Papers* 114 (August 2016): 99–110.

129   The Java Trench had turned out: Alan J. Jamieson, Heather A. Stewart, Johanna N. Weston, Patrick Lahey, and Victor L. Vescovo, "Hadal Biodiversity, Habitats, and Potential Chemosynthesis in the Java Trench, Eastern Indian Ocean," *Frontiers in Marine Science* 9, no. 856992 (March 2022), https:// doi:org/10.3389/fmars.2022.856992.

130   unlimited dives to full ocean depth: The *Limiting Factor*'s full-ocean-depth certification—signed by DNV-GL inspector and marine engineer Jonathan Struwe, who collaborated extensively with Triton on the sub's design and build—was a major milestone in deep-sea exploration: it was the first time in history that a submersible had received this designation. "Certifying is hard and expensive," Lahey told me. "When you introduce the DNV-GL requirement, it means there are no shortcuts, you know? Everything you do has to be tested. There was an internationally recognized set of standards that we had to comply with. So to me, that's the most significant thing, because ultimately it separates this sub from anything that's preceded it."

CHAPTER 5: . . . STAYS IN HADES

132   "Let everything happen to you": Rainer Maria Rilke, *Rilke's Book of Hours: Love Poems to God,* trans. Anita Barrows and Joanna Macy (New York: Riverhead Books, 1996), 119.

134   "It is a chance to find harmony": Joe MacInnis, *Underwater Man* (New York: Dodd, Mead & Company, 1974), 13.

135   Sublimnos had captivated: For more on Sublimnos and its impact, see: Caitlin Stall-Paquet, "This Undersea Explorer from Toronto Helped Inspire Blockbusters like *Titanic* and *Avatar*," torontolife.com, https://torontolife.com/city /im-an-undersea-explorer-from-toronto-my-work-inspired-blockbusters -like-titanic-and-avatar/ (accessed January 11, 2023).

137   Until evolution changes that equation: Paul H. Yancey, Mackenzie E. Ger-

ringer, Jeffrey C. Drazen, Ashley A. Rowden, and Alan Jamieson, "Marine Fish May Be Biochemically Constrained from Inhabiting the Deepest Ocean Depths," *Proceedings of the National Academy of Sciences* 111, no. 12 (March 2014): 4461–65.

138   A hadal snailfish's life span: M. E. Gerringer, "On the Success of the Hadal Snailfishes," *Integrative Organismal Biology* 1, no. 1 (March 2019): 1–18, https://doi.org/10.1093/iob/obz004.

138   "A beautifully complicated trench": Alan J. Jamieson, Heather A. Stewart, Johanna N. Weston, and Cassandra Bongiovanni, "Hadal Fauna of the South Sandwich Trench, Southern Ocean: Baited Camera Survey from the Five Deeps Expedition," *Deep Sea Research Part II: Topical Studies in Oceanography* 194 (December 2021), https://doi.org/10.1016/j.dsr2.2021.104987.

140   The previous depth record for an octopus: Alan J. Jamieson and Michael Vecchione, "First in Situ Observation of Cephalopoda at Hadal Depths (Octopoda: Opisthoteuthidae: Grimpoteuthis sp.)," *Marine Biology* 167, no. 82 (May 2020), https://doi.org/10.1007/s00227-020-03701-1.

140   When it arrived at the lander: For a video clip of the stalked ascidian, see "Newcastle University Scientist Discovers New Species in Java Trench," YouTube video, 0:43, https://www.youtube.com/watch?v=OXSwk_ikms8 (accessed January 11, 2023).

141   Underneath its smooth skin: To view images, diagrams, and technical specifications of the *Limiting Factor,* as well as a video about the forging of its titanium pressure hull, see https://tritonsubs.com/subs/t36000-2/ (accessed January 11, 2023).

157   "The surface of things": Terence McKenna, *Nature Is Alive and Talking to Us. This Is Not a Metaphor* (n.p., 2020). Pages not numbered.

CHAPTER 6: "THIS IS THE MOTHER OF ALL SHIPWRECKS"

160   "This was also how he learned": Gabriel García Márquez, *Love in the Time of Cholera,* trans. Edith Grossman (New York: Alfred A. Knopf, 1988), 64.

160   four types of sonar: The REMUS 6000 AUV, developed at the Woods Hole Oceanographic Institution, has pencil-beam sonar to avoid collisions with undersea obstacles; dual-frequency side-scan sonar that creates a two-dimensional image of the seafloor (and any objects resting on it); multibeam profiling sonar that creates a three-dimensional model of the seafloor; and sub-bottom profiling sonar that emits a powerful sound beam to reveal objects buried beneath seafloor sediment. To view an animation of REMUS 6000 in action, see https://vimeo.com/325064291.

162   to fund the ongoing War of the Spanish Succession: Carla Rahn Phillips, *The Treasure of the San José: Death at Sea in the War of the Spanish Succession* (Baltimore, MD: The Johns Hopkins University Press, 2007), 2.

163   Fourteen crewmen survived: Ibid., 170.

164   some three million ships linger on the seabed: The United Nations Educational, Scientific and Cultural Organization, *The UNESCO Convention on the Protection of the Underwater Cultural Heritage* (n.d.), 4, https://unesdoc.unesco.org/ark:/48223/pf0000152883 (accessed January 12, 2023).

165   When they successfully: George F. Bass, *Beneath the Seven Seas: Adventures with the Institute of Nautical Archaeology* (London: Thames & Hudson, 2005), 48–54.

165   "It was a real uphill battle": George F. Bass, interview for University of Pennsylvania Museum of Archaeology and Anthropology, September 5, 2014, YouTube video, 7:09, https://www.youtube.com/watch?v=RSlOKzsq_K0.

165   "We were always wet and cold": George F. Bass, *Archaeology Beneath the Sea: My Fifty Years of Diving on Ancient Shipwrecks* (Istanbul, Turkey: Boyut, 2011). Updated edition of 1975 printed edition from Walker & Company, ch. 5, Kindle.

165   Bass viewed the seafloor as the world's greatest museum: Ibid., ch. 29.

166   A seventh-century CE Byzantine ship: Bass, *Beneath the Seven Seas*, 92–97.

166   At another wreck site: Ibid., 106–17.

166   Now known as the Uluburun wreck: Ibid., 34–47.

167   "I was determined to use a submarine": Bass, *Archaeology Beneath the Seas*, ch. 7.

167   "All around the *Titanic*": Robert D. Ballard with Will Hively, *The Eternal Darkness: A Personal History of Deep-Sea Exploration* (Princeton, NJ: Princeton University Press, 2000), 253.

167   "unmanned tethered eyeballs": Ibid., 228.

168   there were historical shipwrecks everywhere: Fredrik Søreide, *Ships from the Depths: Deepwater Archaeology* (College Station, TX: Texas A&M University Press, 2011), 70–73.

168   crenellated by scars from the Storegga Slide: Petter Bryn, Kjell Berg, Carl F. Forsberg, Anders Solheim, and Tore J. Kvalstad, "Explaining the Storegga Slide," *Marine and Petroleum Geology* 22, nos. 1–2 (January 2005): 11–19.

168   walloped northern Europe: Astrid J. Nyland, James Walker, and Graeme Warren, "Evidence of the Storegga Tsunami 8200 BP? An Archaeological Review of Impact After a Large-Scale Marine Event in Mesolithic Northern Europe," *Frontiers in Earth Science* 9, no. 767460 (December 2021), https://doi.org/10.3389/feart.2021.767460.

168   the ROV built specifically: Søreide, *Ships from the Depths*, 74.

169   the ten-million-dollar bill: Ibid., 84.

169   The archaeologists identified it: Ibid., 83.

170   in the Java Sea alone, three British warships: Travis M. Andrews, "Several World War II Warships Mysteriously Disappear from Watery Grave at the Site of Battle of Java Sea," *Washington Post*, November 18, 2016. See also Oliver Holmes, Monica Ulmanu, and Simon Roberts, "The World's Biggest Grave Robbery: Asia's Disappearing WWII Shipwrecks," *The Guardian*, November 2, 2017.

170   And this metal piracy: Martijn R. Manders, "The Issues with Large Metal Wrecks from the 20th Century," in *Heritage Underwater at Risk: Threats, Challenges, Solutions*, eds. Albert Hafner, Hakan Oniz, Lucy Semaan, and Christopher J. Underwood (Paris, France: International Council on Monuments and Sites, 2020), 73–76.

171   "provided AUV mission planning": Garry Kozak, "The Discovery of the Capitana San José," *Ocean News and Technology* (August 2019): 24–25.

172   The galleon was the ship: Timothy R. Watson, *The Spanish Treasure Fleets* (Sarasota, FL: Pineapple Press, 1994), 57–59.

175   An American treasure-hunting outfit: Sea Search Armada's grievances against the Colombian government—including the company's demand for seventeen billion dollars in "compensatory damages"—are detailed in a colorful memorandum opinion from the U.S. District Court for the District of Columbia, issued on October 24, 2011, https://law.justia.com/cases/federal /district-courts/district-of-columbia/dcdce/1:2010cv02083/145469/19/ (accessed January 13, 2023).

176   The treasure hunters claimed: The tale of Sea Search Armada (and its many lawsuits) is a complicated one, and I've chosen not to include it in any detail because, while its contractual dispute with the Colombian government has been taken up by the courts, its claim to have found the *San José* is unsupported. And even if Sea Search Armada found *something* on the seafloor—perhaps even pieces of another ship—by the company's own admission that debris was located in waters significantly shallower and bathymetrically different than the deep resting place of the *San José*.

176   When asked about the treasure's provenance: John Colapinto, "Secrets of the Deep," *New Yorker*, April 7, 2008. Also: Tom Brown, "Spain Rejects U.S. Treasure-Hunters' Shipwreck Claim," *Reuters*, April 18, 2008, reuters.com, https://www.reuters.com/article/us-spain-treasure-spain-rejects-u-s -treasure-hunters-shipwreck-claim-idUSN1832990720080418 (accessed January 13, 2023).

176   "This is not money": Álvaro de Cozár, "Odyssey Treasure Heads Back to Spain," *El País*, February 24, 2012, elpais.com, https://english.elpais .com/elpais/2012/02/24/inenglish/1330107486_223846.html (accessed January 13, 2023).

177   In a recent paper about the mechanism: Tony Freeth, David Higgon, Aris Dacanalis, Lindsay MacDonald, Myrto Georgakopoulou, and Adam Wojcik, "A Model of the Cosmos in the Ancient Greek Antikythera Mechanism," *Nature Scientific Reports* 11, no. 5821 (March 2021), https://doi.org/10.1038 /s41598-021-84310-w.

179   "Not a single splinter": Richard Emblin, "Galleon San José's Treasure Will Not Finance Salvage, Claims VP Ramirez," *City Paper*, October 10, 2019, thecitypaperbogota.com, https://thecitypaperbogota.com/news/galleon-san -joses-treasure-will-not-finance-salvage-claims-vp-ramirez/ (accessed January 13, 2023).

180   "In itself gold is of no greater value": Bass, *Archaeology Beneath the Sea*, ch. 21, Kindle.

CHAPTER 7: THE END OF THE BEGINNING

182   "The ocean is a place": Rachel Carson, "Undersea," in *Lost Woods: The Discovered Writing of Rachel Carson*, ed. Linda Lear (Boston: Beacon Press, 1998), 6.

184   "This is the Tank Room": For a video tour through the Tank Room, see "What Lies Beneath," Natural History Museum, broadcast on June 13, 2018, online video, 31:20, https://www.nhm.ac.uk/discover/what-lies-beneath.html.

185   These ten appendages: Roger Hanlon, Mike Vecchione, and Louise Allcock, *Octopus, Squid, and Cuttlefish: A Visual, Scientific Guide to the Ocean's Most Advanced Invertebrates* (Chicago, IL: The University of Chicago Press, 2018), 82–83.

186   George Drevar, captain of a schooner called the *Pauline:* A. C. Oudemans, *The Great Sea-Serpent: An Historical and Critical Treatise* (London: Luzac & Co., 1892), 334–35.

186   The skipper, Olivier de Kersauson: Adam Sage, "French Timbers Are Shivered by Sea Monster," *Sunday Times*, January 16, 2003.

186   "No one ever understood the wonder": Alan Watts, *Become What You Are*, ed. Mark Watts (Boston, MA, and London: Shambala, 2003), 61.

187   Before 2012, when a luminous lure: For years, scientists sought to find and film and observe a giant squid in the wild; the woman who got it done was marine biologist Edie Widder. Widder details her adventures among giant squid in her book, *Below the Edge of Darkness: A Memoir of Exploring Light and Life in the Deep Sea* (New York: Random House, 2021).

187   While preserving Archie: For a more detailed description of Archie's journey from the Falkland Islands abyss to a glass tank in London, see Jonathan Ablett, "The Giant Squid, *Architeuthis dux* Steenstrup, 1857 (Mollusca: Cephalopoda): The Making of an Iconic Specimen," *NatSCA News* 23 (January 2012): 16–20.

187   A giant squid's body is infused: Steve O'Shea, "Architeuthis Buoyancy and Feeding," tonmo.com, https://tonmo.com/articles/architeuthis-buoyancy-and-feeding.24/ (accessed January 15, 2023).

187   Both species are cryptic: Douglas Long, "Super Colossal," deepseanews.com, https://deepseanews.com/2015/04/super-colossal/ (accessed February 20, 2023).

187   "In no department of zoological science": H. G. Wells, *The Sea Raiders*, in *Best Science Fiction Stories of H. G. Wells* (New York: Dover Publications, 1966), 280.

188   "Their forms are horrible": Olaus Magnus, quoted in "Historical Account of a Giant Squid," teara.govt.nz, https://teara.govt.nz/en/document/7922/historical-account-of-a-giant-squid (accessed January 15, 2023).

189   The expedition's sonar efforts: Cassandra Bongiovanni, Heather A. Stewart, and Alan J. Jamieson, "High-Resolution Multibeam Sonar Bathymetry of the Deepest Place in Each Ocean," *Geoscience Data Journal* 9, no. 1 (June 2022): 108–23.

192   It's more like a bowl: To view the bathymetry of the Molloy Hole and its surrounding terrain, see "Fly Through: Molloy Hole, Caladan Oceanic 2019 Field Season," YouTube video, 1:24, https://www.youtube.com/watch?v=PSW2KkOXp2k.

192   Major ocean currents converge: Tore Hattermann, Pål Erik Isachsen, Wilken-Jon von Appen, Jon Albretsen, and Arild Sundfjord, "Eddy-Driven Recirculation of Atlantic Water in Fram Strait," *Geophysical Research Letters* 43, no. 17 (April 2016): 3406–14.

192   It's the planet's tallest, mightiest waterfall: Denmark Strait, worldatlas.com, https://www.worldatlas.com/straits/denmark-strait.html (accessed January 15, 2023).

194   In an article about personal subs: Catherine Nixey, "Do You Know a Good Submarine-Maker?," *1843 Magazine*, November 18, 2019.

194   James Bond himself: Aston Martin, "Project Neptune: Triton and Aston Martin," press release issued September 28, 2017, https://media.astonmartin.com /project-neptune-triton-and-aston-martin/.

CHAPTER 8: YOU ARE NOW ENTERING THE TWILIGHT ZONE

198   "If one dives and returns": William Beebe, *Half Mile Down* (New York: Harcourt, Brace and Company, 1934), 66–67.

199   "The capacity of humanity": James Cameron, Ray Dalio, Peter de Menocal, and Edith Widder, "Illuminating the Abyss: Inspiration, Exploration, and Discovery in the Ocean Twilight Zone," hosted by Tatiana Schlossberg, video panel discussion, 1:01:20, https://www.whoi.edu/multimedia /illuminating-the-abyss/.

200   "No kid ever dreamed": "James Cameron Responds to Robert Ballard on Deep-Sea Exploration," January 14, 2013, mission-blue.org, https:// mission-blue.org/2013/01/james-cameron-responds-to-robert-ballard-on -deep-sea-exploration/ (accessed January 15, 2023).

203   researchers keep revising: Xabier Irigoien, T.A. Klevjer, A. Røstad, U. Martinez, G. Boyra, J. L. Acuña, A. Bode et al., "Large Mesopelagic Fishes Biomass and Trophic Efficiency in the Open Ocean," *Nature Communications* 5, no. 3271 (February 2014), https://doi.org/10.1038/ncomms4271.

203   bristlemouths are the most abundant vertebrates: "Bristlemouth Dominance: How Do We Know?," Woods Hole Oceanographic Institution, Ocean Twilight Zone, twilightzone/whoi.edu, https://twilightzone.whoi.edu/bristle mouth-dominance-how-do-we-know/ (accessed January 15, 2023). See also William J. Broad, "An Ocean Mystery in the Trillions," *New York Times*, June 29, 2015.

204   A dragonfish's teeth: Wudan Yan, "Meet the Deep-Sea Dragonfish. Its Transparent Teeth Are Stronger Than a Piranha's," *New York Times*, June 5, 2019.

204   But in the ocular weirdness Olympics: Monterey Bay Aquarium Research Institute, "Researchers Solve Mystery of Deep-Sea Fish with Tubular Eyes and Transparent Head," news release, February 23, 2009, https://www.mbari .org/news/researchers-solve-mystery-of-deep-sea-fish-with-tubular-eyes -and-transparent-head/.

205   But even Charles Wyville Thomson: Sir Charles Wyville Thomson, *The Voyage of the Challenger: The Atlantic*, vol. 2 (London: Macmillan and Co., 1877), 352. Wyville wrote: "There is every reason to believe that the fauna of deep water is confined principally to two belts, one at and near the surface, and the other on and near the bottom; leaving an intermediate zone in which the larger animal forms, vertebrate and invertebrate, are nearly or entirely absent."

205   scientists who've dedicated their careers: Adrian Martin, Philip Boyd, Ken Buesseler, Ivona Cetinic, Hervé Claustre, Sari Giering, Stephanie Henson et al., "The Oceans' Twilight Zone Must Be Studied Now, Before It Is Too Late," *Nature* 580 (April 2020): 26–28.

205   This nocturnal commute: Veronique LaCapra, "Mission to the Twilight Zone,"

Oceanus: Woods Hole Oceanographic Institution's Journal of Our Ocean Planet 54, no. 1 (Spring 2019): 10–23.

205   some species pursue: Bruce H. Robison, Rob E. Sherlock, Kim R. Reisenbichler, and Paul R. McGill, "Running the Gauntlet: Assessing Threats to Vertical Migrators," *Frontiers in Marine Science* 7, no. 64 (February 2020), https://doi.org/10.3389/fmars.2020.00064.

205   These tiny beasts sink a mighty: The Ocean Twilight Zone Project, *The Ocean Twilight Zone's Role in Climate Change* (Woods Hole, MA: Woods Hole Oceanographic Institution, 2022), 5–9, https://twilightzone.whoi.edu/wp-content/uploads/2022/02/The-Ocean-Twilight-Zones-Role-in-Climate-Change.pdf.

208   the soul-stirring blue: Beebe, 110.

208   Trying to describe its sensory effect: Ibid., 135.

210   bioluminescence is a central fact of life: Michael Actil, *Luminous Creatures: The History and Science of Light Production in Living Organisms* (Montreal, Quebec, and Kingston, Ontario: McGill–Queen's University Press, 2018), xiv–xvi.

210   In the deep, light is a strategy: Bruce H. Robison, "Deep Pelagic Biology," *Journal of Experimental Marine Biology and Ecology* 300, no. 1–2 (March 2004): 253–72.

210   The *Colobonema* jellyfish: Bruce H. Robison, "Light in the Ocean's Midwaters," *Scientific American*, July 1, 1995.

211   Twilight zone creatures have as many: Steven H. D. Haddock, Mark A. Moline, and James F. Case, "Bioluminescence in the Sea," *Annual Review of Marine Science* 2 (January 2010): 443–93.

212   The box jellyfish: Michael Hopkin, "Box Jellyfish Show a Keen Eye," *Nature*, May 11, 2005, https://doi.org/10.1038/news050509-7.

212   it can reverse its life cycle: Emily Osterloff, "Immortal Jellyfish: The Secret to Cheating Death," nhm.ac.uk, https://www.nhm.ac.uk/discover/immortal-jellyfish-secret-to-cheating-death.html (accessed January 15, 2023).

215   They're *ultra*black: Alexander L. Davis, Kate N. Thomas, Freya E. Goetz, Bruce H. Robison et al., "Ultra-Black Camouflage in Deep-Sea Fishes," *Current Biology* 30, no. 17 (September 2020): 3470–76.

218   "When once it has been seen": Beebe, 175.

218   Show the machine world: Dag Standal and Eduardo Grimaldo, "Institutional Nuts and Bolts for a Mesopelagic Fishery in Norway," *Marine Policy* 119, no. 104043 (May 2020), https://doi.org/10.1016/j.marpol.2020.104043.

219   Without its carbon-pumping services: The Ocean Twilight Zone Project, *Value Beyond View: Illuminating the Human Benefits of the Ocean Twilight Zone* (Woods Hole, MA: Woods Hole Oceanographic Institution, 2022), https://twilightzone.whoi.edu/value-beyond-view/.

      See also Madeline Drexler, "The Ocean Twilight Zone's Crucial Carbon Pump," Woods Hole Oceanographic Institution, whoi.edu, https://www.whoi.edu/news-insights/content/the-ocean-twilight-zones-crucial-carbon-pump/ (accessed January 15, 2023).

219   Norway and Pakistan have already issued: Glen Wright (IDDRI), Kristina M. Gjerde (IUCN), Aria Finkelstein (Massachusetts Institute of Technology),

and Duncan Currie (GlobeLaw), "Fishing in the Twilight Zone: Illuminating Governance Challenges at the Next Fisheries Frontier," IDDRI, November 2020 study, iddri.org, https://www.iddri.org/en/publications-and-events /study/fishing-twilight-zone-illuminating-governance-challenges-next.

CHAPTER 9: SELLING THE ABYSS

224   American geologist Bruce Heezen: Bruce C. Heezen and Charles D. Hollister, *The Face of the Deep* (New York, London, and Toronto: Oxford University Press, 1971), 423.

224   In 1965, another American geologist: John L. Mero, *The Mineral Resources of the Sea* (Amsterdam, London, and New York: Elsevier Scientific Publishing Company, 1965), 1.

224   Mero wanted to suck the: Ibid., 114.

224   "With nuclear explosives": Ibid., 86.

224   Noting that sea squirts: Ibid., 52.

224   Industrialists gasped at Mero's claims: Ibid., 277–79.

224   Mero envisioned: Ibid., 252.

225   "It is a treasure hunter's dream": Tom Kizzia, "Deep Sea Billions," *Washington Post*, August 2, 1981.

225   This effort would result: Michael Lodge, "The International Seabed Authority and Deep Seabed Mining," United Nations, un.org, https://www.un.org/en /chronicle/article/international-seabed-authority-and-deep-seabed-mining (accessed January 17, 2023).

226   Legally, the ISA was mandated: On the ISA's website (isa.org.jm), the answer to the question "What are the ISA's mandate and policies?" states that it must "promote the orderly, safe and responsible management and development of the resources of the deep seabed area for the benefit of mankind as a whole. In doing so, ISA has the duty to adopt appropriate rules, regulations and procedures to ensure the effective protection of the marine environment from harmful effects . . ." https://www.isa.org.jm/index.php/frequently -asked-questions-faqs (accessed January 17, 2023).

226   the corporate players formed consortiums: Yajuan Kang and Shaojun Liu, "The Development History and Latest Progress of Deep-Sea Polymetallic Nodule Mining Technology," *Minerals* 11, no. 10 (October 2021): 1132, https://doi .org/10.3390/min11101132.

226   a Zamboni-like nodule-collecting machine: Ted Brockett, "Ocean Mining: Lessons from the Past," *Ocean News & Technology* (May 2020): 14–17.

227   The nodule-forming process: Xiaohong Wang, Ute Schloßmacher, Matthias Wiens, Heinz C. Schröder, and Werner E. G. Müller, "Biogenic Origin of Polymetallic Nodules from the Clarion-Clipperton Zone in the Eastern Pacific Ocean: Electron Microscopic and EDX Evidence," *Marine Biotechnology* 11, no. 1 (February 2009): 99–108.

227   Mero believed that the nodules: Mero, 278.

228   And people do care about it: Laura Kaikkonen and Ingrid van Putten, "We May Not Know Much About the Deep Sea, But Do We Care About Mining It?," *People and Nature* 3, no. 4 (August 2021): 843–60.

228   But now, the technologies: Takuma Watari, Keisuke Nansai, and Kenichi Naka-jima, "Review of Critical Metal Dynamics to 2050 for 48 Elements," *Resources, Conservation and Recycling* 155 (April 2020), https://doi.org/10.1016/j.resconrec.2019.104669.

229   as one marine scientist put it: Canadian marine biologist and emeritus profes-sor at the University of Victoria Verena Tunnicliffe, quoted in Damian Car-rington, "Is Deep Sea Mining Vital for a Greener Future—Even If It Destroys Ecosystems?," *The Guardian,* June 4, 2017.

229   Despite its legal obligation: International Seabed Authority, "Protection of the Seabed Environment," isa.org.jm, https://isa.org.jm/files/files/documents/eng4.pdf (accessed January 17, 2023).

229   lists heavily to the side: Jack Lo Lau, "Sandor Mulsow: 'The ISA Is Not Fit to Regulate Any Activity in the Oceans,'" *China Dialogue,* chinadialogueocean.net, https://chinadialogueocean.net/en/governance/19905-sandor-mulsow-isa-not-suited-to-regulate-oceans/ (accessed January 17, 2023).

229   an opaque group of forty-one: Todd Woody, "Seabed Mining: The 30 People Who Could Decide the Fate of the Deep Ocean," *Oceans Deeply,* deeply.thenew humanitarian.org/oceans, https://deeply.thenewhumanitarian.org/oceans/articles/2017/09/06/seabed-mining-the-24-people-who-could-decide-the-fate-of-the-deep-ocean (accessed January 17, 2023).

229   "absolutism and dogmatism": Michael Lodge at the "CIL International Law Year in Review 2021 Conference–Session 6," June 28, 2021, YouTube video, 40:42, https://www.youtube.com/watch?v=-HvVOI4aBLQ.

229   "an essential component": Michael Lodge, "New Developments in Deep Sea-bed Mining," speech in Hamburg, Germany, September 25, 2018, isa.org.jm, https://isa.org.jm/files/documents/EN/SG-Stats/25_September_2018.pdf. Quote appears on page 3.

229   To date, the ISA has granted: International Seabed Authority, *Secretary-General Annual Report 2022* (June 2022): 72–74. The report notes on page 72, figure 9: "Map of Regions Being Explored for Mineral Resources in the Area," that mining exploration contracts have been issued for regions in "the Clarion-Clipperton Zone, the Indian Ocean, the Mid-Atlantic Ridge, the South Atlan-tic Ocean and the Western Pacific Ocean."

229   Any member state that pays a $500,000 fee: Jonathan Watts, "Race to the Bottom: The Disastrous, Blindfolded Rush to Mine the Deep Sea," *The Guardian,* theguardian.com, https://www.theguardian.com/environment/2021/sep/27/race-to-the-bottom-the-disastrous-blindfolded-rush-to-mine-the-deep-sea (accessed January 17, 2023).

230   The sponsoring nation receives: Duncan Currie, "Seabed Mining: Legal Risks, Responsibilities and Liabilities for Sponsoring States," Deepsea Conservation Coalition, savethehighseas.org, https://www.savethehighseas.org/wp-content/uploads/2020/10/Seabed-Mining-Liability-Factsheet_DSCC_July2020.pdf (accessed January 17, 2023).

230   Ultimately the ISA stands: "Deep-Sea Mining: Who Stands to Benefit," *Deep-Sea Mining Fact Sheet 6* (February 2022): 7, https://savethehighseas.org/wp-content/uploads/2022/07/DSCC_FactSheet7_DSM_WhoBenefits_4pp_web.pdf.

230 a jarring conflict of interest: The fact that the ISA will be participating (however legally and for whatever reason) in an industry for which it is also the regulator has raised widespread concerns about conflict of interest, a charge the organization denies. In a rebuttal to the *Los Angeles Times*, a lawyer for the ISA wrote: "There is no conflict within the mandate of the ISA. The ISA, through its constituent organs, fulfills its mandate to manage the regulation of seabed activities in the context of the aim of protecting the marine environment. There is no 'inherent conflict of interest' within that mandate." The ISA lawyer added that "All regulators, domestic and international, and in whatever sector or context, are required to balance competing interests. This is the very function of and why regulation is required. To conflate the effective regulation of diverse interests as a conflict of interest fails to understand that role."

230 develop its own nodule mining concession: United Nations Convention on the Law of the Sea, "Article 170: The Enterprise," 94, un.org/depts./los /convention_agreements/texts/unclo/unclos_e.pdf.

230 the ISA and its delegates are still negotiating: International Seabed Authority, "The Mining Code," https://www.isa.org.jm/mining-code (accessed January 21, 2023).

230 But the exploration phase: International Seabed Authority, "Environmental Impact Assessments," https://www.isa.org.jm/minerals/environmental -impact-assessments (accessed January 21, 2023).

230 But the mining contractors are self-reporting: "Deep-Sea Mining: Is the International Seabed Authority Fit for Purpose?" *Deep-Sea Mining Fact Sheet 7* (February 2022): 3, https://savethehighseas.org/wp-content/uploads /2022/03/DSCC_FactSheet7_DSM_ISA_4pp_28Feb22.pdf.

230 One part of their impact: Tom Peacock, "Mining the Deep Sea," TEDxMIT talk, January 23, 2020, YouTube video, 24:07, https://www.youtube.com /watch?v=6TpibSTjOww.

232 "When you're down there": Claudia Dreifus, "Deep in the Sea, Imagining the Cradle of Life on Earth," *New York Times*, October 16, 2007.

232 A chance to conduct baseline studies: Cindy Lee Van Dover, "Tighten Regulations on Deep-Sea Mining," *Nature* 470 (February 2011): 31–33.

232 "We can't just stick our heads": Elham Shabahat, "'Antithetical to Science': When Deep-Sea Research Meets Mining Interests," Mongabay, October 4, 2021, news.mongabay.com, https://news.mongabay.com/2021/10/antithet ical-to-science-when-deep-sea-research-meets-mining-interests/ (accessed March 23, 2023).

232 But a vent field's vertical structure: Porter Hoagland, Stace Beaulieu, Maurice A. Tivey, Roderick G. Eggert, Christopher German, Lyle Glowka, and Jian Lin, "Deep-Sea Mining of Seafloor Massive Sulfides," *Marine Policy* 34, no. 3 (May 2010): 728–32.

233 This slurry would be a mix: Nautilus Minerals Niugini Limited, *Solwara 1 Project: Environmental Impact Study, Volume A, Main Report* (September 2008): Section 9.6.2, yumpu.com, https://www.yumpu.com/en/document /read/38646617/environmental-impact-statement-nautilus-cares-nautilus -minerals (accessed January 17, 2023). Among other disclosures, this section

states, "Sessile fauna in the path of mining will be unavoidably entrained in the ore stream and pumped to the surface, resulting in an unavoidable loss of these animals."

233   Then the remains: Ibid., section 9.4.3, "Suspended Sediment and Plume Formation."

233   Nautilus would extract two million tons: Richard Steiner, *Independent Review of the Environmental Impact Statement for the Proposed Nautilus Minerals Solwara 1 Seabed Mining Project, Papua New Guinea* (January 2009), 3–4.

233   Nautilus had obtained: Ibid., 3.

234   stock was worth a penny: On December 4, 2019, Nautilus Minerals' closing stock price was $0.0100. Historical data, Yahoo Finance, https://finance.yahoo.com, https://finance.yahoo.com/quote/NUSMF/history (accessed January 13, 2023).

234   The company needed such a gargantuan ship: Amanda Stutt, "Nautilus Minerals Officially Sinks, Shares Still Trading," mining.com, https://www.mining.com/nautilus-minerals-officially-sinks-shares-still-trading/ (accessed January 17, 2023).

235   The group noted: Cedric Patjole, "Deep Sea Mining to Have Zero Impact: Nautilus," Loop Papua New Guinea, looppng.com, https://www.looppng.com/business/deep-sea-mining-have-zero-impact-nautilus-67238 (accessed January 17, 2023).

235   "the actual impact": Nautilus Minerals Inc., *Annual Information Form for the Fiscal Year Ended December 31, 2014* (March 2015), 59.

235   The list of hazards: Steiner, *Independent Review*, 2–5.

235   "an exciting opportunity": David Shukman, "Agreement Reached on Deep Sea Mining," BBC News, April 25, 2014, bbc.com/news. https://www.bbc.com/news/science-environment-27158883 (accessed February 16, 2023).

235   The enterprise was awash: Nautilus Minerals Inc., *Annual Information Form for the Fiscal Year Ended December 31, 2017* (March 2018), 11–12.

236   former Nautilus executives and investors: Justin Scheck, Eliot Brown, and Ben Foldy, "Environmental Investing Frenzy Stretches Meaning of 'Green,' " *Wall Street Journal*, June 24, 2021.

236   To nobody's surprise: Deep Sea Mining Campaign, London Mining Network, and Mining Watch Canada, *Why the Rush? Seabed Mining in the Pacific Ocean* (July 2019), 9.

236   Michael Lodge appeared in DeepGreen's promotional video and photographs: Deep Sea Mining Campaign, *Why the Rush?*, cover. DeepGreen's promotional video—which the company removed from Vimeo after criticism of Lodge's appearance in it—can be viewed on the *Los Angeles Times* website at https://www.latimes.com/00000180-0150-dd9c-a5ca-ff75c88e0000-123 (accessed January 21, 2023).

237   On a *60 Minutes* segment: *60 Minutes*, "Into the Deep," producer: Heather Abbott, November 17, 2019, https://www.cbsnews.com/news/rare-earth-elements-u-s-on-sidelines-in-race-for-metals-sitting-on-ocean-floor-60-minutes-60-minutes-2019-11-17/ (accessed January 21, 2023).

237   "I'm doing this for the planet": Barron quoted in Scheck, *Wall Street Journal*.

237   Also, for an 8.1 percent stake: The Metals Company, *2021 Annual Report*

(Vancouver, British Columbia, April 2022): 160, https://investors.metals.co/financials/annual-reports (accessed March 28, 2023).

237   The TMC stock debuted: For a chart of the Metals Company's stock performance from its debut on September 10, 2021, to the present, see https://finance.yahoo.com/quote/TMC/history/.

237   Shareholders joined a class-action suit: The Rosen Law Firm, Class Action Complaint for the Violation of Federal Securities Laws (New York: October 2021): 6. "Making materially false and misleading statements" is one overarching allegation made by the lawyers in this suit. The complaint also alleges other wrongdoings by the Metals Company (and its chairman and CEO, Gerard Barron), including that the company "overpaid on licenses to potential undisclosed insiders," "artificially inflated exploration expenses by more than 100% to mislead investors about the scale of its operations," and has a "history of affiliating with bad actors." Barron and TMC are also alleged to have "significantly downplayed the environmental risks of deep-sea mining polymetallic nodules . . ." https://www.dandodiary.com/wp-content/uploads/sites/893/2021/10/TMC-The-Metals-Company-complaint.pdf (accessed January 21, 2023).

237   the Metals Company sued: Ortenca Aliaj, "Deep Sea Mining Group Left in the Lurch After $200m Disappears," *Financial Times*, October 20, 2021.

237   "Deep sea mining is not needed": The Pacific Blue Line, "Drawing the Pacific Blue Line." pacificblueline.org (accessed January 21, 2023). See also Deep Sea Mining Campaign, "Investors Take Flight from the Metals Company: Pacific Island Countries Should Do the Same," media release, September 23, 2021, https://dsm-campaign.org/wp-content/uploads/2022/03/Blue-Line-TMC-Media-Statement.pdf.

237   called for a moratorium: The International Union for Conservation of Nature, "Protection of Deep-Sea Ecosystems and Biodiversity Through a Moratorium on Seabed Mining," news release no. 069, September 22, 2021, https://www.iucncongress2020.org/motion/069.

237   A petition signed by hundreds: "Marine Expert Statement Calling for a Pause to Deep-Sea Mining," https://www.seabedminingsciencestatement.org/.

238   And it's debatable: K.A. Miller, K. Brigden, D. Santillo, D. Currie, P. Johnston, and K. F. Thompson, "Challenging the Need for Deep Seabed Mining from the Perspective of Metal Demand, Biodiversity, Ecosystems Service, and Benefit Sharing," *Frontiers in Marine Science* 8, no. 706161 (July 2021), https://10.3389/fmars.2021.706161.

238   protect the deep from harmful effects: Lisa A. Levin, *Significant, Serious and Sobering: Defining Serious Harm and Harmful Effects from Seabed Mining* (PowerPoint presentation, Deep-Ocean Stewardship Initiative Workshop, Scripps Oceanographic Institute, March 2014), https://www.isa.org.jm/files/documents/EN/Workshops/2017/Berlin/PPT/Levin.pdf.

238   the waters above will also be affected: Jeffrey C. Drazen, Craig R. Smith, Kristina M. Gjerde, Steven H. D. Haddock, Glenn S. Carter, C. Anela Choy et al., "Midwater Ecosystems Must Be Considered When Evaluating Environmental Risks of Deep-Sea Mining," *Proceedings of the National Academy of Sciences* 117, no. 30 (July 2020), https://doi.org/10.1073/pnas.2011914117.

238   "It is frightfully clear": Steven H. D. Haddock and C. Anela Choy, "Treasure and Turmoil in the Deep Sea," *New York Times*, August 14, 2020.

239   In a brassy move: The Metals Company, *2021 Annual Report*, 37. See also International Seabed Authority, "Nauru Requests the President of ISA Council to Complete Adoption of Rules, Regulations and Procedures Necessary to Facilitate the Approval of Plans of Work for Exploitation in the Area," press release, June 29, 2021, https://www.isa.org.jm/news/nauru-requests-president-isa -council-complete-adoption-rules-regulations-and-procedures.

239   It was an ultimatum: Gerard Barron has made it clear in public statements and actions that Nauru is acting on behalf of its corporate sponsor. At the February 2019 ISA meeting in Kingston, Jamaica, Barron spoke for Nauru as its delegate—a flagrant display of influence. In "A Mining Startup's Rush for Underwater Metals Comes with Deep Risks," in *Bloomberg* (June 23, 2021), journalist Todd Woody writes, "Nauru broke UN protocol by ceding its seat to Barron . . . allowing an executive to address the organization's policymaking body." In another article written by Woody, Barron discusses his company's need to move fast: "We want the mining code finalized. . . . We as a contractor hope to be in operation in 2023 and we're going out on a limb because of the extensive work and money we're investing now." Of the ISA's two-year trigger rule, which was invoked by Nauru in June 2021, Barron had previously said, "It's something that's consistently under review—it's not off the table, that's for sure." Todd Woody, "Covid-19 Throws Seabed Mining Negotiations Off Track," *China Dialogue Ocean*, May 7, 2020, https://chinadialogueocean.net /en/governance/13685-covid-19-could-throw-seabed-mining-negotiations -off-track/ (accessed January 22, 2023).

239   "one of the scariest things": Jack Hitt, "The Billion-Dollar Shack," *New York Times Magazine*, December 10, 2010, https://www.nytimes.com/2000/12/10 /magazine/the-billion-dollar-shack.html (accessed January 22, 2023). Hitt discusses his article, and Nauru, with host Ira Glass in December 5, 2003, "The Middle of Nowhere," *This American Life*, podcast, MP3 audio, 30:00, https://www.thisamericanlife.org/253/the-middle-of-nowhere.

239   "long history of": Helen Hughes, "Tough Love Key to Nauru's Future," *Global Policy Forum*, January 22, 2008, https://archive.globalpolicy.org/nations /micro/2008/0122nauru.htm#author (accessed January 22, 2023).

240   Nauru had decided to become a crook itself: U.S. House of Representatives Committee on Banking and Financial Services Hearing on Money Laundering transcript (Washington, DC: March 2000): 61, http://commdocs.house.gov /committees/bank/hba63312.000/hba63312_0f.htm.

240   Nauru also sold passports: Anne Davies and Ben Doherty, "Corruption, Incompetence and a Musical: Nauru's Cursed History," *The Guardian*, September 3, 2018, https://www.theguardian.com/world/2018/sep/04 /corruption-incompetence-and-a-musical-naurus-riches-to-rags-tale (accessed January 22, 2023).

240   considered letting its phone lines: Hitt, "The Billion-Dollar Shack."

240   Nauru closed its borders: Helen Davidson, "Australia Jointly Responsible for Nauru's Draconian Media Policy, Documents Reveal," *The Guard-*

ian, October 3, 2018, https://www.theguardian.com/australia-news/2018/oct/04/australia-jointly-responsible-for-naurus-draconian-media-policy-documents-reveal (accessed January 22, 2023).

240   "Nauru has proudly taken": The Government of the Republic of Nauru, "Nauru Requests the International Seabed Authority Council to Adopt Rules and Regulations Within Two Years," FAQ on Two Year Notice, http://naurugov.nr/government/departments/department-of-foreign-affairs-and-trade/faqs-on-2-year-notice.aspx.

240   "We'll create a bit of a storm": Greg Stone Interview with DeepGreen CEO Gerard Barron, June 10, 2019, in *The Sea Has Many Voices* podcast, YouTube, 1:06:00, https://www.youtube.com/watch?v=TIgfH8rW49Q.

241   Three miles down in the CCZ: P. P. E. Weaver, J. Aguzzi, R. E. Boschen-Rose, A. Colaço, H. de Stigter et al., "Assessing Plume Impacts Caused by Polymetallic Nodule Mining Vehicles," *Marine Policy* 139 (March 2022), https://doi.org/10.1016/j.marpol.2022.105011.

241   Fifty percent of the CCZ's creatures: Tanja Stratmann, Karline Soetaert, Daniel Kersken, and Dick van Oevelen, "Polymetallic Nodules Are Essential for Food-Web Integrity of a Prospective Deep-Seabed Mining Area in Pacific Abyssal Plains," *Nature Scientific Reports* 11, no. 12238 (June 2021), https://doiorg/10.1038/s41598-021-91703-4.

241   "the abundance of life": DeepGreen Metals Inc., "Open Letter to Brands Calling for a Ban on Seafloor Minerals," https://oceanminingintel.com/insights/deepgreen-metals-inc-open-letter-to-brands-on-the-benefits-of-seafloor-nodules.

241   "Despite life likely originating": The Metals Company, *2021 Impact Report* (Vancouver, BC, May 2022): 30, https://metals.co/tmcs-inaugural-impact-report/.

241   The terrestrial biosphere: Yinon M. Bar-On and Ron Milo, "The Biomass Composition of the Oceans: A Blueprint of Our Blue Planet," *Cell* 179, no. 7 (December 12, 2019): 1451–54. Bar-On and Milo note that "the disparity between the biomass of primary producers on land and in the ocean is the huge mass of supportive woody tissues in land plants. . . . On land, being taller gives greater access to sunlight and makes a plant a better competitor. . . . In the ocean, it is not necessary to grow tall in order to reach the light."

242   What's more significant is biodiversity: Holly J. Niner, Jeff A. Ardron, Elva G. Escobar, Matthew Gianni, Aline Jaeckel, Daniel O. B. Jones et al., "Deep-Sea Mining with No Net Loss of Biodiversity—an Impossible Aim," *Frontiers in Marine Science* 5, no. 53 (March 2018), https://doi.org/10.3389/fmars.2018.00053.

242   the more scientists explore the CCZ: Diva J. Amon, Amanda F. Ziegler, Thomas G. Dahlgren, Adrian G. Glover, Aurélie Goineau et al., "Insights into the Abundance and Diversity of Abyssal Megafauna in a Polymetallic-Nodule Region in the Eastern Clarion-Clipperton Zone," *Nature Scientific Reports* 6, no. 30492 (July 2016), https://doi.org/10.1038/srep30492.

242   likely made by beaked whales: Leigh Marsh, Veerle A. I. Huvenne, and Daniel O. B. Jones, "Geomorphological Evidence of Large Vertebrates Interacting with the Seafloor at Abyssal Depths in a Region Designated for

Deep-Sea Mining," *Royal Society Open Science* 5, no. 8 (August 2018), https://doi .org/10.1098/rsos.180286.

243 Recently, researchers analyzed: Tristan Cordier, Inés Barrenechea Angeles, Nicolas Henry, Franck Lejzerowicz, Cédric Berney, Raphaël Morard et al., "Patterns of Eukaryotic Diversity from the Surface to the Deep-Ocean Sediment," *Science Advances* 8, no. 5 (February 2022), https://doi.10.1126/sciadv .abj9309.

243 "We're talking about small animals": Patrick Pester, "The Deep Seafloor Is Filled with Entire Branches of Life Yet to Be Discovered," *Live Science*, February 5, 2022, https://www.livescience.com/deep-ocean-floor-teeming-with -unknown-life (accessed January 22, 2023).

243 "Just as scientists have discovered": Marine Biological Laboratory, "Ocean Microbe Census Discovers Diverse World of Rare Bacteria," *Science Daily*, September 2, 2006, https://www.sciencedaily.com/releases/2006/08/06082 9081744.htm (accessed January 22, 2023).

243 the genetic resources: MacKenzie Elmer and Lauren Fimbres Wood, "Deep Sea Anti-Cancer Drug Discovered by Scripps Scientists Enters Final Phase of Clinical Trials," Scripps Institution of Oceanography, April 7, 2020, https:// scripps.ucsd.edu/news/deep-sea-anti-cancer-drug-discovered-scripps -scientists-enters-final-phase-clinical-trials (accessed January 22, 2023). See also Stephanie Stone, "Hope for New Drugs Arises from the Sea," *Scientific American*, August 1, 2022, https://www.scientificamerican.com/article /hope-for-new-drugs-arises-from-the-sea/ (accessed January 22, 2023).

243 Razing the seafloor will disrupt life: Beth N. Orcutt, James A. Bradley, William J. Brazelton, Emily R. Estes, Jacqueline M. Goordial, Julie A. Huber et al., "Impacts of Deep-Sea Mining on Microbial Ecosystem Services," *Limnology and Oceanography* 65, no. 7 (July 2020): 1489–1510.

244 "The sediments are": Rachel Carson, *The Sea Around Us* (Oxford, UK: Oxford University Press, 1989), 76.

244 "Given the significant volume": The Metals Company, *2021 Annual Report*, 40.

244 large swaths of seafloor: Craig R. Smith, Malcolm R. Clark, Erica Goetze, Adrian G. Glover, and Kerry L. Howell, "Biodiversity, Connectivity and Ecosystem Function Across the Clarion-Clipperton Zone: A Regional Synthesis for an Area Targeted for Nodule Mining," *Frontiers in Marine Science* 8 (December 2021), https://doi.org/10.3389/fmars.2021.797516.

244 scientists reeled in disbelief: An international group of marine scientists signed a protest letter to the ISA and Michael Lodge, written by Gretchen Früh-Green. It ends with this paragraph: "We should study and conserve the deep sea, not open it up to destructive activities. We strongly encourage the ISA to consider the input of the international science community when evaluating future requests for deep-sea mineral exploitation. These unique hydrothermal vent sites are irreplaceable, and their vulnerability to nearby exploration, let alone seabed mining, is entirely unknown."

245 in an episode about mining: Alan Jamieson and Thom Linley, "Deep-Sea Mining Special," December 10, 2020, in *The Deep-Sea Podcast*, podcast, MP3 audio, 1:07:38, https://www.armatusoceanic.com/podcast/006-deep-sea -mining-special.

246   "This is not supposing": David Edward Johnson, "Protecting the Lost City Hydrothermal Vent System: All Is Not Lost, or Is It?," *Marine Policy* 107 (September 2019), https://doi.org/10.1016/j.marpol.2019.103593. Johnson points out that Poland's mining exploration area includes vent fields along the Mid-Atlantic Ridge that do contain metal sulfide deposits, and that Lost City doesn't have to be mined to be harmed: "A principal threat to Lost City ecosystem is indirect impact of deep-sea mining from possible plumes and discharges, if exploitation for polymetallic sulfides is sanctioned at adjacent or neighboring sites."

246   even the rules laid down: In a 2020 "Information Note," Jiyhun Lee, the ISA's director of the office of environmental management and mineral resources, wrote: "Under Article 145 of the Convention, ISA shall adopt appropriate rules, regulations and procedure for *inter alia*, the prevention, reduction and control of pollution and other hazards to the marine environment, and the protection and conservation of the natural resources of the Area and the prevention of damage to the flora and fauna of the marine environment." Lee goes on to state that, in assessing the "environmental implications of activities in the Area," the ISA's Legal and Technical Commission must take into account "the views of recognized experts." Yet the ISA is moving ahead *despite* the views of experts. In April 2022, thirty-one deep-sea scientists and policymakers published a paper that outlines, in rigorous and methodical terms, the mammoth gaps that remain in our understanding of deep-sea mining's environmental impact. "Given that the minimum level of knowledge needed has not been gathered for any exploration region or resource yet, this proposed plan aligns with the increasing calls for slowing the transition from exploration to exploitation," the authors wrote. They recommended ten to thirty years of further study. (To date, more than seven hundred marine science experts have signed a public statement in favor of a moratorium.)

Jiyhun Lee, "Information Note on the Work of the International Seabed Authority Relating to Environmental Impact Assessments," https://ec.europa.eu/newsroom/mare/redirection/document/64790 (accessed January 21, 2023).

Diva J. Amon, Sabine Gollner, Telmo Morato, Craig R. Smith, Chong Chen, Sabine Christiansen et al., "Assessment of Scientific Gaps Related to the Effective Environmental Management of Deep-Seabed Mining," *Marine Policy* 138 (April 2022), https://doi.org/10.1016/j.marpol.2022.105006.

247   *The New York Times* and the *Los Angeles Times:* The two articles referenced here are Todd Woody, "A Gold Rush in the Deep Sea Raises Questions About the Authority Charged with Protecting It," *Los Angeles Times*, April 19, 2022, and Eric Lipton, "Secret Data, Tiny Islands and a Quest for Treasure on the Ocean Floor," *New York Times*, August 29, 2022.

CHAPTER 10: KAMAʻEHUAKANALOA

253   At the summit we could descend: David A. Clague, Jennifer B. Paduan, David W. Caress, Craig L. Moyer, Brian T. Glazer, and Dana R. Yoerger, "Struc-

ture of Loʻihi Seamount, Hawaii, and Lava Flow Morphology from High-Resolution Mapping," *Frontiers in Earth Science* 26 (March 2019), https://doi.org/10.3389/feart.2019.00058.

255   According to his web page: https://www.soest.hawaii.edu/oceanography/glazer/Brian_T._Glazer/Research.html.

257   In 1954, it had appeared: K. O. Emery, "Submarine Topography South of Hawaii," *Pacific Science* 9 (July 1955): 286–91.

260   The men had also been clad: Chris Gebhardt, "50 Years On, Reminders of Apollo 1 Beckon a Safer Future," nasaspaceflight.com, https://www.nasaspaceflight.com/2017/01/50-years-on-apollo-1-safer-future/ (accessed January 17, 2023).

266   "When, at any time in our earthly life": William Beebe, *Half Mile Down* (New York: Harcourt, Brace and Company, 1934), 133–34.

271   "More and more complete severance": Ibid., 210.

EPILOGUE: THE DEEP FUTURE

275   surveying every corner of it: Vescovo achieved his goal of surveying every part of the Challenger Deep, including the Eastern Pool, the Western Pool, and the Central Pool. Data from the ship's sonar, the sub, and the landers has been scientifically analyzed, resulting in this definitive paper: Samuel F. Greenaway, Kathryn D. Sullivan, S. Harper Umfress, Alice B. Beittel, and Karl D. Wagner, "Revised Depth of the Challenger Deep from Submersible Transects; Including a General Method for Precise, Pressure-Derived Depths in the Ocean," *Deep Sea Research Part I: Oceanographic Research Papers* 178 (December 2021), https://doi.org/10.1016/j.dsr.2021.103644.

276   The world's two deepest shipwrecks: In March 2021, Vescovo and Triton's Shane Eigler dived 21,180 feet to the wreck of the USS *Johnston*, a U.S. Navy destroyer than sank in the Philippine Sea during a World War II battle in 1944. The following year, Vescovo—diving with French sonar specialist Jeremie Morizet—found a World War II wreck that lay even deeper in the Philippine Sea, at 22,621 feet: the USS *Samuel B. Roberts*, known as the "Sammy B." For more details on the USS *Johnston* dives, see Caladan Oceanic, "Submersible Crew Completes the World's Deepest Shipwreck Dive (USS *Johnston*)," news release on March 31, 2021, https://caladanoceanic.com/wp-content/uploads/2021/04/Samar-media-release-3-Final-Version-Rev1.pdf.

276   "As NASA and SpaceX": Christian Davenport, "As NASA and SpaceX Prepare to Fly Another Crew to the Space Station, Engineers are Fixing a Leaky Toilet on the Spacecraft," *Washington Post*, October 29, 2021.

277   the stealthy, red-lit Mesobot: Veronique LaCapra, "Mesobot, Follow That Jellyfish: New Robot Will Track Animals in the Twilight Zone," *Oceanus* (Spring 2019): 38–39.

277   Nereid Under-Ice: Woods Hole Oceanographic Institution, "HROV Nereid Under Ice," whoi.edu, https://www.whoi.edu/what-we-do/explore/under water-vehicles/hybrid-vehicles/nereid-under-ice/ (accessed January 17, 2023).

277 Orpheus and Eurydice: Woods Hole Oceanographic Institution, "Orpheus," whoi.edu, https://www.whoi.edu/what-we-do/explore/underwater-vehicles/auvs/orpheus/ (accessed January 17, 2023).

277 humanoid mer-bots with stereoscopic eyes: Taylor Kubota, "Stanford's OceanOneK Connects Human's Sight and Touch to the Deep Sea," news.stanford.edu, https://news.stanford.edu/2022/07/20/oceanonek-connects-humans-sight-touch-deep-sea/ (accessed January 17, 2023).

277 a new Colombian president: "Gustavo Petro Raises the Rescue of the Galleon San José with a New Research Vessel," earlybulletins.news, https://early bulletins.news/world/193781.html (accessed January 17, 2023).

278 The *Alvin* would soon return: Hannah Piecuch, "Overhaul to Take Alvin to Greater Extremes," *Oceanus,* whoi.edu/oceanus, https://www.whoi.edu/oceanus/feature/overhaul-to-take-alvin-to-greater-extremes/ (accessed January 17, 2023).

279 China's new three-man submersible: Bo Yang, "Manned Submersibles—Deep-Sea Scientific Research and Exploitation of Marine Resources," *Bulletin of Chinese Academy of Sciences* 36:5, Article 18 (May 2021).

279 "Dear friends": Chen Yu, "The Deep-Sea Dream," chinatoday.com, http://www.chinatoday.com.cn/ctenglish/2018/ln/202102/t20210226_800237469.html. (accessed January 4, 2023).

279 a nesting colony of *sixty million* icefish: Autun Purser, Laura Hehemann, Lilian Boehringer, Andreas Rogge, Moritz Holtappels, and Frank Wenzhoefer, "A Vast Icefish Breeding Colony Discovered in the Antarctic," *Current Biology* 32, no. 4 (February 2022), https://doi.org/10.1016/j.cub.2021.12.022.

279 The sponges were feeding: T. M. Morganti, B. M. Slaby, A. de Kluijver, K. Busch, U. Hentschel, J. J. Middleburg, H. Grotheer et al., "Giant Sponge Grounds of Central Arctic Seamounts Are Associated with Extinct Seep Life," *Nature Communications* 13, no. 638 (February 2022), https://doi.org/10.1038/s41467-022-28129-7.

280 *Plasticus* was discovered: Johanna N. J. Weston, Priscilla Carrillo-Barragan, Thomas D. Linley, William D. K. Reid, and Alan J. Jamieson, "New Species of Eurythenes from Hadal Depths of the Mariana Trench, Pacific Ocean (Crustacea: Amphipoda)," *Zootaxa* 4748, no. 1 (March 2020): 163–68.

280 the hadal trenches are thick with every toxin: Alan J. Jamieson, Tamas Malkocs, Stuart B. Piertney, Toyonobu Fujii, and Zulin Zhang, "Bioaccumulation of Persistent Organic Pollutants in the Deepest Ocean Fauna," *Nature Ecology & Evolution* 1, no. 0051 (February 2017), https://doi.org/10.1038/s41559-016-0051.

282 now the volcano's official name: United States Geological Survey Hawaiian Volcano Observatory, "Kamaʻehuakanaloa—The Volcano Formerly Known as Loʻihi Seamount," usgs.gov, https://www.usgs.gov/observatories/hvo/news/volcano-watch-kamaehuakanaloa-volcano-formerly-known-loihi-seamount (accessed January 15, 2023).

Bibliography

Anctil, Michel. *Luminous Creatures: The History and Science of Light Production in Living Organisms* (Montreal, Quebec, and Kingston, Ontario: McGill–Queen's University Press, 2018).

Asma, Stephen T. *On Monsters: An Unnatural History of Our Worst Fears* (Oxford, UK: Oxford University Press, 2009).

Atwater, Brian F., Musumi-Rokkaku Satoko, Satake Kenji, Tsuji Yoshinobu, Ueda Kazue, and David K. Yamaguchi. *The Orphan Tsunami of 1700: Japanese Clues to a Parent Earthquake in North America* (Seattle, WA: University of Washington Press, 2015).

Baker, Maria, Eva Ramirez-Llodra, and Paul Taylor, eds. *Natural Capital and Exploitation of the Deep Ocean* (Oxford, UK: Oxford University Press, 2020).

Ballard, Robert D. *The Eternal Darkness: A Personal History of Deep-Sea Exploration* (Princeton, NJ: Princeton University Press, 2000).

Bass, George F. *Archaeology Beneath the Sea: My Fifty Years of Diving on Ancient Shipwrecks* (Istanbul, Turkey: Boyut, 2011). Updated Kindle edition of 1975 printed edition from Walker & Company, 1975.

Bass, George F., ed. *Beneath the Seas: Adventures with the Institute of Nautical Archaeology* (London: Thames & Hudson, 2005).

Beebe, William. *Half Mile Down* (New York: Harcourt, Brace and Company, 1934).

Burke, Edmund. *A Philosophical Enquiry into the Sublime and Beautiful* (Oxford, UK: Oxford University Press, 2015).

Busby, Frank R. *Manned Submersibles* (Washington, DC: Office of the Oceanographer of the Navy, 1976).

Campbell, Lord George. *Log-Letters from the Challenger* (London: Macmillan and Co., 1877).

Carson, Rachel. *The Sea Around Us* (Oxford, UK: Oxford University Press, 1989).

———. *Lost Woods: The Discovered Writing of Rachel Carson*, ed. Linda Lear (Boston, MA: Beacon Press, 1998).

Chapman, Roger. *No Time on Our Side* (New York: W. W. Norton & Company, 1975).

Corfield, Richard. *The Silent Landscape: The Scientific Voyage of HMS Challenger* (Washington, DC: The Joseph Henry Press, 2003).

Delgado, James P. *War at Sea: A Shipwrecked History from Antiquity to the Twentieth Century* (Oxford, UK: Oxford University Press, 2019).

Delgado, James, Terry Kerby, Hans K. Van Tilburg, Steven Price, Ole Varmer, Maximilian D. Cremer, and Russell Matthews. *The Lost Submarines of Pearl Harbor: The Rediscovery and Archaeology of Japan's Top-Secret Midget Submarines of World War II* (College Station, TX: Texas A&M University Press, 2016).

Eiseley, Loren. *The Star Thrower* (San Diego, CA, and New York: Harcourt Brace & Company, 1978).

Ellis, Richard. *The Search for the Giant Squid: The Biology and Mythology of the World's Most Elusive Sea Creature* (New York: Penguin Books, 1998).

————. *Monsters of the Sea* (Guilford, CT: Lyons Press, 2006).

Forbes, Edward. *The Natural History of the European Seas*, ed. Robert Godwin-Austen (London: John Van Voorst, 1859).

Gershwin, Lisa-Ann. *Jellyfish: A Natural History* (Chicago, IL: University of Chicago Press, 2016).

Gould, Carol Grant. *The Remarkable Life of William Beebe, Explorer and Naturalist* (Washington, DC: Shearwater Books, 2004).

Gould, Richard A. *Archaeology and the Social History of Ships*, 2nd ed. (Cambridge, UK: Cambridge University Press, 2011).

Hamilton-Paterson, James. *Three Miles Down: A Hunt for Sunken Treasure* (London: Jonathan Cape, 1998).

Hand, Kevin Peter. *Alien Oceans: The Search for Life in the Depths of Space* (Princeton, NJ: Princeton University Press, 2020).

Hanlon, Roger, Mike Vecchione, and Louise Allcock. *Octopus, Squid, and Cuttlefish: A Visual, Scientific Guide to the Ocean's Most Advanced Invertebrates* (Chicago: University of Chicago Press, 2018)

Heezen, Bruce C., and Charles D. Hollister. *The Face of the Deep* (New York, London, and Toronto: Oxford University Press, 1971).

Homer. *The Odyssey*. Translated by Stephen Mitchell (New York: Atria Books, 2013).

Jamieson, Alan. *The Hadal Zone: Life in the Deepest Oceans* (Cambridge, UK: Cambridge University Press, 2015).

Johannesson, Kurt. *The Renaissance of the Goths in Sixteenth-Century Sweden: Johannes and Olaus Magnus as Politicians and Historians*. Translated by James Larson (Berkeley, CA: University of California Press, 1991).

Karson, Jeffrey A., Deborah S. Kelley, Daniel J. Fornari, Michael R. Perfit, and Timothy M. Shank, *Discovering the Deep: A Photographic Atlas of the Seafloor and Ocean Crust* (Cambridge, UK: Cambridge University Press, 2015).

Leroi, Armand Marie. *The Lagoon: How Aristotle Invented Science* (New York: Penguin Books, 2014).

Lyons, Sherrie Lynne. *Species, Serpents, Spirits, and Skulls* (Albany, NY: State University of New York Press, 2009).

Macdougall, Doug. *Endless Novelties of Extraordinary Interest: The Voyage of H.M.S. Challenger and the Birth of Modern Oceanography* (New Haven, CT, and London: Yale University Press, 2019).

MacInnis, Joe. *Underwater Man* (New York: Dodd, Mead & Company, 1974).

————. *The Breadalbane Adventure* (Montreal: Optimum Publishing International, 1982).

Magnus, Olaus. *A Description of the Northern Peoples,* vol. 3. Translated by Peter Fisher and Humphrey Higgens; edited by Peter Foote (London: The Hakluyt Society, 1998).

Matsen, Brad. *Descent: The Heroic Discovery of the Abyss* (New York: Pantheon Books, 2005).

Mero, John L. *The Mineral Resources of the Sea* (Amsterdam, London, and New York: Elsevier Scientific Publishing Company, 1965).

Moseley, H. N. *Notes by a Naturalist: An Account of Observations Made During the Voyage of H.M.S. Challenger* (New York: G. P. Putnam's Sons, 1892).

Munktell, Ing-Marie. *Museum Gustavianum: A Window to the Surrounding World* (Uppsala, Sweden: Uppsala University, 2015).

Oudemans, A. C. *The Great Sea-Serpent: An Historical and Critical Treatise* (London: Luzac & Co., 1892).

Piccard, Jacques, and Robert S. Dietz. *Seven Miles Down: The Story of the Bathyscaph Trieste* (New York: G. P. Putnam's Sons, 1961).

Rahn Phillips, Carla. *The Treasure of the San José: Death at Sea in the War of the Spanish Succession* (Baltimore, MD: Johns Hopkins University Press, 2007).

Rehbock, Philip F., ed. *At Sea with the Scientifics: The Challenger Letters of Joseph Matkin* (Honolulu: University of Hawaii Press, 1992).

Robison, Bruce, and Judith Connor. *The Deep Sea* (Monterey, CA: Monterey Bay Aquarium Press, 1999).

Rozwadowski, Helen. *Fathoming the Ocean: The Discovery and Exploration of the Deep Sea* (Cambridge, MA, and London: Belknap Press of Harvard University Press, 2005).

———. *Vast Expanses: A History of the Oceans* (London: Reaktion Books, 2018).

Sars, Georg Ossian. *On Some Remarkable Forms of Animal Life from the Great Depths Off the Norwegian Coast* (Oslo: Brøgger & Christie, 1872).

Schlee, Susan. *The Edge of an Unfamiliar World: A History of Oceanography* (New York: E. P. Dutton & Co., 1973).

Søreide, Fredrik. *Ships from the Depths: Deepwater Archaeology* (College Station, TX: Texas A&M University Press, 2011).

Spry, R. N. *The Cruise of Her Majesty's Ship Challenger* (London: Sampson Low, Marston, Searle, & Rivington, 1876).

Thomson, Charles Wyville. *The Depths of the Sea* (London: Macmillan and Company, 1873).

Thomson, Sir Charles Wyville. *The Voyage of the Challenger: The Atlantic* (London: Macmillan and Co., 1877).

Wallich, G. C. *The North-Atlantic Sea-Bed: A Diary of the Voyage on Board H.M.S. Bulldog, in 1860* (London: John Van Voorst, 1862).

Watson, Timothy R. *The Spanish Treasure Fleets* (Sarasota, FL: Pineapple Press, 1994).

Widder, Edith. *Below the Edge of Darkness: A Memoir of Exploring Light and Life in the Deep Sea* (New York: Random House, 2021).

Wild, John James. *At Anchor: A Narrative of Experiences Afloat and Ashore During the Voyage of H.M.S. Challenger from 1872 to 1876* (London: Marcus Ward and Co., 1878).

Wilson, Edward O. *Biophilia: The Human Bond with Other Species* (Cambridge, MA, and London: Harvard University Press, 1984).

Van Dover, Cindy Lee. *Deep-Ocean Journeys: Discovering New Life at the Bottom of the Sea* (Reading, MA: Helix Books, 1996).

Van Duzer, Chet. *Sea Monsters on Medieval and Renaissance Maps* (London: The British Library, 2013).

Young, Josh. *Expedition Deep Ocean: The First Descent to the Bottom of All Five of the World's Oceans* (New York and London: Pegasus Books, 2020).

Resources

ORGANIZATIONS

Caladan Oceanic: https://caladanoceanic.com
Deep-Ocean Stewardship Initiative: https://www.dosi-project.org/
Deep Sea Conservation Coalition: https://savethehighseas.org/
EYOS Expeditions: https://www.eyos-expeditions.com/
Five Deeps Expedition: https://fivedeeps.com
Inkfish: https://ink.fish/
Marine Technology Society: https://www.mtsociety.org/
Minderoo–UWA Deep-Sea Research Center: https://www.uwa.edu.au/oceans
    -institute/Research/Deep-Sea-Research-Centre
Mission Blue: https://missionblue.org/
Monterey Bay Aquarium Research Institute: https://www.mbari.org/
Nautilus Live: Ocean Expedition Trust: https://nautiluslive.org/
OceanX: https://oceanx.org/
NOAA Ocean Exploration: https://www.oceanexplorer.noaa.gov/
Schmidt Ocean Institute: https://schmidtocean.org/
Scripps Institution of Oceanography: https://scripps.ucsd.edu/
Triton Submarines: https://tritonsubs.com
University of Washington School of Oceanography: https://www.ocean
    .washington.edu/
Woods Hole Oceanographic Institution: https://www.whoi.edu/

OTHER RESOURCES

Along with the organizations listed above, I recommend the following to those who would like to dive deeper:

The Deep-Sea Podcast

Alan Jamieson and Thom Linley (with regular contributor Captain Don Walsh) host a monthly show they describe as "a punk take on a science podcast about everything deep sea." It's available widely on podcast platforms, or at https://armatusoceanic.com/the-deepsea-podcast.

The Regional Cabled Array

This educational portal offers a closer look at the world's most advanced deep-sea observatory and the images and data it's gathering from the Pacific Northwest seafloor—including live video from a hydrothermal vent on Axial Volcano: https://interactiveoceans.washington.edu/about/regional-cabled-array/.

Monterey Bay Aquarium, "Into the Deep: Exploring Our Undiscovered Ocean"

As of April 2022, anyone who would like to meet a siphonophore, blood-belly comb jelly, giant isopod, and other deep-sea creatures in person can do so at the Monterey Bay Aquarium in Monterey, California. Working with deep-sea scientists from the Monterey Bay Aquarium Research Institute (MBARI), the aquarists have re-created deep-sea conditions in highly specialized tanks, offering an unprecedented glimpse of the extraordinary life below: https://www.monterey bayaquarium.org/visit/exhibits/into-the-deep.

The Pagoo *Submersible*

In spring 2023, Triton purchased the *Pagoo*, a ten-person sub previously owned by the late Microsoft cofounder, Paul Allen. (A passionate ocean advocate and explorer, Allen kept the *Pagoo* on his private yacht, the *Octopus*.) Beginning in late 2023 or early 2024, anyone who would like to witness the deep's twilight zone will be able to book a seat on this sub, which dives to 1,200 feet. Triton's plans for the *Pagoo* will be in place soon, and a site in the Caribbean announced as its home base. For more information and updates: https://tritonsubs.com /pagoo/

Illustration Credits

INSERT 1

PAGE 1 Uppsala University Library collections. PAGE 2 *Top:* Ralph White / COR-BIS / via Getty Images. *Bottom:* U.S. Naval History and Heritage Command. PAGE 3 *Top:* Steve Nicklas / NOS / NGS. *Bottom:* Five Deeps Expedition. PAGE 4 *Top:* Courtesy of Terry Kerby. *Bottom:* Courtesy of Terry Kerby / HURL. PAGE 5 *Top*: Courtesy of Terry Kerby / HURL. *Bottom:* Courtesy of Terry Kerby / HURL. PAGE 6 *Top:* Courtesy of Deborah Kelley, University of Washington. *Center:* Deborah Kelley / University of Washington, NSF-OOI-WHOI; V19. *Bottom:* NSF-OOI / UW / CSSF: ROPOS Dive R1757. PAGE 7 *Top left:* Susan Casey. *Bottom left:* Courtesy of Deborah Kelley / University of Washington. *Top right:* University of Washington. *Center right:* NSF / OOI / UW / ISS; R1838; V15. *Bottom right:* UW / NSF / OOI / WHOI; V19. PAGE 8 *Top:* Deborah Kelley and Mitchell Elend, University of Washington; URI-ROV *Hercules,* and NOAA Ocean Exploration. *Bottom:* Deborah Kelley, University of Washington; URI-ROV Hercules, IFE, URI-IAO, Lost City Science Party, and NOAA Ocean Exploration.

INSERT 2

PAGE 1 *Top:* Reeve Jolliffe / Five Deeps Expedition. *Bottom right:* Nick Verola / Caladan Oceanic. *Bottom left:* Nick Verola / Caladan Oceanic. PAGE 2 *Top:* Reeve Jolliffe / Five Deeps Expedition. *Bottom:* © Atlantic Productions / Discovery, from the Caladan Oceanic Five Deeps Expedition. Photo by Tamara Stubbs. PAGE 3 *Top:* © Atlantic Productions / Discovery, from the Caladan Oceanic Five Deeps Expedition. *Center left:* © Atlantic Productions / Discovery, from the Caladan Oceanic Five Deeps Expedition. Photo by Joe MacInnis. *Center right:* © Atlantic Productions / Discovery, from the Caladan Oceanic Five Deeps Expedition. Photo by Joe MacInnis. *Bottom:* © Alan Jamieson and Thomas Linley. PAGE 4 *Top:* © Alan Jamieson and Thomas Linley. *Center:* © Alan Jamieson and Thomas Linley. *Bottom:* © Atlantic Productions / Discovery, from the Caladan Oceanic Five Deeps Expedition. PAGE 5 *Top:* © Alan Jamieson / Caladan Oceanic. *Center:* © Alan Jamieson / Caladan Oceanic. *Bottom:* Alan Jamieson. PAGE 6 *Top left:* EYOS Expeditions. *Center left:* © Atlantic Productions / Discovery, from the Caladan Oceanic Five Deeps Expedition. Photo by Susan Casey. *Bottom left:* © Atlantic Productions / Discovery, from the Caladan Oceanic Five Deeps Expedition. Photo by Joe MacInnis. *Top right:* © Atlantic Productions / Discovery, from the Caladan Oceanic Five Deeps Expedition. Photo by Joe MacInnis. *Second from top right:* © Atlantic Productions / Discovery, from the Caladan Oceanic Five Deeps Expedition. Photo by Joe MacInnis. *Second from bottom right:* © Atlantic Productions / Discovery, from the Caladan Oceanic Five Deeps Expedition. Photo by Joe MacInnis. *Bottom right:* © Atlantic Productions / Discovery, from the Caladan Oceanic Five Deeps Expedition. Photo by Joe MacInnis. PAGE 7 *Top left:* Reeve Jolliffe / Five Deeps Expedition. *Bottom left:* Reeve Jolliffe / Five Deeps Expedition. *Second from top right:* Susan Casey. *Bottom right:* © Alan Jamieson. PAGE 8 *Left:* © Alan Jamieson / Caladan Oceanic / Minderoo Foundation. *Top right:* Victor Vescovo / Caladan Oceanic. *Bottom right:* © Alan Jamieson / Caladan Oceanic.

INSERT 3

PAGE 1 *Top left:* Courtesy of Roger Dooley. *Top right:* Courtesy of Roger Dooley. *Center left:* Courtesy of Roger Dooley. *Bottom left:* Courtesy of Roger Dooley. *Bottom right:* Courtesy of Roger Dooley. PAGE 2 *Top:* Paul Caiger / Woods Hole Oceanographic Institution. *Center:* Paul Caiger / Woods Hole Oceanographic Institution. *Bottom:* Paul Caiger / Woods Hole Oceanographic Institution. PAGE 3 Top: Paul Caiger. *Center:* Paul Caiger / Woods Hole Oceanographic Institution. *Bottom:* Paul Caiger. PAGE 4 *Top:* Susan Casey. *Bottom left:* Courtesy of Buck Taylor. *Bottom right:* Susan Casey. PAGE 5 *Top:* © 2004 MBARI. *Center:* Courtesy of the NOAA Office of Ocean Exploration and Research. *Bottom:* Courtesy of NOAA Office of Ocean Exploration and Research. PAGE 6 *Top:* © Alan Jamieson / Caladan Oceanic / Minderoo Foundation. *Bottom left:* © Alan Jamieson / Caladan Oceanic / Minderoo Foundation. *Bottom right:* © Alan Jamieson / Caladan Oceanic / Minderoo Foundation. PAGE 7 *Top left:* © Todd Brown. *Top right:* Courtesy of NOAA Office of Ocean Exploration and Research. *Center:* Courtesy of Craig Smith and Diva Amon, ABYSSLINE Project. *Bottom:* © Alan Jamieson / Caladan Oceanic / Minderoo Foundation. PAGE 8 *Top left:* Susan Casey. *Bottom left:* Susan Casey. *Top right:* Courtesy of the U.S. Geological Survey. *Bottom right:* Courtesy of NOAA Office of Ocean Exploration and Research.

About the Author

Susan Casey is the author of *New York Times* bestsellers *Voices in the Ocean*, *The Wave*, and *The Devil's Teeth* and is the former editor in chief of *O, The Oprah Magazine*. She is a National Magazine Award–winning journalist whose work has been featured in the *Best American Science and Nature Writing*, *Best American Sports Writing*, and *Best American Magazine Writing* anthologies, and has appeared in *Esquire*, *Sports Illustrated*, *Fortune*, and *Outside*.

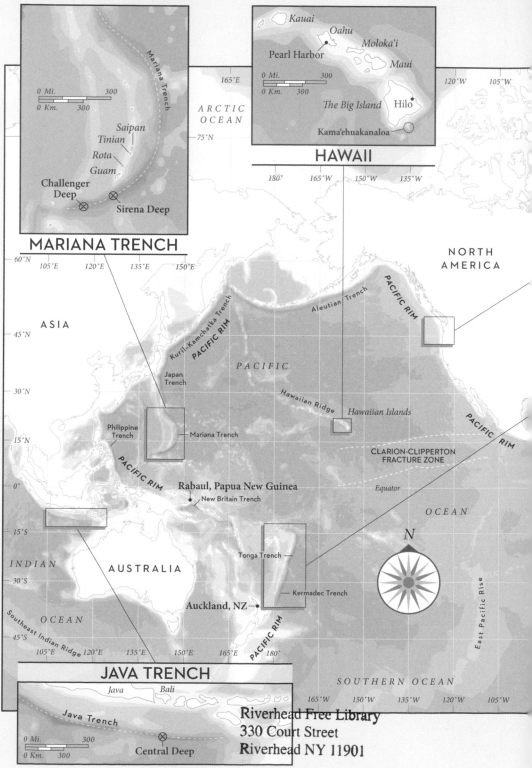

MARIANA TRENCH

Saipan
Tinian
Rota
Guam
Challenger
Deep
Sirena Deep

0 Mi. 300
0 Km. 300

HAWAII

Kauai
Oahu
Moloka'i
Pearl Harbor
Maui
The Big Island
Hilo
Kama'ehuakanaloa

0 Mi. 300
0 Km. 300

ARCTIC
OCEAN

165°E
75°N
120°W 105°W

NORTH
AMERICA

60°N
105°E 120°E 135°E 150°E

180° 165°W 150°W 135°W

ASIA

Aleutian Trench
PACIFIC RIM

45°N

Kuril-Kamchatka Trench
PACIFIC RIM

PACIFIC

30°N

Japan
Trench

Hawaiian Ridge
Hawaiian Islands

PACIFIC RIM

15°N

Philippine
Trench

Mariana Trench

CLARION-CLIPPERTON
FRACTURE ZONE

0°

Rabaul, Papua New Guinea
New Britain Trench

Equator

OCEAN

PACIFIC RIM

N

15°S

Tonga Trench

INDIAN

30°S

AUSTRALIA

Kermadec Trench

East Pacific Rise

OCEAN

45°S

Southeast Indian Ridge
105°E 120°E 135°E 150°E 165°E 180°

Auckland, NZ

PACIFIC RIM

JAVA TRENCH

Java Bali

Java Trench

Central Deep

0 Mi. 300
0 Km. 300

165°W 150°W 135°W 120°W 105°W

SOUTHERN OCEAN

Riverhead Free Library
330 Court Street
Riverhead NY 11901